Data Structures and Program Design Using C

Second Edition

Data Structures and Program Design Using C

A Self-Teaching Introduction

Second Edition

Dr. Dheeraj Malhotra
Dr. Neha Malhotra

Mercury Learning and Information
Boston, Massachusetts

MERCURY LEARNING AND INFORMATION
121 High Street, 3rd Floor
Boston, MA 02110
info@merclearning.com

D. Malhotra and N. Malhotra. *Data Structures and Program Design Using C: A Self-Teaching Introduction, Second Edition*
ISBN: 978-1-5015-2430-1

Library of Congress Control Number: 2025946490
242526321 This book is printed on acid-free paper in the United States of America.

Dedicated to our loving parents and beloved students

CONTENTS

PREFACE

Data structures are the building blocks of computer science. The objective of this text is to emphasize fundamentals of data structures as an introductory subject. It is designed for beginners (students or professionals) who would like to learn the basics of data structures and their implementation using the C programming language. With this focus in mind and feedback of the first edition, we present second edition of this title, well supported with real world analogies throughout text to enable a quick understanding of the technical concepts and to help in identifying appropriate data structures to solve specific, practical problems. This book will serve the purpose of a text / reference book and will be of immense help especially to undergraduate or graduate students of various courses in information technology, engineering, computer applications, and information sciences.

Key Features:

- *Practical Applications:* Real world analogies as practical applications are given throughout the text to easily understand and connect the fundamentals of data structures with day to day, real-world scenarios. This approach, in turn, will assist the reader in developing the capability to easily identify the most appropriate and efficient data structure for solving a specific problem.
- *Frequently Asked Questions:* Frequently asked theoretical/practical questions are integrated throughout the content of the book, within related topics to assist readers in grasping the subject.
- *Algorithms and Programs:* To better understand the fundamentals of data structures at a generic level-followed by its specific implementation in C, syntax independent algorithms as well as implemented programs in C are discussed throughout the book. This presentation will assist the reader in getting both algorithms and their corresponding implementation within a single book.

- *Numerical and Conceptual Exercises:* To assist the reader in developing a strong foundation of the subject, more number of numerical and conceptual problems are included throughout the text.
- *Multiple Choice Questions:* To assist students for placement-oriented exams in various IT areas, several exercises are suitably chosen and are given in an MCQ format.

Dr. Dheeraj Malhotra
Dr. Neha Malhotra

Acknowledgments

We are humbled and profoundly grateful to the Almighty God for the gift of life, family, opportunities, and education. His unwavering blessings have guided us throughout this journey and played a pivotal role in the overwhelming success of the first edition of this book. We extend our heartfelt thanks to our readers, whose enthusiastic support led to its adoption as a reference text in the BCA and MCA curricula of the esteemed GGS IP University.

We express our sincere appreciation to the leadership of our parent institution, Vivekananda Institute of Professional Studies (affiliated with GGS IP University) especially Dr. S.C. Vats (Chairman, VIPS), Sh. Suneet Vats (Vice Chairman, VIPS), Sh. Vineet Vats (Vice Chairman, VIPS), and Prof. Deepali Kamthania (Dean, VSIT). Their continued encouragement and belief in our work have been a source of inspiration and honor.

We are deeply indebted to our mentors—Dr. O.P. Rishi (University of Kota), Dr. Sushil Chandra (DRDO, Government of India), and Dr. Udyan Ghose (GGS IP University)—for their invaluable guidance and motivation throughout the development of various books authored and published by us with MLI.

Our sincere thanks go to Ms. Stuti Suthar (SAP Labs), Mr. Sahil Pathak (TMB), Mr. Deepanshu Gupta (Tech Mahindra Ltd.), Ms. Aditi Vats (VIPS, GGS IPU), and our students and readers for their constructive feedback on the first edition. Their insights have significantly contributed to the refinement of this manuscript.

We are also grateful to David Pallai, Jennifer Blaney, and the entire team at Mercury Learning and Information, De Gruyter Brill, for their enthusiastic collaboration and support throughout this project.

Our heartfelt gratitude goes to our parents, siblings, and family members, whose unwavering support uplifted us in moments of both triumph and challenge.

Finally, we remain inspired by our readers across the USA, Canada, and India. Their continued appreciation and positive feedback on our five authored titles—*Data Structures using C/C++/Java/Python* and *C++ Programming*, published with MLI and De Gruyter Brill—have been instrumental in shaping and enhancing this second edition.

Dr. Dheeraj Malhotra
Dr. Neha Malhotra

CHAPTER 1

Introduction to Data Structures

INTRODUCTION

A *data structure* is an efficient way of storing and organizing data elements in computer memory. *Data* means a value or a collection of values. *Structure* refers to a method of organizing the data. The mathematical or logical representation of data in memory is referred to as a data structure. The objective of a data structure is to store, retrieve, and update the data efficiently. A data structure can be referred to as a group of elements grouped under one name. The group of data elements is called *members*, and they can be of different types. Data structures are used in almost every program and software system. There are various kinds of data structures that are suited for different types of applications. Data structures are the building blocks of a program. For a program to run efficiently, a programmer must choose appropriate data structures. A data structure is a crucial part of data management. As the name suggests, data management is a task that includes different activities such as the collection of data, the organization of data into structures, and much more. Some examples where data structures are used include stacks, queues, arrays, binary trees, linked lists, hash tables, and so forth.

A data structure helps us understand the relationship of one element to another element and organize it within the memory. It is a mathematical or logical representation or organization of data in memory. Data structures are extensively applied in the following areas:

- Compiler design
- *Database management systems (DBMSs)*
- Artificial intelligence
- Network and numerical analysis
- Statistical analysis packages
- Graphics
- *Operating systems (OSs)*
- Simulations

As we can see in the previous list, there are many applications in which different data structures are used for their operations. Some data structures sacrifice speed for the efficient utilization of memory, while others sacrifice memory utilization and result in faster speed. In today's world, programmers aim not just to build a program, but instead, to build an effective program. As previously discussed, for a program to be efficient, a programmer must choose appropriate data structures. Hence, data structures are classified into various types. Now, let us discuss and learn about different types of data structures.

Frequently Asked Questions

Q1. Define the term "data structure."

Answer.

A data structure is an organization of data in a computer's memory or disk storage. In other words, a logical or mathematical model of a particular organization of data is called a data structure. A data structure in computer science is also a way of storing data in a computer so that it can be used efficiently. An appropriate data structure allows a variety of important operations to be performed using both resources (that is, memory space and execution time) efficiently.

TYPES OF DATA STRUCTURES

Data structures are classified into various types.

Linear and Non-Linear Data Structures

A *linear data structure* is one in which the data elements are stored in a linear, or sequential, order (that is, data is stored in consecutive memory locations). A linear data structure can be represented in two ways; either it is represented by a linear relationship between various elements utilizing consecutive memory locations, as in the case of arrays, or it may be represented by a linear relationship between the elements utilizing links from one element to another, as in the case of linked lists. Examples of linear data structures include arrays, linked lists, stacks, queues, and so on.

A *non-linear data structure* is one in which the data is not stored in any sequential order or consecutive memory locations. The data elements in this structure are represented by a hierarchical order. Examples of non-linear data structures include graphs, trees, and so forth.

Static and Dynamic Data Structures

A *static data structure* is a collection of data in memory that is fixed in size and cannot be changed during runtime. The memory size must be known in advance, as the memory cannot be reallocated later in a program. One example is an *array*.

A *dynamic data structure* is a collection of data in which memory can be reallocated during the execution of a program. The programmer can add or remove elements according to their need. Examples include linked lists, graphs, trees, and so on.

Homogeneous and Non-Homogeneous Data Structures

A *homogeneous data structure* is one that contains data elements of the same type, such as arrays.

A *non-homogeneous data structure* contains data elements of different types, such as structures.

Primitive and Non-Primitive Data Structures

Primitive data structures are the fundamental data structures or predefined data structures that are supported by a programming language. Examples of primitive data structure types are short, integer, long, float, double, char, pointers, and so forth.

Non-primitive data structures are comparatively more complicated data structures that are created using primitive data structures. Examples of non-primitive data structures are arrays, files, linked lists, stacks, queues, and so on. The classification of different data structures is shown below in Figure 1.1.

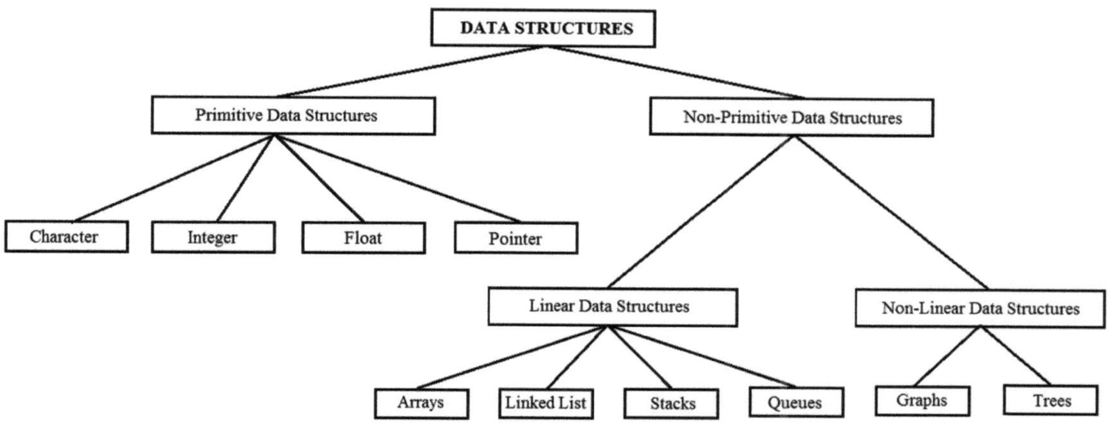

FIGURE 1.1 Classification of different data structures.

We know that C supports various data structures. So, we will now introduce all these data structures, and they will be discussed in detail in the upcoming chapters.

Frequently Asked Questions

Q2. Write the difference between primitive data structures and non-primitive data structures.

Answer.

Primitive data structures are typically directly operated upon by machine-level instructions; that is, the fundamental data types (such as `int`, `float`, `char`, and so on) are known as primitive data structures.

Data structures that are not fundamental are called non-primitive data structures.

Frequently Asked Questions

Q3. Explain the difference between linear and non-linear data structures.

Answer.

The main difference between linear and non-linear data structures lies in the way in which data elements are organized. In the linear data structure, elements are organized sequentially, and therefore, they are easy to implement in a computer's memory. In® non-linear data structures, a data element can be attached to several other data elements to represent specific relationships existing among them.

Arrays

An *array* is a collection of homogeneous (similar) types of data elements in contiguous memory. An array is a linear data structure because all elements of an array are stored in linear order. The various elements of the array are referenced by their index value, also known as the subscript. In C, an array is declared using the following syntax:

Syntax – <Data type> array name [size];

The elements are stored in the array as shown in Figure 1.2.

1st element	2nd element	3rd element	4th element	5th element	6th element
array [0]	array [1]	array [2]	array [3]	array [4]	array [5]

FIGURE 1.2 Memory representation of an array.

Arrays are used for storing a large amount of data of a similar type. They have various advantages and limitations.

These are the advantages of using arrays:

- Elements are stored in adjacent memory locations; hence, searching is very fast, as any element can be easily accessed.
- Arrays do not support dynamic memory allocation, so all the memory management is done by the compiler.

These are some of the limitations of using arrays:

- The insertion and deletion of elements in arrays is complicated and very time-consuming, as it requires the shifting of elements.
- Arrays are static; hence, the size must be known in advance.
- Elements in the array are stored in consecutive memory locations, which may or may not be available.

Queues

A *queue* is a linear collection of data elements in which the element inserted first will be the element that is taken out first; that is, a queue is a *FIFO* (short for *first in, first out*) data structure. A queue is a popular linear data structure in which the first element is inserted from one end, called the *rear* end (also called the tail end), and the deletion of the element takes place from the other end, called the *front* end (also called the head).

Practical Application:

For a simple illustration of a queue, there is a line of people standing at the bus stop and waiting for the bus. Therefore, the first person standing in the line will get on the bus first.

In a computer's memory, queues can be implemented using arrays or linked lists. Figure 1.3 shows the array implementation of a queue. Every queue has `FRONT` and `REAR` variables, which point to the position from where insertion and deletion are done, respectively.

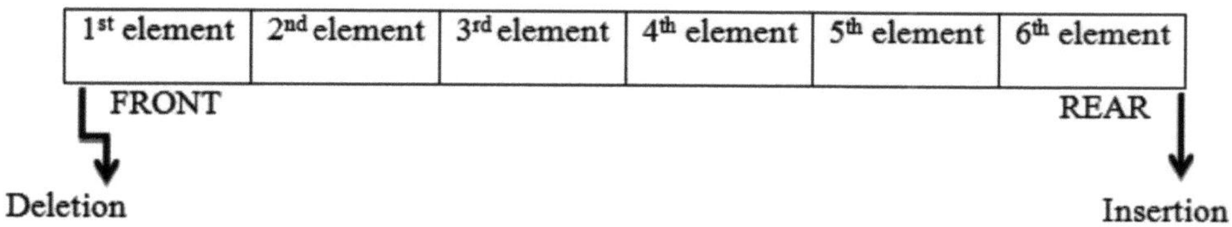

FIGURE 1.3 Memory representation of a queue.

Stacks

A *stack* is a linear collection of data elements in which insertion and deletion take place only at the top of the stack. A stack is a *last-in, first-out* (*LIFO*) data structure because the last element pushed onto the stack will be the first element to be deleted from the stack. Three operations can be performed on the stack, which include `PUSH`, `POP`, and `PEEP` operations. The `PUSH` operation inputs an element into the top of the stack, while the `POP` operation removes an element from the stack. The `PEEP` operation returns the value of the topmost element in the stack without deleting it from the stack. Every stack has a `TOP` variable, which is associated with it. The `TOP` pointer stores the address of the topmost element in the stack and is the position where insertion and deletion take place.

Practical Application:

A real-life example of a stack is if there is a pile of plates arranged on a table. A person will pick up the first plate from the top of the stack.

In a computer's memory, stacks can be implemented using arrays or linked lists. Figure 1.4 shows the array implementation of a stack.

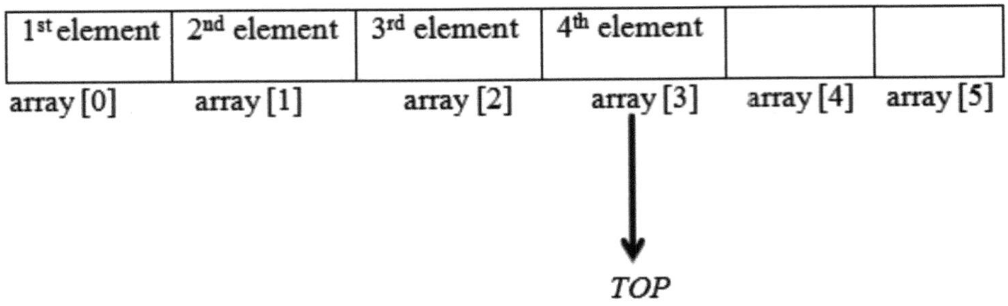

FIGURE 1.4 Memory representation of a stack.

Linked List

The major drawback of an array is that the size or the number of elements must be known in advance. This drawback gave rise to the new concept of a linked list. A *linked list* is a linear collection of data elements. These data elements are called *nodes*, which point to the next node using pointers. A linked list is a sequence of nodes in which each node contains one or more data fields and a pointer that points to the next node. Linked lists are also dynamic; that is, memory is allocated as and when required. Figure 1.5 shows the implementation of a linked list.

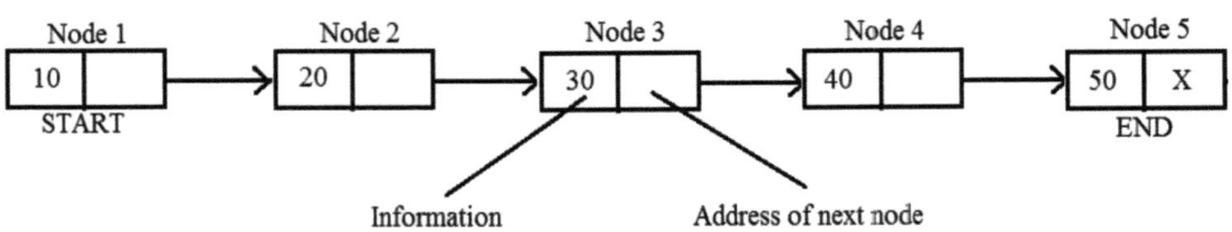

FIGURE 1.5 Memory representation of a linked list.

In Figure 1.5, we have made a linked list in which each node is divided into two slots:

- The first slot contains the information/data.
- The second slot contains the address of the next node.

Practical Application:

A simple real-life example is a train; here, each coach is connected to its previous and next coach (except the first and last coach).

The address part of the last node stores a special value called NULL, which denotes the end of the linked list. The advantage of a linked list over arrays is that it is now easier to insert and delete data elements, as we don't have to shift each time. On the contrary, searching for an element has become difficult. More time is required to search for an element, and it also requires a large amount of memory space. Hence, linked lists are used where a collection of data elements is required, and the number of data elements in the collection is not known to us in advance.

Frequently Asked Questions

Q4. Define the term "linked list."

Answer.

A linked list or one-way list is a linear collection of data elements called nodes, where pointers give the linear order. It is a popular dynamic data structure. The nodes in the linked list are not stored in consecutive memory locations. For every data item in a node of the linked list, there is an associated pointer that gives the address location of the next node in the linked list.

Trees

A *tree* is a popular non-linear data structure in which the data elements or the nodes are represented in a hierarchical order. Here, one of the nodes is shown as the root node of the tree, and the remaining nodes are partitioned into two disjoint sets such that each set is a part of a subtree. A tree makes the searching process very easy, and its recursive programming makes a program optimized and easy to understand.

A *binary tree* is the simplest form of a tree. A binary tree consists of a root node and two subtrees known as the *left subtree* and the *right subtree*, where both subtrees are also binary trees. An AVL tree is a height balanced binary tree where height difference of left and right subtree can't be more than one to avoid skewed structure. Each node in a binary tree consists of three parts. The extreme left part stores a pointer that points to the left subtree, the middle part consists of the data element, and the extreme right part stores a pointer that points to the right subtree. The root is the topmost element of the tree. When there are no nodes in a tree, that is, when `ROOT = NULL`, then it is called an *empty tree*.

For example, consider a binary tree where R is the root node of the tree, as shown in Figure 1.6. LEFT and RIGHT are the left and right subtrees of R, respectively. A is designated as the root node of the tree. Nodes B and C are the left and right child of A, respectively. Nodes B, D, E, and G constitute the left subtree of the root. Similarly, nodes C, F, H, and I constitute the right subtree of the root.

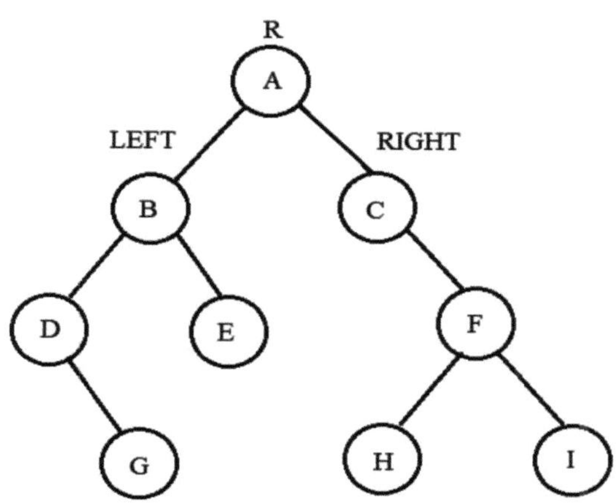

FIGURE 1.6 A binary tree.

Here are some advantages of a tree:

- The searching process is very fast in trees.
- The insertion and deletion of elements have become easier compared to other data structures.

Frequently Asked Questions

Q5. Define a binary tree.

Answer.

A binary tree is a hierarchical data structure in which each node has, at most, two children, that is, a left and right child. In a binary tree, the degree of each node can be, at most, two. Binary trees are used to implement binary search trees, which are used for efficient searching and sorting. A binary tree is a popular subtype of a k-ary tree, where k is 2.

Graphs

A *graph* is a general tree with no parent-child relationship. It is a non-linear data structure that consists of vertices, also called nodes, and the edges that connect those vertices to one another. In a graph, any complex relationship can exist. A graph, G, may be defined as a finite set of V vertices and E edges. Therefore, G = (V, E), where V is the set of vertices and E is the set of edges. Graphs are used in various applications of mathematics and computer science. Unlike a root node in trees, graphs don't have root nodes; rather, the nodes can be connected to any node in the graph. Two nodes are termed as *neighbors* when they are connected via an edge.

Practical Application:

A real-life example of a graph can be seen in workstations where several computers are joined to one another via network connections.

For example, consider a graph, G, with six vertices and eight edges, as shown in Figure 1.7. Here, Q and Z are neighbors of P. Similarly, R and T are neighbors of S.

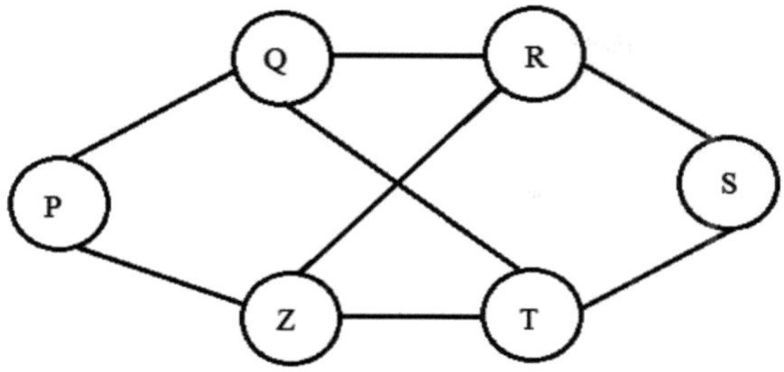

FIGURE 1.7 A graph.

OPERATIONS ON DATA STRUCTURES

Here, we will discuss various operations that are performed on data structures:

- *Creation* – This is the process of creating a data structure. Declaration and initialization of the data structure are done here. It is the first operation.
- *Insertion* – This is the process of adding new data elements to the data structure, for example, adding the details of an employee who has recently joined an organization.
- *Deletion* – This is the process of removing a particular data element from the given collection of data elements, for example, removing the name of an employee who has left the company.
- *Updating* – This is the process of modifying the data elements of a data structure. For example, if the address of a student has changed, then it should be updated.
- *Searching* – This is used to find the location of a particular data element or all the data elements with the help of a given key, for example, finding the names of people who live in New York.
- *Sorting* – This is the process of arranging the data elements in some order, that is, either ascending or descending order. An example is arranging the names of students in a class in alphabetical order.
- *Merging* – This is the process of combining the data elements of two different lists to form a single list of data elements.
- *Traversal* – This is the process of accessing each data element exactly once so that it can be processed. An example is to print the names of all the students in a class.
- *Destruction* – This is the process of deleting the entire data structure. It is the last operation in the data structure.

ALGORITHMS

An *algorithm* is a systematic set of instructions combined to solve a complex problem. It is a step-by-step sequence of instructions, each of which has a clear meaning and can be executed in a minimum amount of effort in a finite time. In general, an algorithm is a blueprint for writing a program to solve a problem. Once we have a blueprint of the solution, we can easily implement it in any high-level language like C, C++, and so forth. It solves the problem in a finite number of steps. An algorithm written in a programming language is known as a *program*. A computer is a machine with no brain or intelligence. Therefore, the computer must be instructed to perform a given task in unambiguous steps. Hence, a programmer must define their problem in the form of an algorithm written in English. Such an algorithm should have the following features:

- An algorithm should be simple and concise.
- It should be efficient and effective.
- It should be free of ambiguity; that is, the logic must be clear.

Similarly, an algorithm must have the following characteristics:

- *Input* – It reads the data of the given problem.
- *Output* – The desired result must be produced.
- *Process/definiteness* – Each step or instruction must be unambiguous.
- *Effectiveness* – Each step should be accurate and concise. The desired result should be produced within a finite time.
- *Finiteness* – The number of steps should be finite.

Developing an Algorithm

To develop an algorithm, some steps are suggested:

1. Define or understand the problem.
2. Identify the result or output of the problem.
3. Identify the inputs required by the problem and choose the best input.
4. Design the logic from the given inputs to get the desired output.
5. Test the algorithm for different inputs.
6. Repeat the above steps till it produces the desired result for all the inputs.

APPROACHES FOR DESIGNING AN ALGORITHM

A complicated algorithm is divided into smaller units, which are called *modules*. Then, these modules are further divided into sub-modules. In this way, a complex algorithm can easily be solved. The process of dividing an algorithm into modules is called *modularization*. There are two popular approaches for designing an algorithm:

- Top-down approach
- Bottom-up approach

Now, let us understand both approaches:

- *Top-down approach* – A top-down approach states that the complex/complicated problem/algorithm should be divided into a smaller number of one or more modules. These smaller modules are further divided into one or more sub-modules. This process of decomposition is repeated until we achieve the desired output of module complexity. A top-down approach starts from the topmost module, and the modules are incremented accordingly till a level is reached where we don't require any more sub-modules, that is, the desired level of complexity is achieved. This is shown in Figure 1.8.

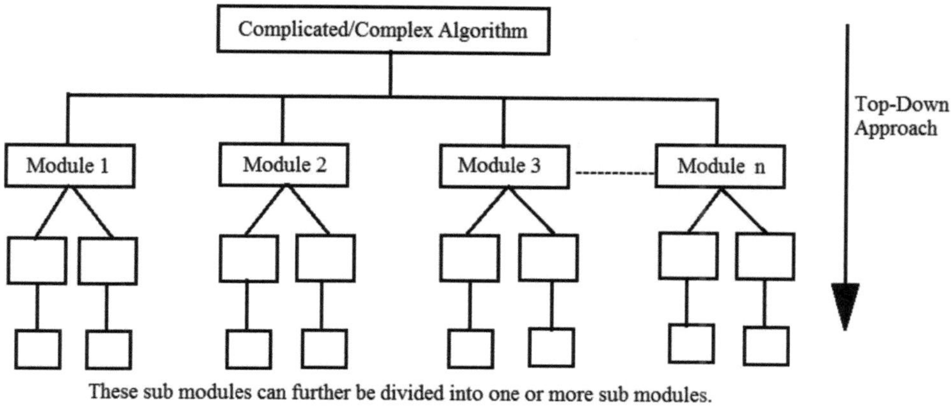

FIGURE 1.8 Top-down approach.

- *Bottom-up approach* – A bottom-up algorithm design approach is the opposite of a top-down approach. In this kind of approach, we first start with designing the basic modules and proceed further toward designing the high-level modules. The sub-modules are grouped together to form a module of a higher level. Similarly, all high-level modules are grouped to form higher-level modules. Thus, this process of combining the sub-modules is repeated until we obtain the desired output of the algorithm. This is shown in Figure 1.9.

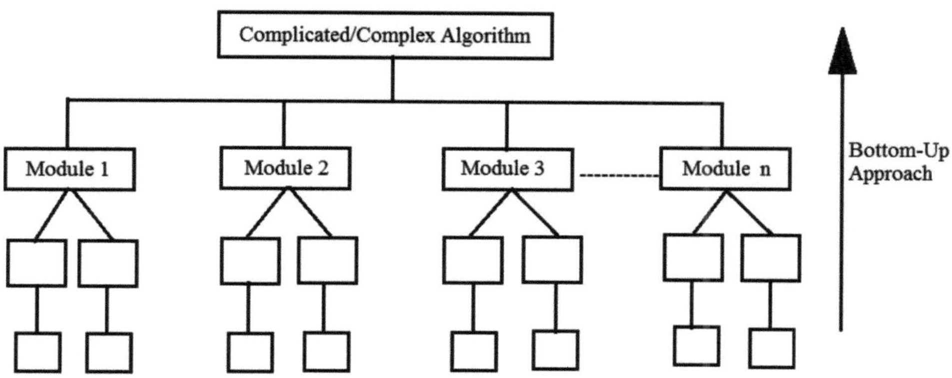

FIGURE 1.9 Bottom-up approach.

ANALYZING AN ALGORITHM

An algorithm can be analyzed by two factors: *space* and *time*. We aim to develop an algorithm that makes the best use of both these resources. Analyzing an algorithm measures the efficiency of the algorithm. The efficiency of the algorithm is measured in terms of speed and time complexity. The complexity of an algorithm is a function that measures the space and time used by an algorithm in terms of input size.

The *time complexity* of an algorithm is the amount of time it takes for the algorithm to run the program completely. It is the running time of the program. The time complexity of an algorithm depends on the input size. The time complexity is commonly represented by using big O notation. For example, the time complexity of a linear search is O(n).

The *space complexity* of an algorithm is the amount of memory space required to run the program completely. The space complexity of an algorithm depends on the input size.

Time complexity is categorized into three types:

- *Best-case running time* – The performance of the algorithm will be best under optimal conditions. For example, the best case for a binary search occurs when the desired element is the middle element of the list. Another example can be of sorting; that is, if the elements are already sorted in a list, then the algorithm will execute in the best time.
- *Average-case running time* – This denotes the behavior of an algorithm when the input is randomly drawn from a given collection or distribution. It is an estimate of the running time for "average" input. It is usually assumed that all inputs of a given size are likely to occur with equal probability.
- *Worst-case running time* – The behavior of the algorithm in this case concerns the worst possible case of an input instance. The worst-case running time of an algorithm is an upper bound on the running time for any input. For example, the worst case for a linear search occurs when the desired element is the last element in the list or the element does not exist in the list.

Frequently Asked Questions

Q6. Define time complexity.

Answer.

Time complexity is a measure that evaluates the number of operations performed by a given algorithm as a function of the size of the input. It is the approximation of the number of steps necessary to execute an algorithm. It is commonly represented with asymptotic notation, that is, O(g) notation, also known as big O notation, where g is the function of the size of the input data.

Time-Space Trade-Off

In computer science, the *time-space trade-off* is a way of solving a particular problem, either in less time and more memory space or in more time and less memory space. If we talk in practical terms, designing such an algorithm in which we can save both space and time is a challenging task. So, we can use more than one algorithm to solve a problem. One may require less time, and the other may require less memory space to execute. Therefore, we sacrifice one thing for the other. Hence, there exists a time-space or time-memory trade-off between algorithms. This time-space trade-off gives the programmer a rational choice from an informed point of view.

So, if time is a big concern for a programmer, then they might choose a program that takes less or the minimum time to execute. On the other hand, if space is a prime concern for a programmer, then they might choose a program that takes less memory space to execute at the cost of more time.

ABSTRACT DATA TYPES

An *abstract data type* (*ADT*) is a popular mathematical model of the data objects that define a data type along with various functions that operate on these objects. To understand the meaning of an ADT, we will simply break the term into two parts, that is, "data type" and "abstract." The data type of a variable is a collection of values that a variable can take. There are various data types in C, including integer, float, character, long, double, and so on. When we talk about the term "abstract" in the context of data structures, it means apart from a detailed specification. It can be considered as a description of the data in a structure with a list of operations to be executed on the data within the structure. Thus, an ADT is the specification of a data type that specifies the mathematical and logical model of the data type. For example, when we use stacks and queues, then at that point in time, our prime concern is only with the data type and the operations to be performed on those structures. We are not worried about how the data will be stored in memory. We don't bother about how `push()` and `pop()` operations work. We just know that we have two functions available to us, so we have to use them for insertion and deletion operations.

BIG O NOTATION

The performance of an algorithm (that is, time and space requirements) can be easily compared with other competitive algorithms using asymptotic notations such as the big O notation, the Omega notation, and the Theta notation. The algorithmic complexity can be easily approximated using asymptotic notations by simply ignoring the implementation-dependent factors. For instance, we can compare various available sorting algorithms using big O notation or any other asymptotic notation.

Big O notation is one of the popular analysis characterization schemes, since it provides an upper bound on the complexity of an algorithm. In big O, $O(g)$ is representative of the class of all functions that grow no faster than g. Therefore, if $f(n) = O(g(n))$, then $f(n) <= c(g(n))$ for all $n > n_0$, where n_0 represents a threshold and c represents a constant.

An algorithm with $O(1)$ complexity is referred to as a constant computing time algorithm. Similarly, an algorithm with $O(n)$ complexity is referred to as a linear algorithm, $O(n^2)$ for quadratic algorithms, $O(2^n)$ for exponential time algorithms, $O(n^k)$ for polynomial time algorithms, and $O(log\ n)$ for logarithmic time algorithms.

An algorithm with complexity of the order of $O(log_2 n)$ is considered one of the best algorithms, while an algorithm with complexity of the order of $O(2^n)$ is considered the worst algorithm. The complexity of computations or the number of iterations required in various types of functions may be compared as follows:

$$O(log_2 n) < O(n) < O\ (n\ log_2 n) < O(n^2) < O(\ n^3) < O(2^n)$$

SUMMARY

- A data structure determines a way of storing and organizing the data elements in the computer memory. Data means a value or a collection of values. Structure refers to a way of organizing the data. The mathematical or logical representation of data in memory is referred to as a data structure.
- Data structures are classified into various types, which include linear and non-linear data structures, primitive and non-primitive data structures, static and dynamic data structures, and homogeneous and non-homogeneous data structures.
- A linear data structure is one in which the data elements are stored in a linear or sequential order; that is, data is stored in consecutive memory locations. A non-linear data structure is one in which the data is not stored in any sequential order or consecutive memory locations.
- A static data structure is a collection of data in memory that is fixed in size and cannot be changed during runtime. A dynamic data structure is a collection of data in which memory can be reallocated during the execution of a program.
- Primitive data structures are fundamental data structures or predefined data structures that are supported by a programming language. Non-primitive data structures are comparatively more complicated data structures that are created using primitive data structures.
- A homogeneous data structure is one that contains all data elements of the same type. A non-homogeneous data structure contains data elements of different types.
- An array is a collection of *homogeneous* (similar) types of data elements in *contiguous* memory.
- A queue is a linear collection of data elements in which the element inserted first will be the element taken out first, that is, a FIFO data structure. A queue is a linear data structure in which the first element is inserted from one end, called the rear end, and the deletion of the element takes place from the other end, called the front end.
- A linked list is a sequence of nodes in which each node contains one or more data fields and a pointer that points to the next node.
- A stack is a linear collection of data elements in which insertion and deletion take place only at one end, called the top of the stack. A stack is a LIFO data structure, because the last element added to the top of the stack will be the first element to be deleted from the top of the stack.
- A tree is a non-linear data structure in which the data elements or the nodes are represented in a hierarchical order. Here, an initial node is designated as the root node of the tree, and the remaining nodes are partitioned into two disjoint sets such that each set is a part of a subtree.
- A binary tree is the simplest form of a tree. A binary tree consists of a root node and two subtrees known as the left subtree and right subtree, where both the subtrees are also binary trees.
- A graph is a general tree with no parent-child relationship. It is a non-linear data structure that consists of vertices or nodes and the edges that connect those vertices with one another.
- An algorithm is a systematic set of instructions combined to solve a complex problem. It is a step-by-step sequence of instructions, each of which has a clear meaning and can be executed in a minimum amount of effort in a finite time.
- The process of dividing an algorithm into modules is called modularization.

- The time complexity of an algorithm is described as the amount of time taken by an algorithm to run the program completely. It is the running time of the program.
- The space complexity of an algorithm is the amount of memory space required to run the program completely.
- An ADT is a mathematical model of the data objects that define a data type as well as the functions to operate on these objects.
- Big O notation is one of the most popular analysis characterization schemes, since it provides an upper bound on the complexity of an algorithm.

EXERCISES

1. What do you understand by a good program?
2. Explain the classification of data structures.
3. What is an algorithm? Discuss the characteristics of an algorithm.
4. What are the various operations that can be performed on the data structures? Explain each of them with an example.
5. Differentiate an array from a linked list.
6. Explain the terms *time complexity* and *space complexity*.
7. What do you understand by the complexity of an algorithm?
8. Write a short note on graphs.
9. What is the process of modularization?
10. Differentiate between stacks and queues with examples.
11. What is meant by abstract data types? Explain in detail.
12. Discuss the worst-case, best-case, and average-case time complexity of an algorithm.
13. Write a brief note on trees.
14. Explain how you can develop an algorithm to solve a complex problem.
15. Explain the time-memory trade-off in detail.

MULTIPLE CHOICE QUESTIONS

1. Which of these data structures is a FIFO data structure?
   - **A.** Array
   - **B.** Stack
   - **C.** Queue
   - **D.** Linked list
2. What is the maximum number of children a binary tree can have?
   - **A.** 0
   - **B.** 2
   - **C.** 1
   - **D.** 3

3. Which of the following data structures uses dynamic memory allocation?
   - A. Graphs
   - B. Linked lists
   - C. Trees
   - D. All of these
4. In a queue, deletion is always done from the ________.
   - A. Front end
   - B. Rear end
   - C. Middle
   - D. None of these
5. Which data structure is used to represent complex relationships between the nodes?
   - A. Linked lists
   - B. Trees
   - C. Stacks
   - D. Graphs
6. Which of the following is an example of a heterogeneous data structure?
   - A. Array
   - B. Structure
   - C. Linked list
   - D. None of these
7. In a stack, insertion and deletion take place from the ________.
   - A. Bottom
   - B. Middle
   - C. Top
   - D. All of these
8. Which of the following is not part of the ADT description?
   - A. Operations
   - B. Data
   - C. Both (a) and (b)
   - D. None of the above
9. Which of the following data structures allows deletion at both ends of the list but insertion at one end only?
   - A. Stack
   - B. Input-restricted dequeue
   - C. Output-restricted dequeue
   - D. Priority queue

10. Which of the following data structures is a linear type?
    - A. Trees
    - B. Graphs
    - C. Queues
    - D. None of the above
11. Which one of the following is beneficial when the data is stored and has to be retrieved in reverse order?
    - A. Stack
    - B. Linked list
    - C. Queue
    - D. All of the above
12. A binary search tree whose left and right subtrees differ in height by 1 at most is a ________.
    - A. Red black tree
    - B. M-way search tree
    - C. AVL tree
    - D. None of the above
13. The operation of processing each element in the list is called ________.
    - A. Traversal
    - B. Merging
    - C. Inserting
    - D. Sorting
14. Which of the following are the two primary measures of the efficiency of an algorithm?
    - A. Data and time
    - B. Data and space
    - C. Time and space
    - D. Time and complexity
15. Which one of the following cases does not exist/occur in complexity theory?
    - A. Average case
    - B. Worst case
    - C. Best case
    - D. Minimal case

CHAPTER 2

Introduction to the C Language

INTRODUCTION

C is a high-level programming language that was developed by Denis Ritchie, a scientist at Bell Labs, in the early 1970s. C is a powerful language, and it is widely used with the Unix operating system, which is also an extensive and complex operating system. C is a user-friendly language in which simple English words like `printf`, `scanf`, `if-else`, and so on are used as statements. Initially, C was developed for writing system software, but as we can see nowadays, it has become a common language such that a variety of different software programs are written using the C language. The C language is often called a middle-level language, as it not only provides different data types or data structures that are needed by a programmer, but it can also access the computer hardware with the help of specially designed functions and declarations. Many programmers are using the C language to program all tasks because of its growing popularity. C is a case-sensitive language; that is, it distinguishes between uppercase letters and lowercase letters. Hence, the C language is becoming popular because of the following:

- One reason for its popularity is its portability; that is, a program written in the C language on one machine or computer can easily be transferred to another machine with minimal modification or with no modifications to it. Hence, it is a platform-independent language.
- Programs written in the C language are very efficient and fast.
- C supports various other programming languages, so in the future, it will be easier for us to understand programs written in C.
- C is a structured programming language; that is, it makes the user think of a problem in terms of blocks or functions.
- C is more powerful and flexible than any other high-level language, as it uses/offers many more functions than any of the other programming languages.
- The most significant advantage of the C language is that it is available to anyone and can be used on different types of personal computers.
- Programs written in C are small and efficient.
- C is a general-purpose language.

Frequently Asked Questions

Q1. List some important features of C.

Answer.

- *Flexibility* – C is a powerful and flexible language. It is used for projects as an operating system, a word processor, for graphics and spreadsheets, and even as a compiler for other languages. As a result, a large variety of C compilers and accessories are available.
- *Portability* – C is a portable language, which means that a C program written on one computer system can be run on another system with little or no modification.
- *Compactness* – C code should be written in routines called functions. These functions can be reused in other applications or programs. By using pieces of information, we can create useful and reusable code.

HEADER FILES IN C

We all know that when we load a C compiler, some supporting files are stored on the disk. During programming, if we want those files or functions, first we have to include those functions or files in our program to use them. Thus, these files will be loaded from the disk into memory, where these files are called to perform the desired operations. These files are known as *header files*. For example, if we want to use the `printf()` and `scanf()` functions, then both of these functions are defined in the `stdio.h` header file. There are various other functions provided by C compilers that are included in standard header files. Header files can be included by either of the following methods:

```
# include "header file"
# include <header file>
```

The following are some of the standard header files:

- `stdio.h` – This includes standard input and output functions. Examples are `printf()`, `scanf()`, `gets()`, `puts()`, etc.
- `conio.h` – This is used for clearing the screen or holding the screen. Examples include `getch()`, `clrscr()`, and `getche()`.
- `string.h` – This includes string handling functions. Examples include `strcpy()`, `strcmp()`, `strlen()`, and `strcat()`.
- `alloc.h` – This is used for dynamic memory allocation. Examples include `malloc()`, `calloc()`, and `free()`.
- `math.h` – This includes all the mathematical functions. Examples include `isalpha()`, `isdigit()`, `sin()`, and `cos()`.

Frequently Asked Questions

Q2. What are header files?

Answer.

In the C programming language, header files are a source code that contains standard definitions and data structures that all programmers may use as and when required. These are also called "include files," because the `#include` statement is used to incorporate them into C programs. Header files use the a `.h` extension.

MAIN FUNCTION

The `main()` function is a part of every C program. A program cannot execute without the `main()` function. The `main()` function is not terminated by any semicolon. The C language permits various forms of the `main()` function:

- `main()`
- `void main()`
- `int main()`

The empty parentheses indicate that the function is without arguments. The keywords `void` and `int` before the `main()` function mean that the function does not return any information and the function returns integer values to the operating system, respectively. Remember that when `int` is used before `main()`, then the last statement must have a `return 0` statement.

INPUT AND OUTPUT METHODS

The most common operations in a C program are to accept input values from a standard input device and to display the data produced by the program on a standard output device. One way of assigning the values is by inputting the values from the user at the beginning of the program, for example, `int x = 5`. The other way of assigning the values is by inputting the values from the user at runtime. Thus, there are various functions in C that are used for assigning values at runtime. One of the commonly used functions is the `scanf()` function.

The `scanf()` function is used to read any kind of data from the keyboard. The general syntax of the `scanf()` function is:

```
scanf("specifier/ control string", arguments) ;
```

The control string or the specifier specifies the type of data that has to be entered from the keyboard. The arguments are the addresses of the variables, which are stored in the memory locations. Each control string starts with a `%`. Various control strings are as follows:

Control String	Description
`%d, %i`	Integer values
`%u`	Unsigned integer values
`%e, %f, %g`	Floating type values
`%c`	Character values
`%s`	String of characters

The `scanf()` function ignores any blank spaces entered by the user. This function returns the number of inputs that are successfully scanned and stored. The `scanf()` function is used as follows:

```
int number ;
scanf("%d", &number) ;
```

Now, there is another function that is used for displaying the data entered by the user. This function is the `printf()` function.

The `printf()` function is used to display the data required by the user and print the values of the variables on the screen. The general syntax for the `printf()` function is:

```
printf("specifier/ control string", arguments) ;
```

The control strings and arguments were discussed previously.

The `printf()` function is used as follows:

```
printf("Hello World!!") ;
```

CHARACTER SET USED IN C

The character set allowed in C consists of the following characters:

- *Alphabet* – It includes uppercase as well as lowercase letters in English, i.e., {A, B, C, ..., Z} and {a, b, c, ..., z}.
- *Digits* – It includes decimal digits, i.e., {0, 1, 2, ..., 9}.
- *White spaces* – It includes spaces, enters, and tabs.
- *Special characters* – It consists of special symbols, which include {, !, ?, #, <, >, (,), %, ", &, ^, *, <<, >>, [,], +, =, /, −, _, :, ;, and }.

C TOKENS

C tokens help us to write a program in C. C supports various types of tokens, as shown in Figure 2.1:

- Keywords
- Identifiers
- Constants
- Variables

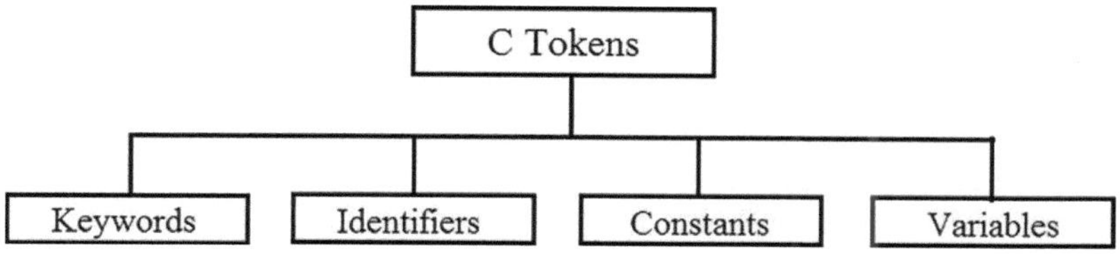

FIGURE 2.1. Various C tokens.

Now, let us discuss all of them.

Keywords in C are the reserved words that have a special meaning. They are written in lowercase. Keywords cannot be used as identifiers. Examples include `auto`, `int`, `float`, `char`, `break`, `switch`, `continue`, `double`, `long`, `long double`, `short`, `unsigned`, `signed`, `while`, `for`, `else`, `void`, and so forth.

An *identifier* is a name that is given to a constant, variable, function, or array. The rules that are used to define identifiers are as follows:

- An identifier can have letters, digits, or underscores.
- It should not start with a digit.
- It can start with an underscore or a letter.
- An identifier cannot have special symbols.
- A keyword cannot be used as an identifier.

For example:

Acceptable Identifiers	Unacceptable Identifiers
`a345_`	`au to`
`c_65`	`12d`
`average`	`n 3_`

Constants are the fixed values in C that can never be changed. These are used to define fixed values in a program. For example, the value of pi is always fixed. A constant can be of any basic data type, such as an integer constant, a character constant, or a float constant:

- *Integer constant* – This is a constant to which only the integer values are assigned, for example, `const int area = 1000;`
- *Character constant* – This is a constant to which only the character values are assigned, for example, `const char Malhotra = 'A';`
- *Float constant* – This is a constant to which only the real or floating type values are assigned, for example, `const float pi = 3.1427;`

A *variable* is a name that is used to refer to some memory location. While working with a variable, we refer to the address of the memory where the data is stored. C supports two types of variables, character variables and numeric variables:

- *Numeric variables* – These are used to store integer or float type values.
- *Character variables* – In these variables, single characters are enclosed in single quotes.

Frequently Asked Questions

Q3. Write any five keywords in the C language.

Answer.

- `if`
- `switch`
- `do`
- `while`
- `for`

Frequently Asked Questions

Q4. Which of the following are invalid identifiers and why?

- **`static`**
- **`January 2`**
- **`VIPS`**
- **`2VSIT`**

Answer.

- `static` is a keyword, and as per convention, one should not use a keyword as an identifier, so `static` is an invalid identifier.
- `January 2` is an invalid identifier as an identifier can't have spaces.
- `VIPS` is a valid identifier.
- `2VSIT` is an invalid identifier as the identifier cannot start with a numeral.

DATA TYPES IN C

Data types are the special keywords that define the type of data and the amount of data a variable holds. Data types are categorized into three categories, as shown in Figure 2.2:

- Basic (primary) data types
- Modified data types
- Derived data types

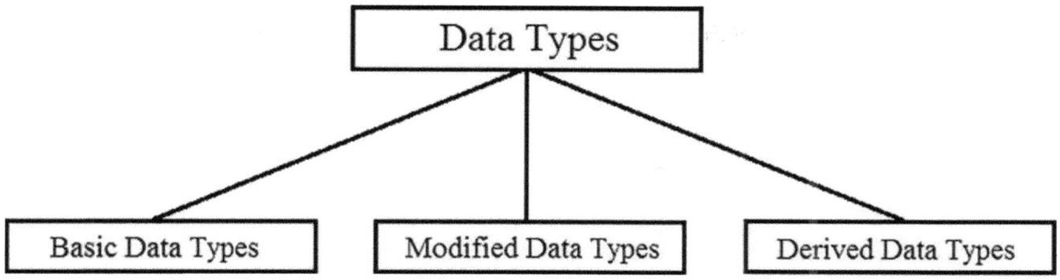

FIGURE 2.2 Categorization of data types.

Basic data types are the primary data types available in C, as shown in the following table.

Data Type	Bytes (in Memory)	Range
`int`	2	–32,768 to 32,767
`char`	1	–128 to 127
`double`	8	-1.7×10^{-308} to $1.7 \times 10^{+308}$
`float`	4	-3.4×10^{-38} to $3.4 \times 10^{+38}$

Modified data types are those data types that are created by altering the basic data types with the help of some keywords:

- *Integer* – `unsigned int` (positive values only), `signed int` (positive as well as negative values), `short int`, `long int`, `signed long int`, `unsigned short int`, `signed short int`, and `unsigned long int`
- *Character* – `unsigned char` and `signed char`
- *Double* – `long double`

Further classification is shown in the following table.

Data Type	Bytes (in Memory)	Range
`unsigned int`	2	0 to 65535
`signed int`	2	-32,768 to 32,767
`short int`	2	-32,768 to 32,767
`long int`	4	-2147483648 to 2147483647
`unsigned short int`	2	0 to 65535
`signed short int`	2	-32,768 to 32,767
`unsigned long int`	4	0 to 4294967294
`signed long int`	4	2147483648 to 2147483647
`unsigned char`	1	0 – 255
`signed char`	1	-128 to 127
`long double`	10	-3.4×10^{-493} to $1.1 \times 10^{+493}$

Derived data types are those data types that are created using basic as well as modified data types. These include arrays, structures, unions, enumerations, and so on.

OPERATORS IN C

Operators in C are used to perform some specific operations between different variables and constants. C supports a variety of operators, which are given as follows:

- Arithmetic operators
- Logical operators
- Assignment operators
- Relational operators/comparison operators
- Condition operators/ternary operators
- Bitwise operators
- Comma operators
- Unary operators/increment and decrement operators
- `sizeof` operators

Now, let us discuss all of these operators.

Arithmetic Operators

Arithmetic operators are those operators that are used in mathematical computation or calculation. The valid arithmetic operators in C are given in the following table.

Let `x` and `y` be the two variables.

Operator	Operation	Example
`+`	Addition	`x + y`
`-`	Subtraction	`x - y`
`*`	Multiplication	`x * y`
`%`	Remainder/modulus	`x % y`
`/`	Division	`x / y`

Logical Operators

C supports three types of logical operators, which are given as follows:

Operator	Description	Example
`!`	Logical NOT	`!x, !y`
`&&`	Logical AND	`x && y`
`||`	Logical OR	`x || y`

Logical NOT is a unary operator. This operator takes a single expression, and it inverts the result such that true becomes false, and vice versa. The truth table for logical NOT is given as follows:

x	y	!x	!y
0	1	1	0
1	0	0	1

Logical AND is a binary operator. Hence, it takes two inputs or expressions. If both the inputs are true, then the whole expression is true. If both or even any one of the inputs is false, then the whole expression will be false. The truth table for logical AND is given as follows:

X	Y	X && Y
0	0	0
0	1	0
1	0	0
1	1	1

Logical OR is also a binary operator; that is, it also takes two expressions. If both the inputs are false, then the output is false. If a single input or both of the inputs are true, then the output will be true. The truth table for logical OR is given as follows:

X	Y	X || Y
0	0	0
0	1	1
1	0	1
1	1	1

Assignment Operators

Assignment operators are ones that are responsible for assigning values to the variables. These operators are always evaluated from right to left. C supports various assignment operators, which are given in the following table:

Operators	Example
=	x = 5 y = 8
+=	x += y :- x = x + y
-=	x -= y :- x = x - y
*=	x *= y :- x = x * y
%=	x %= y :- x = x % y
/=	x /= y :- x = x / y

Relational Operators

Relational operators are used for comparison between two values or expressions. They are also known as *comparison operators*. These operators are always evaluated from left to right. The various relational operators used in C are as follows:

Operators	Description	Example
>	Greater than	`x > y`
<	Less than	`x < y`
==	Equal to	`x == y`
>=	Greater than or equal to	`x >= y`
<=	Less than or equal to	`x <= y`
!=	Not equal to	`x != y`

Conditional Operators

The *conditional operator* is also known as a *ternary operator* regarding input; it accepts three operands. The syntax of this operator is as follows:

```
(Expression 1) ?(Expression 2) :(Expression 3) ;
```

Here, expression 1 is evaluated first. If expression 1 is true, then expression 2 is evaluated and expression 2 will be the answer of this whole expression; otherwise, expression 3 is evaluated and expression 3 will be the answer of this whole expression. Conditional operators can be used to find the larger of two numbers:

```
Greatest = (x > y) ?x : y ;
```

Here, if `x > y` is true, then `x` is greater than `y`; that is, (greatest = `x`), else `y` is greater than `x`, that is, (greatest = `y`).

Bitwise Operators

Bitwise operators are special operators that are used to perform operations at the bit level. C supports various types of bitwise operators, which include the following:

Operator	Description
<<	Right shift
>>	Left shift
&	Bitwise AND
|	Bitwise OR
^	Bitwise XOR

Comma Operators

The *comma operator* is used to chain together some expressions. First, it evaluates the first expression and discards its value. It then evaluates the second expression, and the calculated result of this expression is returned as the final output. The expressions separated by a comma operator are evaluated in a sequence starting from left to right. It has the lowest precedence of all the operators in C. For example, the following statement is given:

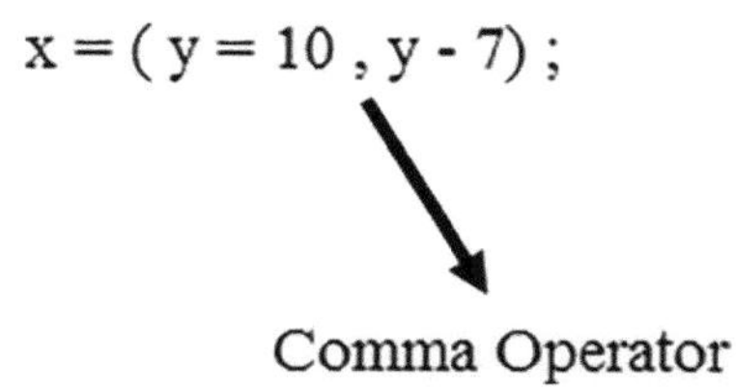

Hence, in the previous example, `y` has been assigned a value of `10`. Now, `y` is decremented by `7`. So, the value of the expression `(y - 7)` will be `3` and is assigned to `x`. Thus, the final value of `x = 3`.

Unary Operators

A *unary operator* is one that requires only a single operand to work. C supports two unary operators, which are the increment (++) and decrement (--) operators. These operators are used to increase or decrease the value of a variable by one, respectively. There are two variants of increment and decrement operators, which are postfix and prefix. In a postfix expression, the operator is applied after the operand. On the other hand, in a prefix expression, the operator is applied before the operand.

Operator	**Postfix**	**Prefix**
Increment (++)	x++	++x
Decrement (--)	--x	--x

Remember, x++ is not the same as ++x; in x++, the value is returned first, and then the value is incremented. In ++x, the value is returned after it is incremented. Similarly, `x--` is not the same as `--x`. It is true that both these operators increment or decrement the value by one. For example, `b = a ++` is equivalent to:

- `b = a`
- `a = a + 1`

Similarly, `b = -- a` is equivalent to:

- `a = a - 1`
- `b = a`

sizeof Operators

The `sizeof` *operator* is a unary operator that returns the size of an object, a variable, or a data type in bytes. It is used to determine the amount of memory storage a variable or a data type will take. It is written in the following manner: the keyword `sizeof` is followed by a variable/expression. A `sizeof` operator has the same precedence as that of unary operators. For example, if we have

```
int x = 100,
```

then the answer = `sizeof(x) = 2.`

Therefore, the final answer will be `2` as `x` is an integer, so it takes 2 bytes of storage space.

Frequently Asked Questions

Q5. What do you mean by the ternary operator?

Answer.

The conditional operator consists of two symbols, the question mark (`?`) and the colon (`:`). The syntax for the ternary operator is as follows:

```
exp1 ? exp2 : exp3
```

The ternary operator works as follows:

Expression 1 is evaluated first. If the expression is true, then expression 2 is evaluated and its value becomes the value of the expression. If expression 1 is false, expression 3 is evaluated and its value becomes the value of the expression. Note that only one of the expressions is evaluated.

DECISION CONTROL STATEMENTS IN C

Whenever we talk of a program written in the C language, we know that a C program will always execute sequentially, that is, line by line. Initially, the first line will be executed. Then, the second line will execute after the execution of the first line, and so on. Control statements are those that enable a programmer to execute a particular block of code and specify the order in which the various instructions of code are required to be executed. It determines the flow of control. Control statements define how the control is transferred to other parts of a program. Hence, they are also called *decision control statements*. A decision control statement is one that helps us jump from one point in a program to another. A decision control statement is executed in C using the following:

- `if` statement
- `if-else` statement
- Nested `if-else` statement
- Switch `case` statement

Now, let us discuss all of them.

if Statement

An `if` statement is a bidirectional control statement that is used to test a condition and take one of the possible actions. It is the simplest decision control statement and is used very frequently in decision-making. The general syntax of an `if` statement is as follows, and as shown in Figure 2.3:

```
if (condition)
 {
 Statement Block of if;   //If condition is true, execute the
statements of if.
 }
Statements Block under if;
```

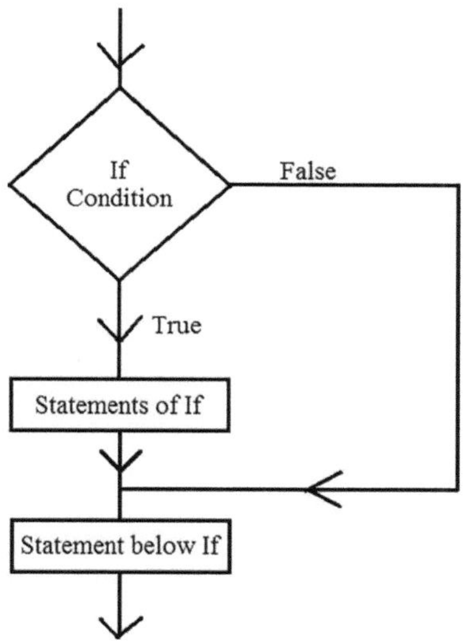

FIGURE 2.3 if statement flow diagram.

An `if` statement will check the condition, and if the condition is true, then the set of statements in the `if` block will be executed; otherwise, the set of statements below the `if` block will be executed. The `if` block can have one or multiple statements enclosed within curly brackets. The `else` block is optional in a simple `if` statement because if the condition is false, then the control directly jumps to the next statement. Remember, there is no semicolon after the condition because the condition and the statement should be put as a single statement.

For example:

```
#include <stdio.h>
void main()
{
 int x, y ;
 printf("\n Enter two values: ") ;
```

```
 scanf("%d %d", &x, &y) ;
 if (y > x)
 {
  printf("\n %d is greater, y") ;
 }
getch() ;
}
```

To get the output, enter two values:

```
23
67
67 is greater.
```

if-else Statement

After discussing the usage of the `if` statement, we learned that the `if` statement does nothing when the condition is false. It just passes the control to the next statement. The `if-else` statement takes care of this aspect. The general syntax of the `if-else` statement is as follows, and as shown in Figure 2.4:

```
if (condition)
 {
    Statements X;          //If condition is true, execute the
statements of If.
 }
else
{
     Statements Y;         //If condition is false, execute the
statements of else.
}
```

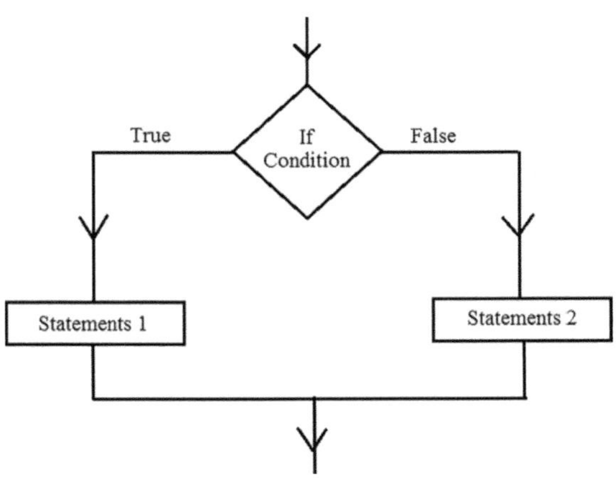

FIGURE 2.4 if-else statement flow diagram.

The `if-else` statement will check the condition. If the condition is true, then the set of `Statements X` is executed, and the `else` block is not executed. Otherwise, if the condition is false, then the set of `Statements Y` is executed and the `if` block is not executed. The `if` or `else` blocks can contain one or multiple statements.

For example:

```
#include <stdio.h>
void main()
{
 int x, y ;
 printf("\n Enter two values: ") ;
 scanf("%d %d", &x, &y) ;
 if (y < x)
 {
  printf("\n %d is smaller", y) ;
 }
 else
 {
  printf("%d is smaller", x) ;
 }
getch() ;
}
```

To get the output, enter two values:

```
100
500
100 is smaller.
```

Nested if-else Statement

The nested `if-else` statement is also known as the `if-else-if` ladder. The `if-else-if` statement works the same as a normal `if` statement. The general syntax of the `if-else-if` statement is as follows, and as shown in Figure 2.5:

```
if (condition 1)
{
    Statements 1;         //If condition 1 is true, execute the
statements of If.
 }
else if (condition 2)
{
    Statements 2;        //If condition 2 is true, execute the
statements of else if.
  }
else
 {
 Statements 3;        //If condition 2 is false, execute the
statements of else.
}
```

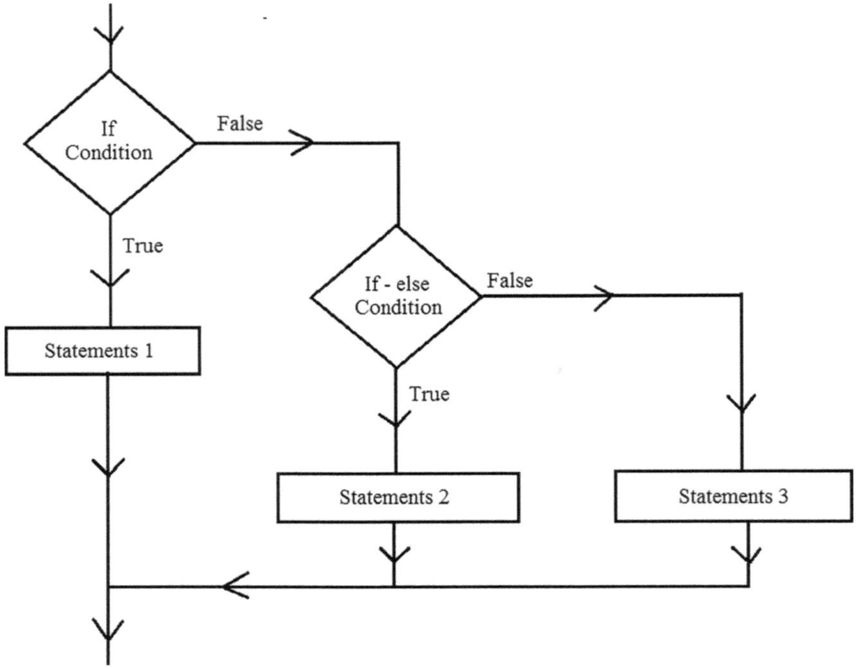

FIGURE 2.5 Nested if-else statement flow diagram.

An `if-else-if` ladder works in the following manner. First, the `if` condition is checked. If the condition is true, then the set of `Statements 1` is executed. If the condition is false, then the `else-if` condition is checked. If the `else-if` condition is true, then the set of `Statements 2` is executed. Otherwise, the set of `Statements 3` is executed. Remember, after the first `if` expression, we can have as many `else-if` branches as are needed, depending on the number of expressions to be tested.

For example:

```
#include <stdio.h>
void main()
{
 int x, y, z ;
 printf("\n Enter three values: ") ;
 scanf("%d %d %d", &x, &y, &z) ;
 if (x > y && x > z)
 {
   printf("\n %d is greater", x) ;
  }
  else if (y > x && y > z)
 {
   printf("%d is greater", y) ;
 }
  else
```

```
  {
    printf("%d is greater", z) ;
   }
getch() ;
}
```

To get the output, enter three values:

```
10
50
30
```

50 is greater.

switch Statement

As we all know, an `if` statement is used to check the given condition and choose one option depending on whether the condition is true or false. If we have several options to choose from, then it will not be a good thing to use `if` statements for each option, as it will become very complex. Hence, to avoid such a problem, a `switch` statement is used. The `switch` statement is a multidirectional conditional control statement. It is a simplified version of an `if-else-if` statement. It selects one option from the number of options available to us. Thus, it is also known as a *selector statement*. Its execution is faster than an `if-else-if` construct. A `switch` statement is comparatively easy to understand and debug. The general syntax of the `switch` statement is as follows, and as shown in Figure 2.6:

```
switch (choice)
{
  case constant 1:
  Statements 1 ;
  break ;

  case constant 2:
  Statements 2 ;
  break ;

  case constant 3:
  Statements 3 ;
  break ;
.
  .
  .
  case constant n:
  Statements n ;
  break ;

  default:
  Statements D ;
}
```

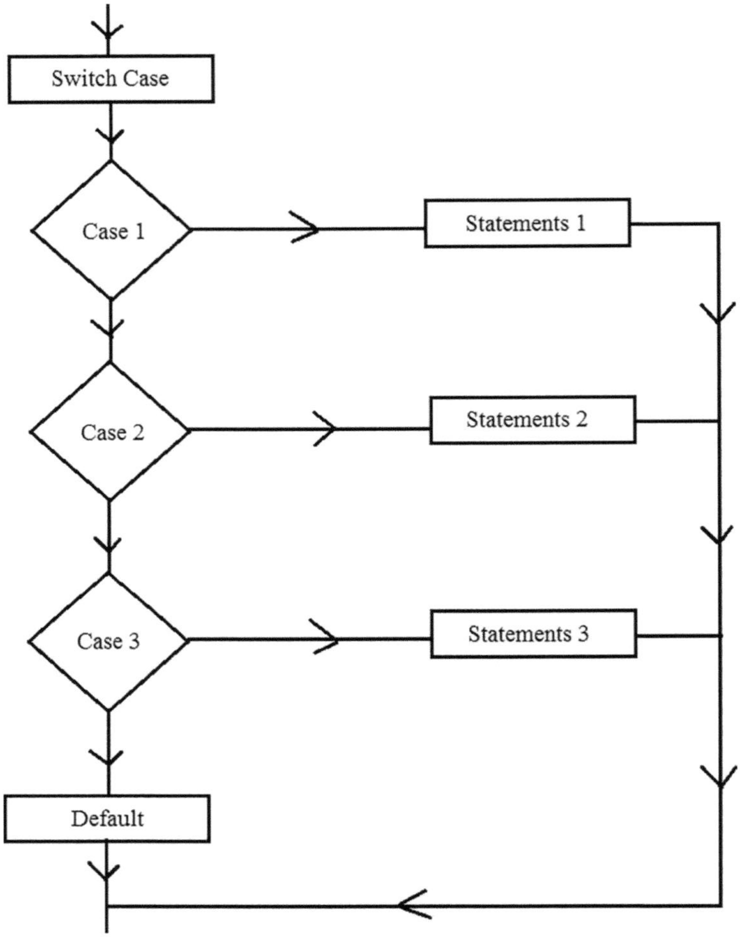

FIGURE 2.6 switch statement flow diagram.

A switch statement works as follows:

1. Initially, the value of the expression is compared with the `case` constants of the `switch` construct.
2. If the value of the expression and the `switch` statement match, then its corresponding block is executed until a break is encountered. Once a break is encountered, the control comes out of the `switch` statement.
3. If there is no match in the `switch` statements, then the set of `default` statements is executed.
4. All the values of the `case` constants must be unique.
5. There can be only one `default` statement in the entire `switch` statement. A `default` statement is optional; if it is not present and there is no match with any of the `case` constants, then no action takes place. The control simply jumps out of the `switch` statement.

For example:

```
# include<stdio.h>
void main()
{
  int choice ;
  printf("\n Enter your choice: ") ;
  scanf("%d", &ch) ;
  switch(ch)
  {
    case 1 :
    printf("First!!") ;
    break ;

    case 2 :
    printf("Second!!") ;
    break ;

    case 3 :
    printf("Third!!") ;
    break ;

    default :
    printf("wrong choice") ;
   }
getch() ;
}
```

To get the output, enter your choice: `2`.

The output is `Second`.

Frequently Asked Questions

Q6. Which one is better—a `switch` case or an `else-if` ladder?

Answer.

- `switch` permits the execution of more than one alternative, whereas an `if` statement does not. Various alternatives in an `if` statement are mutually exclusive, whereas alternatives may or may not be mutually exclusive within a `switch` statement.
- `switch` can only perform equality tests involving integer type or character type constants; an `if` statement, on the other hand, allows for more general comparisons involving other data types as well.

When there are more than three or four conditions, use the `switch` case rather than a long nested `if` statement.

LOOPING STATEMENTS IN C

Looping statements, also known as *iterative statements*, are a set of instructions that are executed repeatedly until a certain condition or expression becomes false. This kind of repetitive execution of the statements in a program is called a *loop*. Loops can be categorized into two categories: pre-deterministic loops and deterministic loops. *Pre-deterministic loops* are ones in which the number of times a loop will execute is known. On the contrary, loops in which it is not known how many times they will execute are called *deterministic loops*. C supports three types of loops, which include the following:

- `while` loop
- `do-while` loop
- `for` loop

Now, let us discuss all these loops in detail.

while Loop

A `while` loop is a loop that is used to repeat a set of one or more instructions/statements until a particular condition becomes false. In a `while` loop, the condition is checked before executing the body of the loop or any statements in the `statements` block. Hence, a `while` loop is also called an *entry control loop*. A `while` loop is a deterministic loop, as the number of times it will execute is known to us. The general syntax of a `while` loop is as follows, and as shown in Figure 2.7:

```
while (condition)
{
  block of statements/ body of loop ;
  increment/ decrement ;
 }
```

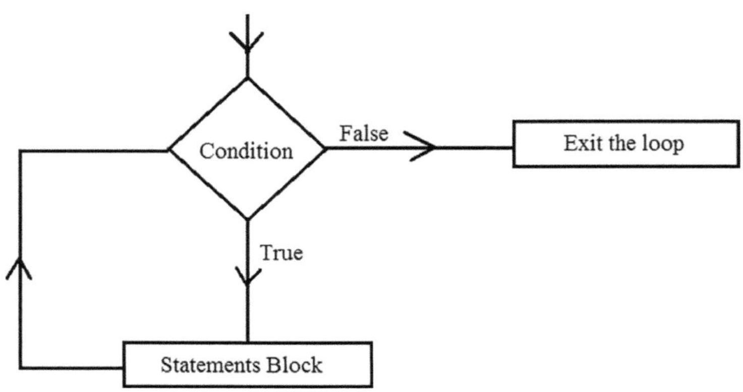

FIGURE 2.7 while loop flow diagram.

A `while` loop is executed as follows:

1. The condition is tested.
2. If the condition is true, then the statement is executed, and step 1 is repeated.
3. If the condition is false, then the loop is terminated, and the control jumps out to execute the rest of the program.

For example:

```
# include<stdio.h>
void main()
{
  inti = 1 ;
  while (i < 10)
 {
   printf("\t %d", i) ;
   i = i + 1 ;
  }
getch() ;
}
```

The output is:

```
1  2  3  4  5  6  7  8  9
```

In the above example, `i` is initialized to `1`, and 1 is less than 10, and therefore the condition is true. Hence, the value of `i` is printed and is incremented by 1. The condition will become false when `i` becomes `10`; thus, at that condition, the loop will end.

do-while Loop

A `do-while` loop is similar to a `while` loop. The only difference is that, unlike a `while` loop in which a condition is checked at the start of the loop, in a `do-while` loop, the condition is checked at the end of the loop. Hence, it is also called an *exit control loop*. This implies that in a `do-while` loop, the statements must be executed at least once, even if the condition is false, because the condition is checked at the end of the loop. The general syntax for a `do-while` loop is as follows, and as shown in Figure 2.8:

```
Do
{
  block of statements/ body of loop ;
  increment/ decrement ;
 } while (condition);
```

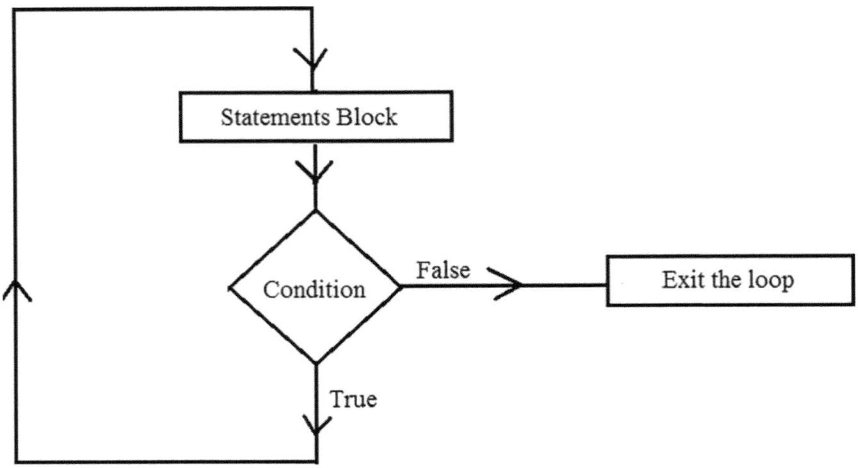

FIGURE 2.8 do-while loop flow diagram.

The `do-while` loop continues to execute until the condition evaluates to false. A `do-while` loop is usually employed in situations where we require the program to be executed at least once, for instance, in menu-driven programs. One of the major disadvantages of using a `do-while` loop is that a `do-while` loop will always execute at least once, even if the condition is false. Therefore, if the user enters some irrelevant data, it will still execute.

For example:

```
# include<stdio.h>
void main()
{
  int i = 0 ;
  do
 {
   printf("\t %d", i) ;
   i = i + 1 ;
   } while (i < 10) ;
getch() ;
}
```

This is the output:

```
0  1  2  3  4  5  6  7  8  9
```

In the above code, `i` is initialized to `0`, so the value of `i` is printed and is incremented by 1. After executing the loop once, the condition will be checked. Now, `i = 1` and the condition is true. Therefore, the loop will execute. The condition will become false when `i` becomes equal to `10`. In that case, the loop will be terminated.

for Loop

A `for` loop is a pre-deterministic loop; that is, it is a count-controlled loop such that the programmer knows in advance how many times the `for` loop is to be executed. In the `for` loop, the loop variable is always initialized exactly once. The general syntax for the `for` loop is as follows, and as shown in Figure 2.9:

```
for (initialization; condition; increment/decrement)
{
  block of statements/ body of loop ;
 }
```

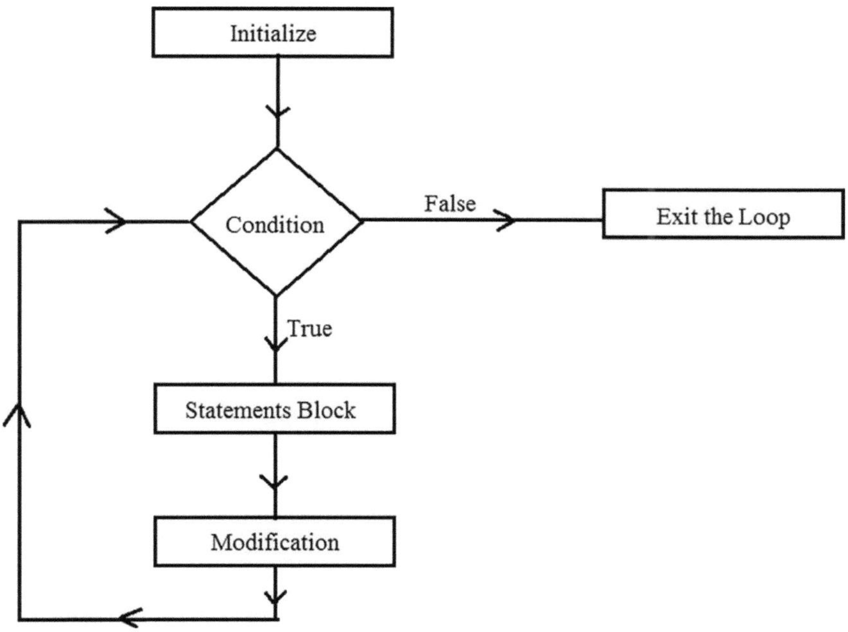

FIGURE 2.9 for loop flow diagram.

In a `for` loop, the condition is always checked at the top of the loop. With every iteration of the loop, the variable and the condition are checked. If the condition is true, then the statements written within the `for` loop are executed; otherwise, the control moves out of the loop and the `for` loop is terminated. As we have seen in the syntax, `initialization` means assigning a particular value to a variable initially. Second, the condition specifies whether the loop will continue to execute or terminate. The condition is checked with every iteration of the loop. *Iteration* means to update the value of a variable either by incrementing it or decrementing it. Also, each section in a `for` loop is separated by a semicolon. So, it is possible that one of the sections may be empty. `for` loops are widely used to execute a particular set of statements a limited number of times.

For example:

```
# include<stdio.h>
void main()
{
  int x  ;
  for (x = 1 ; x <= 10 ; x++)
  {
   printf("\t %d", x) ;
  }
getch() ;
}
```

This is the output:

```
1  2  3  4  5  6  7  8  9  10
```

In the above example, x is a counter variable that is initialized to 1. Now, the condition is checked because 1 is less than 10. Thus, the condition is true, so the value of x is printed. After every iteration, the value of x is incremented and the condition is checked. The condition will become false when i becomes 11, so at that time, the for loop will be terminated and the control will come out of the loop.

BREAK AND CONTINUE STATEMENTS

In C, break statements are used for loops and switch statements. They are used to terminate the execution of the loop. A break statement causes an intermediate exit from the loop in which the statement appears. We have already seen its use in switch statements, as it is used to exit from a switch statement. When a break is encountered, the control jumps out of the loop. The break statement is usually used in a situation in which either there is some error, or if we don't want to execute the rest of the loop. It has a very simple syntax:

```
break;
```

For example:

```
# include<stdio.h>
void main()
{
  int num = 0 ;
  while(num< 5)
  {
    if(num == 2)
    {
      printf("Hello!!") ;
      break ;
     }
```

```
    printf("\n Number = %d", num) ;
    num = num + 1 ;
  }
getch() ;
}
```

Here is the output:

```
Number = 0
Number = 1
Hello!!
```

In the previous code, when the value of `num` is equal to `2`, the `break` statement is executed, and the control jumps out of the `while` loop following the next statement after the `while` loop. Hence, the `break` statement is used to exit from a loop at any point.

As we can see, a `break` statement is used to exit a particular loop, and a `continue` statement is used to do the next iteration of the loop. `continue` statements are also used with loops. Unlike with a `break` statement, the loop does not terminate when a `continue` statement is encountered. A `continue` statement skips the rest of the statements, and the control is transferred to the loop continuation portion of the loop. Therefore, the execution of the loop resumes with the next iteration. The syntax of the continue statement is as follows:

```
continue ;
```

For example:

```
# include<stdio.h>
void main()
{
   int n ;
  for(n =0 ; n <= 8 ; n++)
  {
    if(n == 4)
    continue ;
    printf("\t %d", n) ;
  }
getch() ;
}
```

Here is the output:

```
0  1  2  3  5  6  7  8
```

In the previous code, as soon as the value of `n` becomes equal to `4`, the `continue` statement is executed and the `printf` statement is skipped. The control is transferred to the expression, which increments the value of `n`.

FUNCTIONS IN C

As C programmers, we often experience that the size of our program becomes too large, and its complexity also increases. At that time, it is very difficult for a programmer to read the entire code and also to check for any errors in it. Hence, to overcome this problem, the C language enables us to break the entire program into a smaller number of modules or segments. These modules or segments are called *functions*. A function is a predefined block of code designed to perform a particular task. Functions are used to improve the efficiency of the program. Functions can reduce redundancy and help to understand the code easily. Each function is designed to perform a particular task. Functions are separated into two categories:

- *Library functions* – Library functions are those functions that are predefined in C under various libraries. These are the ready-made functions available in C. These ready-made functions do not require any coding to execute any operation. These functions can be directly used by just including the related header files, for example, `scanf()`, `printf()`, `gets()`, `puts()`, `strcpy()`, and so on.
- *User-defined functions* – Unlike predefined library functions, these functions can be defined by the programmer or user. We can easily create these types of functions. The general form of a user-defined function is as follows:

```
[return type] <Function name> [parameters/ arguments]
              {
                Statements;
                return();
              }
```

In the preceding code, we can see the following:

- `return type` – Return types are used to identify which kind of value is going to be returned by the functions. Return types are the data types. If a function does not return any value, then the return type is void.
- `Function name` – This identifies the name of a function. The name of the function should not be reserved in the C libraries.
- `parameters/ arguments` – These are the variables or values passed with their data types to the functions for performing various operations.
- `Statements` – Statements are the particular steps that are performed by the functions.

There are three things associated with functions, which are as follows:

- *Declaring a function/function declaration* – A function must be declared before it is used. Declaring a function means the compiler must know in advance the number of parameters or arguments and the types of arguments that the function expects to receive, and also the data type that the function will return to the calling program. The general form of declaring a function is [return type] <Function name> [parameters/ arguments], which has already been discussed.

- *Calling a function/function call* – A function call is a call that transfers control to the called function to execute the set of statements in that particular block/function. After calling a function and executing it, the control again jumps back to the calling function. There are two types of functions: a *calling* function and a *called* function. A calling function is one that calls the function, and a called function is one that is called by the calling function. The general syntax for calling a particular function is <function name> (arguments/parameters). Arguments may be passed in the form of expressions to the called function.
- *Defining a function/function definition* – After a function is called, the function must be defined. Defining a function means space is allocated in the memory for that function. The number of arguments and their order must be the same as given in the function declaration. It comprises two parts, that is, the header of the function and the body of the function, where [return type] <Function name> [parameters/ arguments] is the function header and the set of statements is the function body.

Frequently Asked Questions

Q7. Write down the benefits of using functions in C.

Answer.

- *Modular programming* – It facilitates top-down modular programming. The high-level logic of the overall problem is solved first, while the details of each of the lower-level functions are addressed later.
- *Reduction of source code* - The length of the source program can be reduced by using functions at appropriate places.
- *Easier debugging* – It is easy to locate and isolate a faulty function for investigation.
- *Function sharing* – If a program is divided into smaller subprograms, each subprogram can be written by different team members rather than having the whole team work on a single complex program.

Structure of a Multi-Functional Program

The header of all user defined functions are declared before main() function. However function definitions are followed after main() and functions are called within the main() function as demonstrated through the following sample code.

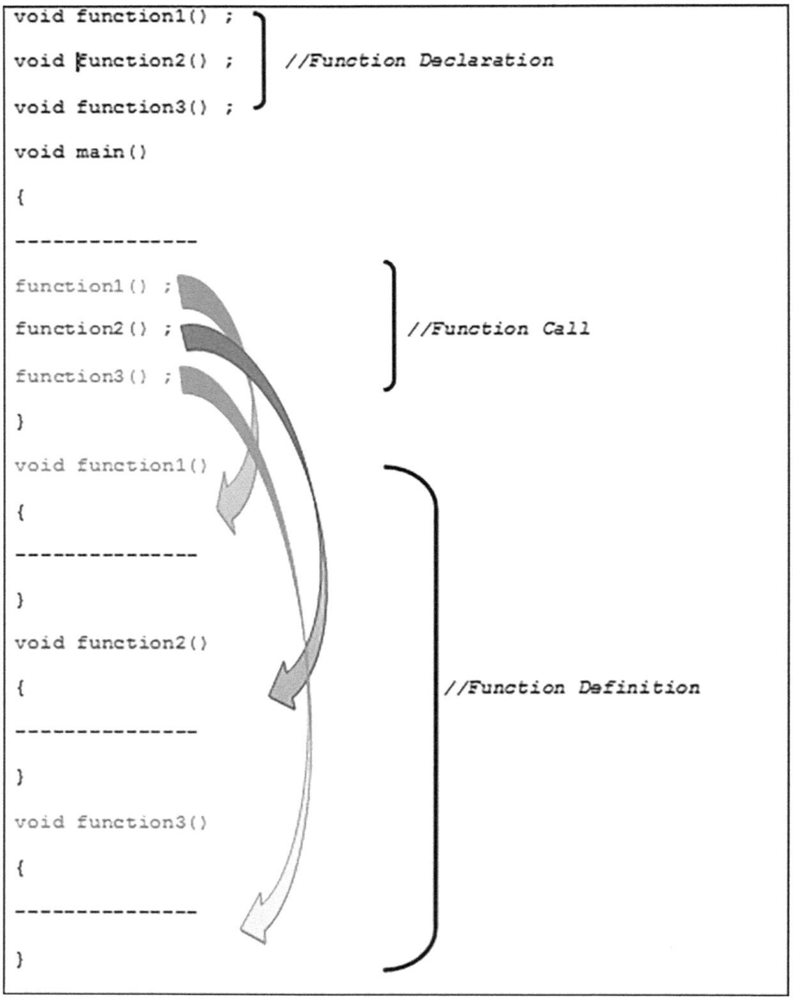

Passing Arguments to Functions

There are two ways to pass arguments or parameters to a function. These two ways are as follows:

- Call by value
- Call by reference

Let us now discuss both in detail.

Call by value is a method in which the values are passed from the calling function to the called function. In this method, the called function creates copies of the actual values of the calling function's argument into its formal parameters. So, in this case, if the called function is supposed to change/modify the actual values of the parameters, then the changes will only be reflected in the called function. These changes will not be reflected in the calling function.

This is because the changes that are made to the variables are not the actual variables, but copies of the actual variables. Hence, this is known as *call by value*. By default, C programming uses the call-by-value method to pass the parameters/arguments.

For example:

```
#include <stdio.h>
void swap(int, int) ;      //Declaration
void main()                        //Main function
{
   int a = 10, b = 20 ;
   swap(a, b) ;                   //Calling function
 }
void swap(int a, int b)
{
  int temp ;
  temp = a ;
  a = b ;
  b = temp ;
printf("\n After swapping a = %d and b = %d", a, b) ;
}
```

Here is the output:

After swapping, a = 20 and b = 10

The major drawback of the call-by-value technique is that a lot of memory space is consumed since a copy of the variables is created. Also, copying data consumes a lot of time in this technique. On the other hand, the biggest advantage of this technique is that any expressions or variables can be passed as arguments.

Call by reference is a method in which the addresses of the variables are passed from the calling function to the called function. In this method, function arguments are declared as references rather than normal variables. So, in this case, any changes made by the called function in the arguments will also be reflected in the calling function. An asterisk (*) is placed after the data type in the argument list to indicate that the parameters are passed by call by reference. In this method, no copies of the actual variables are created. Hence, the changes are also reflected in the calling function.

For example:

```
#include <stdio.h>
void swap(int *, int *) ;          //Declaration
void main()                         //Main function
{
  int a = 10, b = 20 ;
   swap(&a, &b) ;                   //Calling function
 }
void swap(int *a, int *b)
```

```
{
  int temp ;
  temp = *a ;
  *a = *b ;
  *b = temp ;
printf("\n After swapping a = %d and b = %d", a, b) ;
}
```

Here is the output:

After swapping, `a = 20` and `b = 10`

One of the biggest advantages of the call-by-reference technique is that it provides greater time as well as space complexity, as in this method, no copies of data are created. Also, the changes are reflected in the calling function as well. The biggest drawback of this technique is the use of pointers. The use of pointers should be done very carefully, as pointers can point anywhere in our systems. So, only those who are experienced in handling pointers should use this technique.

RECURSION

Recursion is a process performed by a function by calling itself a number of times to perform operations. A recursive function is a function that calls itself and executes the same instruction repeatedly. Recursive programs are fast and well optimized. Recursion is also an application of a stack. Hence, recursion is slower because it must maintain its stack properties.

Recursion makes code small and easy to understand.

For example:

```
#include<stdio.h>
#include<conio.h>
void main()
{
  int fact, n ;
  int factorial(int) ;
  clrscr() ;
  printf("Enter any number:") ;
  scanf("%d", &n) ;
  fact = factorial(n) ;
  printf("Factorial = %d",fact) ;
  getch() ;
}

int factorial(int x)
{
  int f ;
```

```
   if(x == 1 | | x == 0)
   return 1 ;
   else
   f = x * factorial(x - 1) ;
   return f ;
}
```

To get the output:

Enter any number: 6

```
factorial = 720
```

STRUCTURES IN C

A *structure* is a user-defined or custom data type that is used in storing related information, that is, data of different data types. A structure is like an array, but the main difference is that an array contains only information of the same data type. Therefore, a structure is a collection of one or more different variables or data types grouped under a single name. Each variable in a structure is known as a member variable. A structure is declared as follows:

```
Struct structure_name
{
   data type variable name 1;
  data type variable name 2;
   .
   .
   .
Block of statements
};
```

A structure is always declared using the keyword `struct` followed by the name of the structure. `structure_name` is the name of a user-defined data type and will further be used to identify the structure and declare variables/objects of the `struct` type.

For example:

```
struct Student
{
char name[25];
int age ;
float height ;
};
```

In the previous example, the struct `Student` declares a structure of `Student` having three data fields, that is, name, roll number, and class. There are three elements in the previous structure. Hence, all three data types are accessed separately and are not mixed. Now, let us understand how structures are used through a program:

```
#include <stdio.h>
#include <string.h>
struct student
{
int id;
char name[20];
float percentage;
};

int main()
{
struct student record = {0};//Initializing to null
record.id=1;//Accessing member variable
strcpy(record.name, "Ram,");
record.percentage = 90.5;
printf(" Id - %d \n", record.id);
printf(" Name - %s \n", record.name);
printf(" Percentage - %f \n", record.percentage);
return 0;
}
```

Here is the output:

```
Id – 1

Name – Ram

Percentage – 90.500000
```

POINTERS

A *pointer* is a special type of variable that is used to store the address of another variable rather than some simple value. Pointers can be used to access the data stored in memory. Pointers are frequently used with arrays because pointers are more efficient in handling arrays and data tables. A pointer is a variable used to store the address of another variable and is used to perform various operations. Pointers give power as well as flexibility to C programmers. Pointers can directly access memory locations and can easily manipulate addresses.

Practical Application:

An example of a pointer is the address of the home of a human being, which can be used to easily reach the destination.

Here is the declaration of a pointer: `Data type *variable name;`
For example:

```
int *ptr ;
int i = 3 ;
ptr = &i ;
```

Therefore, `ptr` will store the address of the memory location where `i` is stored.

Arrays and Pointers

The concept of arrays is very much bound to the concept of pointers. Now, let us take an example; if we have an array of 10 elements declared as

```
int array [10] = {10,20,30,40,50,60,70,80,90,100} ;
```

The above array will be stored in memory, as shown in Figure 2.10: `102030405060708090100`
`100     102104106108110112114116120`

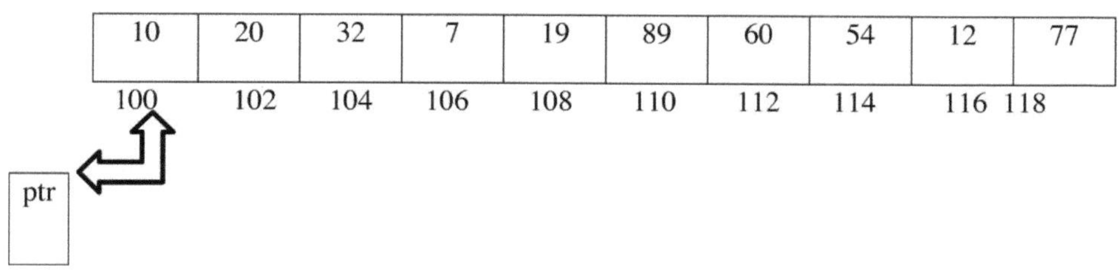

FIGURE 2.10 Memory representation of arr.

Here, the starting address of the array i.e. `100` is the base address of the array. The base address is the address of the first element in the array. Now, we will use a pointer variable to store the address of the first element, and `ptr` will point to the first element of the array.

```
int * ptr;

ptr = &array[0] ;
```

Similarly, we can store the addresses of other elements. If the pointer variable holds the address of the first element in the array, then the address of the successive elements can be calculated by `ptr++`:

```
int * ptr ;
ptr = &array[0] ;
ptr++ ;
printf("The value of second element is %d", *ptr) ;
```

A pointer variable can be prefixed or postfixed with increment or decrement operators. Increment or decrement will depend upon the data type, for example, `int` – 2 bytes, `float` – 4 bytes, `char` – 1 byte, and so on.

Here, the data type identifies the type of data that the pointer points to. An asterisk (`*`) identifies that the variable is going to store and point toward some address. The variable name gives a unique name to the pointer.

Uses of pointers in C include the following:

- Helps in dynamic memory allocation
- Makes use of arrays and strings more efficiently and optimizes the code
- Enhances the speed of the execution of code
- Passes the pointers to functions as arguments
- Helps in returning multiple values from a function

Frequently Asked Questions

Q8. State true or false for the following with reasons:

(i) A pointer is a variable that stores a value.

(ii) * is an address operator.

Answer.

(i) False; a pointer is a variable that is used to store the address of another variable.

(ii) False; * is a dereferencing operator.

Drawbacks of Using Pointers

We all know that pointers are very useful in C, but they are also the most dangerous feature of C. If pointers are not used correctly, they can lead to various problems that are difficult to handle. Some of the drawbacks of pointers are as follows:

- The improper use of pointers can cause the system to crash.
- If a pointer is used to read a memory location and it points to the wrong memory location, then it will be very difficult for the program, as it will produce unexpected results.
- If pointers are not used properly, they can point anywhere in the operating system, which can also cause data corruption.
- Pointers should always be initialized; otherwise, they may contain garbage values, which will cause various problems.
- The debugging process in the case of pointers is a very difficult task.

SUMMARY

- C is a high-level programming language that was developed by Denis Ritchie, a scientist at Bell Labs, in the early 1970s.
- C is a platform that also supports various other languages, such as Java and C++. The C language is often called a middle-level language, as it not only provides different data types or data structures that are needed by a programmer, but it can also access the computer hardware with the help of specially designed functions and declarations.

- During programming, if we want those files or functions, first, we have to include those functions or files in our program to use them. Thus, these files will be loaded from the disk to the memory, where these files are called to perform the desired operations. Hence, these files are known as header files.
- The `main()` function is a function that is a part of every C program. A program cannot execute without the `main()` function.
- The most common operations in a C program are to accept input values from a standard input device and to display the data produced by the program on a standard output device.
- Keywords in C are the reserved words that have a special meaning. They are written in lowercase. Keywords cannot be used as an identifier. An identifier is a name that is given to a constant, variable, function, or array.
- Data types in C are the special keywords that define the type of data and the amount of data a variable holds.
- The four basic data types used in C are `int`, `char`, `float`, and `double`.
- Operators in C are used to perform some specific operations between different variables and constants.
- Arithmetic operators are those operators that are used in mathematical calculations.
- Assignment operators are used for assigning values to the variables. These operators are always evaluated from right to left.
- Relational operators are used for comparison between two values or expressions. They are also known as comparison operators.
- The conditional operator is also known as a ternary operator, because it takes three operands.
- Bitwise operators are special operators that are used to perform operations at the bit level.
- The comma operator is used to chain together some expressions.
- The `sizeof` operator is a unary operator that returns the size of a variable or a data type in bytes.
- A unary operator is one that requires only a single operand to work. C supports two unary operators, which are the increment (++) and decrement (--) operators. These operators are used to increase or decrease the value of a variable by one, respectively.
- Control statements are those that enable a programmer to execute a particular block of code, specifying the order in which the various instructions in a program are required to be executed. It determines the flow of control.
- The `if` statement is a bidirectional control statement that is used to test the condition and take one of the possible actions.
- Nested `if-else` statements are also known as `if-else-if` ladders.
- A `switch` statement is a multidirectional conditional control statement. It is a simplified version of an `if-else-if` statement.
- Looping statements, also known as iterative statements, are a set of instructions that are repeatedly executed until a certain condition or expression becomes false.
- A `while` loop is a loop that is used to repeat a set of one or more instructions/statements until a particular condition becomes false.
- A `do-while` loop is similar to a `while` loop. The only difference is that, unlike a `while` loop in which a condition is checked at the start of the loop, in a `do-while` loop, the condition is checked at the end of the loop.

- A `for` loop is a pre-deterministic loop; that is, it is a count-controlled loop such that the program knows in advance how many times the loop is to be executed.
- A `break` statement causes an intermediate exit from that loop in which the statement appears. A `continue` statement skips the rest of the statements, and the control is transferred to the start of the loop.
- The C language enables us to break the entire program into a smaller number of modules or segments. These modules or segments are called functions. A function is a predefined block of code designed to perform a particular task.
- Call by value is a method in which the values are passed from the calling function to the called function. In this method, the called function creates copies of the actual values of the calling function's argument into its formal parameters.
- Call by reference is a method in which the addresses of the variables are passed from the calling function to the called function. In this method, function arguments are declared as references rather than normal variables.
- A structure is a collection of one or more different variables or data types grouped under a single name.
- A pointer is a special type of variable that is used to store the address of another variable.
- An asterisk (`*`) identifies that the variable is going to store and point toward some address. A variable name gives a unique name to the pointer.

EXERCISES

Theory Questions

1. What are the different characteristics of the C language that make it a very popular language?
2. What is the `main()` function?
3. What are the purposes of the `printf()` and `scanf()` functions?
4. What are the different operators used in C? Discuss all of them in detail.
5. What do we mean by "header files"? Why are header files included in the programs?
6. What are the data types in C? Explain in detail.
7. What do you understand by "C tokens"? Discuss in detail.
8. Define the term "identifiers." What are the various rules for identifying an identifier? Give examples.
9. What do you understand by the conditional operator? Explain with the help of an example.
10. What are the decisional control statements in C?
11. Differentiate between `while` and `do-while` loops. Give examples.
12. What is the difference between simple `if` and `if-else` statements? Explain with the help of an example.
13. Write the syntax of a `for` loop. Can we skip any part in the `for` loop, or not?
14. What do we mean by structure? Give an example.

15. Explain a `switch` case. What are the various advantages of using a `switch` case?
16. Define a function. Why are functions needed?
17. What is the difference between a function definition and a function declaration?
18. Discuss the structure of a program having multiple functions.
19. Differentiate between call by value and call by reference. Give suitable examples.
20. What do we mean by pointers? How are they initialized?
21. What are the various drawbacks of using pointers?
22. What do you understand by iterative statements in C? Briefly discuss all the types.
23. How is a user-defined function different from a predefined function?
24. Explain the difference between `break` and `continue` statements with suitable examples.
25. Write a short note on arrays and pointers.
26. Differentiate between pre-increment and post-increment operators.

Programming Questions

1. Write a program to print your name on the screen.
2. Write a program that reads five integer values and displays them.
3. Write a C program to add two floating-point numbers. Accept the numbers from the user.
4. Write a program to calculate simple interest.
5. Write a program to check whether the given number is even or odd.
6. Write a program to find the largest of three given numbers.
7. Write a menu-driven program performing addition, subtraction, multiplication, and division of two numbers using functions.
8. Write a C program to print the following pattern using a `for` loop:

a)
```
*
**     ***
****
           *****
```

b)
```
a
b c
d e f
g h i j
```

9. Write a program that takes an integer value from the user and prints its corresponding ASCII equivalent.
10. Write a program to find the reverse of a given number using a function.
11. Write a program to check whether a number is divisible by 2 or not using a function.
12. Write a program to add two numbers using pointers.
13. Write a program where both `if` and `else` statements are executed in a program.
14. Write a program to perform a call by value and a call by reference using functions.
15. Write a program to find the factorial of a number using a `for` loop.

16. Write a program to accept a string from the user and to check whether the string is a palindrome or not using a user-defined function.
17. Write a C program to print the numbers from 1 to 10, excluding 5, using a `continue` statement.
18. Write a program to swap two numbers without using the third variable.

MULTIPLE CHOICE QUESTIONS

1. A conditional operator is also called a ternary operator as it has ________ operands.
   - A. 1
   - B. 3
   - C. 2
   - D. 4
2. Which of the following operators is used to declare a pointer?
   - A. `>`
   - B. `<`
   - C. `*`
   - D. `&`
3. Which of the following is a valid identifier in C?
   - A. `a_43`
   - B. `cd bd`
   - C. `apple`
   - D. both (a) and (c)
4. Which operator is used for mathematical computation?
   - A. Assignment operator
   - B. Arithmetic operator
   - C. Bitwise operator
   - D. Relational operator
5. Which of the control strings is associated with floating values?
   - A. `%f`
   - B. `%g`
   - C. `%e`
   - D. All of the above
6. A function declaration identifies a function with its ________.
   - A. Arguments
   - B. Data type of the return value
   - C. Name
   - D. All of the above

7. Which operator is used for comparison between values?
   A. Logical operator
   B. Relational operator
   C. Assignment operator
   D. Unary operator
8. In which of the following loops will a block of statements be executed at least once without checking the condition?
   A. A `for` loop
   B. A `while` loop
   C. A `do-while` loop
   D. All of the above
9. A data structure that is used to store related information together is called a(n)________.
   A. Structure
   B. Array
   C. Linked list
   D. String
10. What is `*(&variable)` equal to?
    A. `&variable`
    B. `*variable`
    C. `&(*variable)`
    D. `Variable`

CHAPTER 3

ARRAYS

INTRODUCTION

We already studied the basics of programming in data structures and C in the previous chapter, in which we aimed to design good programs, where a good program refers to a program that runs correctly and efficiently by occupying less space in memory, and also takes less time to run and execute. Undoubtedly, a program is said to be efficient when it executes with less memory space and also in minimal time. In this chapter, we will learn about the concept of arrays. An array is a user-defined data type that stores related information together. Arrays are discussed in detail in the following sections.

DEFINITION OF AN ARRAY

An *array* is a collection of homogeneous (similar) types of data elements in contiguous memory. An array is a linear data structure because all elements of the array are stored in linear order. Let us take an example in which we have ten students in a class, and we have been asked to store the marks of all ten students, as shown in Figure 3.1; then, we need a data structure known as an array.

36	98	14	74	56	13	7	96	44	82
marks1	marks2	marks3	marks4	marks5	marks6	marks7	marks8	marks9	marks10

FIGURE 3.1 Representation of an array of ten elements.

In the previous example, the data elements are stored in the successive memory locations and are identified by an index number (also known as the subscript), that is, A_i or $A[i]$. A *subscript* is an ordinal number that is used to identify an element of the array. The elements of an array have the same data type, and each element in an array can be accessed using the same name.

Frequently Asked Questions

Q1. What is an array? How can we identify an element in the array?

Answer.
An array is a collection of homogeneous (similar) types of data elements in contiguous memory. An element in an array can be identified by its index number, which is also known as a subscript.

ARRAY DECLARATION

We know that all variables must be declared before they are used in the program. The same concept also holds with array variables. An array must be declared before it is used. During the declaration of an array, the size of the array has to be specified. Declaring an array involves the following specifications:

- *Data type* – The data type means the different kinds of values it can store. The data type can be an integer (`int`), `float`, `char`, or any other valid data type.
- *Array name* – The name refers to the name of the array, which will be used to identify the array.
- *Size* – The size of an array refers to the maximum number of values an array can hold.

This is the syntax: `data_typearray_name [size] ;`

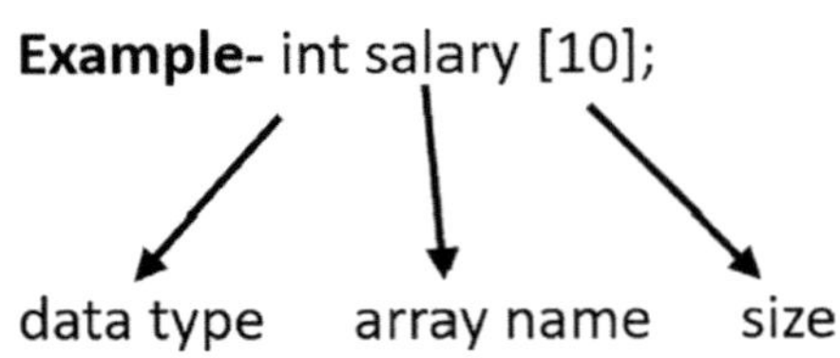

The previous example declares `salary` to be an array that has ten elements. In C, the array index starts from zero. The first element of this array will be stored in `salary[0]`, the second element will be stored in `salary[1]`, and so on. Similarly, the last element will be stored in `salary[9]`. In memory, the array will be shown as in Figure 3.2.

2000	4500	7890	9876	10000	3458	8000	9810	14000	5000
salary[0]	salary[1]	salary[2]	salary[3]	salary[4]	salary[5]	salary[6]	salary[7]	salary[8]	salary[9]

FIGURE 3.2 Memory representation of an array.

Here, `[0]`, `[1]`, `[2]`, ..., `[9]` written in square brackets represent the subscripts that we use to identify a particular element in the array.

ARRAY INITIALIZATION

The initialization of arrays can be done in the following ways:

1. *Initialization at compile time* – Initialization of elements of the array at compile time refers to the same way we initialize the normal or ordinary variables at the time of their declaration. When an array is initialized, there is a need to provide a specific value for every element in the array.
   The general form of initializing arrays is as follows:

```
data_type array_name[size] = { list of values } ;
```

An example of initialization of arrays at compile time is as follows:

```
int age[5] = { 20, 25, 23, 28, 30 } ;
```

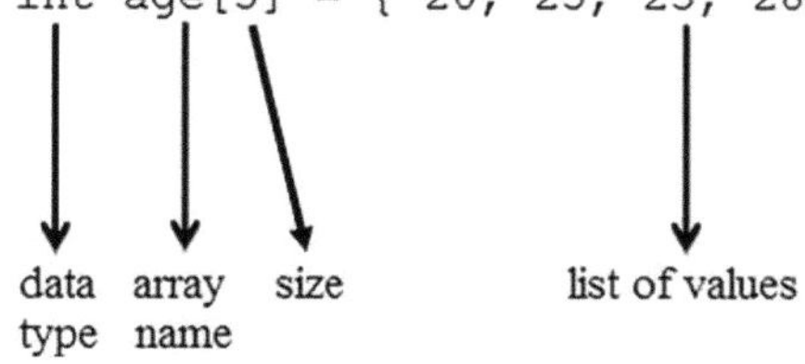

During the initialization of arrays, we may omit the size of the array. For example:

```
int age[ ] = { 25 , 28, 34 } ;
```

In the previous example, the compiler will automatically allocate memory for all the initialized elements of the array. If the number of values is less than the size provided, then such elements will take zeroes as their assigned values. For example:

```
int marks[10] = { 56, 69, 40, 99, 82, 96, 72 } ;
```

Here, the size of the array is `10`, but there are only seven elements; hence, the remaining elements will be considered to be zeroes.

56	69	40	99	82	96	72	0	0	0
marks[0]	marks[1]	marks[2]	marks[3]	marks[4]	marks[5]	marks[6]	marks[7]	marks[8]	marks[9]

FIGURE 3.3 Initialization of array marks[10].

2. *Initialization at runtime* – Initialization of elements of the array at runtime refers to the method of inputting the values from the keyboard. In this method, a `while`, `do-while`, or `for` loop is used to input the values of the array.

```
#include<stdio.h>
void main()
{
    int  i, salary[15] ;
    for( i=0 ; i<15 ; i++ )
    {
          scanf("%d", &salary[i]) ;

    }//End of for loop

}//End of main
```

FIGURE 3.4 Code for inputting the values.

In the previous code, the index `i` is at `0`, and the values will be input for the index values from `0` to `14`, as the array has 15 elements.

CALCULATING THE ADDRESS OF ARRAY ELEMENTS

The address of the elements in the 1D array can be calculated very easily because the array stores all its data elements in contiguous memory locations, storing the base address (address of the first element of the array). Hence, the address of the other data elements can easily be calculated using the base address. The formula to find the address of elements in a 1D array is as follows:

$$\textit{Address of data element, } A[i] = \textit{Base Address } (BA) + w\ (i - \textit{lower bound})$$

where A is the array, i is the index of the element for which the address is to be calculated, BA is the base address of the array A, and w is the size of each element (e.g., the size of `int` is 2 bytes, the size of `char` is 1 byte, etc.).

Frequently Asked Questions

Q2. An array is given `int marks[6] = {34, 53, 87, 100, 98, 65};`. Calculate the address of `marks[3]` if the base address is `3000`.

Answer.

It is given that the base address of the array is `3000`, and we know that the size of an integer is 2 bytes. Hence, we can easily find the address of `marks[3]`.

3000	3002	3004	3006	3008	3010
34	53	87	100	98	65
marks[1]	marks[2]	marks[3]	marks[4]	marks[5]	marks[6]

By putting into the formula:

Address of `marks[3]` $= 3000 + 2(3-1)$

$= 3000 + 2(2)$

Address of `marks[3]` = `3004`

OPERATIONS ON ARRAYS

This section discusses various operations that can be performed on arrays. These operations include:

- Traversing an array
- Inserting an element in an array
- Deleting an element from an array
- Searching for an element in an array
- Merging of two arrays
- Sorting an array

Traversing an Array

Traversing an array means accessing every element in an array exactly once so that it can be processed. Examples are printing all the data elements, performing any process on these elements, and so on. Traversing the elements of the array is a very simple process because of the linear structure of the array (all the elements are stored in contiguous memory locations).

Practical Application:

There is a line of people standing one after the other, and one boy is distributing advertisement pamphlets one by one to each person standing in the line.

For example:

```
#include<stdio.h>
#include<conio.h>
void main()
{
int  i, num[5] ;
    for( i=0 ; i<15 ; i++ )
     {
printf("\n arr[%d] = ", i);
scanf("%d", &num[i]) ;
```

```
 } //End of for loop 1
printf(" The elements of the array are ") ;
     for( i=0 ; i<5 ; i++ )
      {
printf("\t%d", arr[i]) ;

        } //End of for loop 2
} //End of main
```

Here is the output:

```
arr[0] = 5
arr[1] = 8
arr[2] = 16
arr[3] = 1
arr[4] = 7
The elements of the array are 5    8    16    1    7
```

In the previous code, the traversal of the elements of the array is shown. In the first `for` loop, all the elements are inputted into the array. Second, all the elements are traversed and printed in the second `for` loop. Hence, the traversing of an array is done.

Inserting an Element in an Array

Inserting an element in an array refers to the operation of adding an element to the array. In the case of insertion, we assume that there is enough memory space still available in the array. For example, if we have an array that can hold 20 elements and the array contains only 15 elements, then we have space to accommodate five more elements. If the array can hold 15 elements, however, then we will not be able to insert other elements into the array. Insertion in arrays can be done in three ways:

- Insertion at the beginning
- Insertion at a specified position
- Insertion at the end

Now, let us discuss all of these cases in detail:

- *Insertion at the beginning* – In this case, the new element to be inserted is inserted at the beginning of the array. To insert an element at the beginning, all the elements stored in the array must move one place forward to vacate the first position in the array. For example, if an array is declared to hold ten elements and it contains only seven elements, and also if it is given that the new element is to be inserted at the beginning of the array, then all the stored elements must move one place ahead, which is shown as follows:

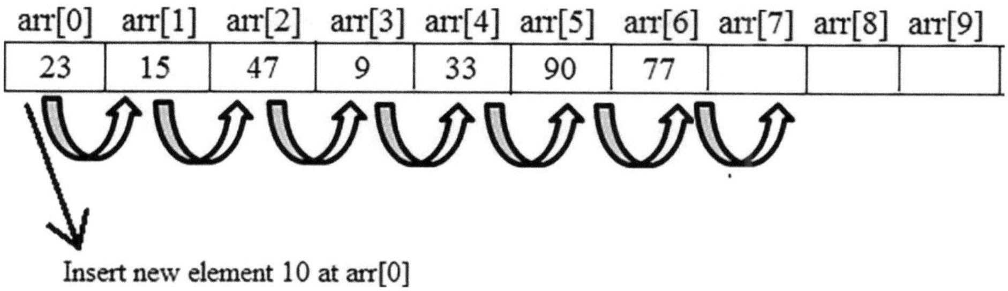

After swapping all the elements and inserting the new element `10` into the array, the new array will be:

arr[0]	arr[1]	arr[2]	arr[3]	arr[4]	arr[5]	arr[6]	arr[7]	arr[8]	arr[9]
10	23	15	47	9	33	90	77		

Let's look at the algorithm for insertion at the beginning.

We assume `ARR` is an array with `N` elements in it. The maximum elements that can be stored in the array is defined by `SIZE`. We should first check if the array has an empty space available to store any element in it or not, and then we proceed with the insertion process:

```
Step 1:START
Step 2:IF N = SIZE,
PRINT OVERFLOW
ELSE
N = N + 1
Step 3:SET I = N
Step 4:Repeat Step 5 while I>=0
Step 5:SET ARR[I+1] = ARR[I]
[ END OF LOOP ]
Step 6:SET ARR[0] = New_Element
Step 7:EXIT
```

- *Insertion at a specified position* – In this case, the new element to be inserted is inserted at the specified location/position, which is entered by the user. In order to insert a new element in the array, the previously stored elements in the array must move one place forward from their current place until the element at the specified position is reached. For example, if an array is declared to hold ten elements and it contains only eight elements, and it is also given that the new element is to be inserted at the fifth position of the array, then the stored elements must move one place ahead, as shown in the following:

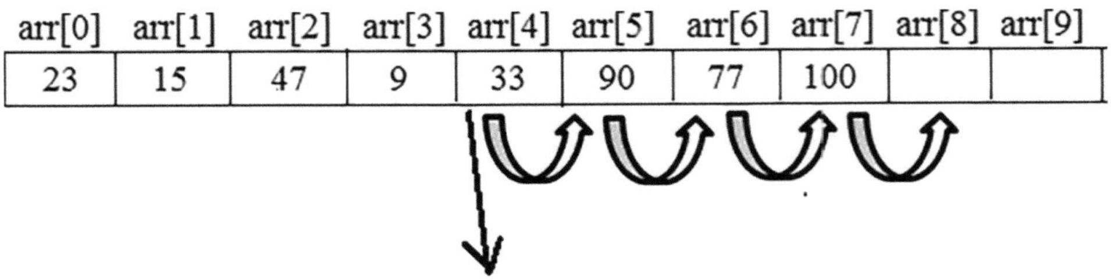

Insert new element 30 at arr[4]

After swapping the elements and inserting a new element `10` into the middle of the array, the new array will be:

arr[0]	arr[1]	arr[2]	arr[3]	arr[4]	arr[5]	arr[6]	arr[7]	arr[8]	arr[9]
23	15	47	9	30	33	90	77	100	

Practical Application:

It is just like if there are people standing in a line and one person just joins the line from the middle, so now every person has to shift one place backward from the middle so that the person can come into the line; hence, it is insertion at the middle.

Let's look at the algorithm for insertion at a specified position.

We assume `ARR` is an array with `N` elements in it. The maximum elements that can be stored in the array is defined by `SIZE`. Let `POS` define the position at which the new element is to be inserted. We should first check if the array has an empty space available to store any element in it or not, and then we proceed with the insertion process:

```
Step 1:START
Step 2: IF N = SIZE,
PRINT OVERFLOW
ELSE
N = N + 1
Step 3:SET I = N
Step 4:Repeat Step 5 while I>=POS
Step 5:SET ARR[I+1] = ARR[I]
[ END OF LOOP ]
Step 6:SET ARR[POS] = New_Element
Step 7:EXIT
```

- *Insertion at the end* – In this case, the new element to be inserted is inserted at the end of the array. So, there is no need for swapping the elements in this case. We are just required to check whether there is enough space available in the array or not. For example, if an array is declared to hold ten elements and it contains only nine elements, then the insertion can take place.

arr[0]	arr[1]	arr[2]	arr[3]	arr[4]	arr[5]	arr[6]	arr[7]	arr[8]	arr[9]
23	15	47	9	30	33	90	77	100	

Insert new element 11 at arr[9]

Now the last element will be inserted at the last position, which is at `arr[9]` and is vacant. Therefore, the new array after insertion will be:

arr[0]	arr[1]	arr[2]	arr[3]	arr[4]	arr[5]	arr[6]	arr[7]	arr[8]	arr[9]
23	15	47	9	30	33	90	77	100	11

Practical Application:

It is just like a normal line where a person comes and joins the line at the end; hence, there is no need for any shifting in this process.

Let's look at the algorithm for insertion at the end.

We assume `ARR` is an array with `N` elements in it. The maximum elements that can be stored in the array is defined by `SIZE`. We should first check if the array has an empty space available to store any element in it or not, and then we proceed with the insertion process:

```
Step 1:START
Step 2:IF N = SIZE,
PRINT OVERFLOW
ELSE
N = N + 1
Step 3:SET ARR[N] = New_Element
Step 4:EXIT
```

Write a menu-driven program to implement insertion in a 1D array, discussing all three cases.

```
#include<stdio.h>
#include<conio.h>
void main()
{
int  i, n, pos, choice, value, arr[10];
clrscr();
printf("Enter the number of elements in array: ") ;
scanf("%d", &n) ;
printf("Enter the elements of array ") ;
 //Accepting the elements of array
for( i=0 ; i<n ; i++ )
     {
scanf("%d", &arr[i]) ;
          }
printf("***MENU***") ;
printf("1. Insertion in beginning ") ;
printf("2. Insertion at specified location ") ;
printf("3. Insertion at end ") ;
printf("Enter your choice: ") ;
scanf("%d", &choice) ;
if(n==10)
{
printf("Overflow error ") ;
exit(0);
 }
else
switch(choice)
{
case 1:
for( i=n-1 ; i>=0 ; i-- )
{
arr[i+1] = arr[i] ;
}
printf("Enter new value: ") ;
scanf("%d", &value) ;
arr[0] =value;
printf(" After insertion array is ") ;
     for( i=0 ; i<=n ; i++ )
      {
printf("\t%d", arr[i]) ;

         }
   break ;
   case 2:
printf("Enter position ") ;
scanf("%d", &pos) ;
   for( i=n-1 ; i>=pos-1 ; i-- )
```

```
    {
arr[i+1] = arr[i] ;
       }
printf("Enter new value: ") ;
scanf("%d", &value) ;
arr[p-1] =value;
printf(" After insertion array is ") ;
     for( i=0 ; i<=n ; i++ )
      {
printf("\t%d", arr[i]) ;

        }
   break ;
   case 3:
printf("Enter new value: ") ;
scanf("%d", &value) ;
arr[n] =value;
printf(" After insertion array is ") ;
     for( i=0 ; i<=n ; i++ )
      {
printf("\t%d", arr[i]) ;

        }
  break ;
  default :
printf("Wrong Choice") ;
  exit(0) ;
  } //End of switch case
getch() ;
} //End of main
```

This is the output:

```
Enter the number of elements
n = 6
arr[0] = 75
arr[1] = 84
arr[2] = 16
arr[3] = 11
arr[4] = 47
arr[5] = 90
****MENU****
Insertion in beginning
Insertion at specified position
Insertion at end
Enter your choice 1
Enter new value 38
arr[0] = 38
After insertion new array is
38    75    84    16    11 47    90
```

```
Enter your choice 2
Enter position 3
Enter new value 65
arr[2] = 65
After insertion new array is
38      75      65      84      16      11      47      90

Enter your choice 3
Enter new value 100
arr[8] = 100
38      75      65      84      16      11      47      90      100
```

Deleting an Element from an Array

Deleting an element from an array refers to the operation of the removal of an element from an array. Deletion in an array can be done in three ways:

- Deletion from the beginning
- Deletion from a specified position
- Deletion from the end

Now, let us discuss all of these cases in detail:

- *Deletion from the beginning* – In this case, the element to be deleted is deleted from the beginning of the array. In order to delete an element from the beginning, all the elements stored in the array must move one place backward in the array. For example, if an array is declared to hold ten elements and it contains only seven elements, and it is also given that the element is to be deleted from the beginning of the array, then all the stored elements must move one place back, as shown in the following:

arr[0]	arr[1]	arr[2]	arr[3]	arr[4]	arr[5]	arr[6]	arr[7]	arr[8]	arr[9]
23	15	47	9	33	90	77			

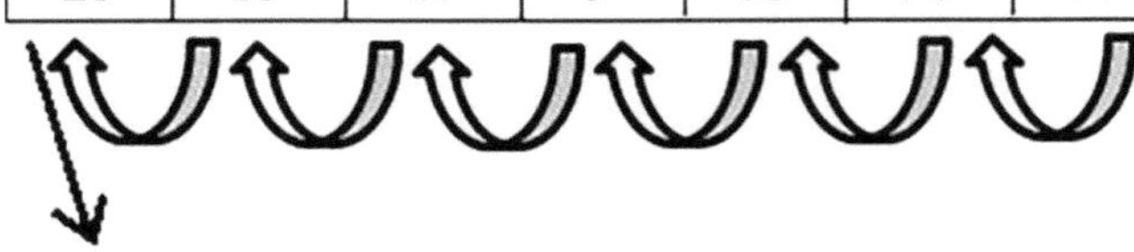

Delete 23 from arr[0]

In order to delete the first element, 23, from the array, we must swap all the stored elements backward so that the first element gets deleted, as shown. After the deletion of 23 from the array, the new array will be:

arr[0]	arr[1]	arr[2]	arr[3]	arr[4]	arr[5]	arr[6]	arr[7]	arr[8]	arr[9]
15	47	9	30	90	77				

Practical Application:

It is just like if there is a pile of books and a person just picks up the book from the end, so now all the books will be shifted one place forward from where they were placed; hence, this is deletion from the beginning.

Let's look at the algorithm for deletion from the beginning.
We assume ARR is an array with N elements in it:

```
Step 1: START
Step 2:SET I = 0
Step 3:Repeat Step 4 while I<N-1
Step 4:SET ARR[I] = ARR[I+1]
[ END OF LOOP ]
Step 5:EXIT
```

- *Deletion from a specified position* – In this case, the element to be deleted is deleted from the specified location/position in the array, which is entered by the user. In order to delete an element from the specified position, the elements stored in the array must move one place backward to their existing place in the array until the element is deleted at the specified position. For example, if an array is declared to hold ten elements and it only contains eight elements, and it is also given that the element is to be deleted from the specified position, which is the fourth position of the array, then the stored elements must move one place back, as shown in the following:

arr[0]	arr[1]	arr[2]	arr[3]	arr[4]	arr[5]	arr[6]	arr[7]	arr[8]	arr[9]
43	22	86	7	13	40	60	75		

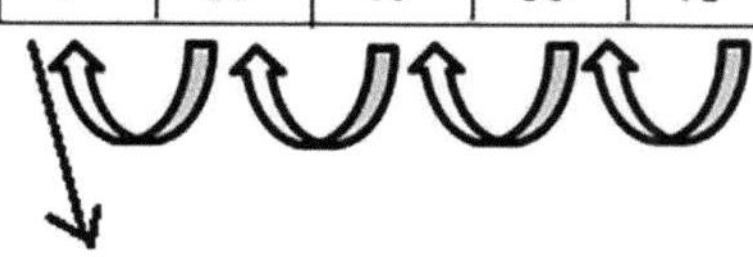

Delete 7 from arr[3]

In order to delete the fourth element, 7, from the array, we must swap the stored elements backward so that the given element gets deleted, as shown. After the deletion of 7 from the array, the new array will be:

arr[0]	arr[1]	arr[2]	arr[3]	arr[4]	arr[5]	arr[6]	arr[7]	arr[8]	arr[9]
43	22	86	13	40	60	75			

Let's look at the algorithm for deletion from a specified position.

We assume ARR is an array with N elements in it. Let POS define the position from which the element is to be deleted:

```
Step 1:START
Step 2:SET I = POS
Step 3:Repeat Step 4 while I<=N-1
Step 4:SET ARR[I+1] = ARR[I]
[ END OF LOOP ]
Step 5:EXIT
```

- *Deletion from the end* – In this case, the deletion is quite simple. Here, we are just required to print all the elements except the last one, as we want to delete the last element. For example, if an array is declared to hold ten elements and it contains only six elements, and it is also given that the element is to be deleted from the end of the array, then the deletion is shown as follows:

arr[0]	arr[1]	arr[2]	arr[3]	arr[4]	arr[5]	arr[6]	arr[7]	arr[8]	arr[9]
43	22	86	13	40	11				

Delete 11 from arr[5]

After deleting `11` from the array, the new array will be:

arr[0]	arr[1]	arr[2]	arr[3]	arr[4]	arr[5]	arr[6]	arr[7]	arr[8]	arr[9]
43	22	86	13	40					

Practical Application:

It is just like if there is a pile of books and a person just picks the first book from the pile; hence, we can say that one book is deleted or removed from the pile, and therefore it is a deletion from the end.

Let's look at the algorithm for deletion from the end.
We assume `ARR` is an array with `N` elements in it:

```
Step 1:START
Step 2:SET N = N-1
Step 3:Repeat Step 4 for I=0 to N
Step 4:Print ARR[I]
[ END OF LOOP ]
Step 5:EXIT
```

Write a menu-driven program to implement deletion in a 1D array, discussing all three cases:

```
#include<stdio.h>
#include<conio.h>
void main()
{
int  i, n, pos, choice, value, arr[10];
clrscr();
printf("Enter the number of elements in array: ") ;
scanf("%d", &n) ;
printf("Enter the elements of array ") ;
 //Accepting the elements of array
 for( i=0 ; i<n ; i++ )
     {
scanf("%d", &arr[i]) ;
         }
printf("***MENU***") ;
printf("1. Deletion from beginning ") ;
printf("2. Deletion from specified location ") ;
printf("3. Deletion from end ") ;
printf("Enter your choice: ") ;
scanf("%d", &choice) ;
 if(n==10)
 {
printf("Overflow error ") ;
   exit(0);
   }
 else
  switch(choice)
  {
   case 1:
    for( i=0 ; i<n-1 ; i++ )
    {
arr[i] = arr[i+1] ;
       }
printf(" After deletion array is ") ;
     for( i=0 ; i<n-1 ; i++ )
      {
printf("\t%d", arr[i]) ;
       }
   break ;
   case 2:
printf("Enter position ") ;
scanf("%d", &pos) ;
   for( i=pos-1 ; i<n-1 ; i++ )
   {
arr[i] = arr[i+1] ;
       }
printf(" After deletion array is ") ;
     for( i=0 ; i<n-1 ; i++ )
      {
printf("\t%d", arr[i]) ;
        }
```

```
    break ;
    case 3:
     n = n-1 ;
printf(" After deletion array is ") ;
     for( i=0 ; i<n ; i++ )
       {
printf("\t%d", arr[i]) ;

         }
  break ;
  default :
printf("Wrong Choice") ;
  exit(0) ;
  } //End of switch case
getch() ;
} //End of main
```

Here is the output:

```
Enter the number of elements
n = 7
arr[0] = 25
arr[1] = 81
arr[2] = 66
arr[3] = 21
arr[4] = 43
arr[5] = 20
arr[6] = 39
****MENU****
Deletion  from beginning
Deletion from specified position
Deletion from end

Enter your choice 1
After deletion new array is
81     66     21     43      20      39

Enter your choice 2
Enter position 3
After deletion new array is
81     66     43     20     39

Enter your choice 3
After deletion new array is
81   66    43    20
```

Searching for an Element in an Array

Searching for an element in an array means finding whether a particular value exists in an array or not. If that particular value is found, then the search is said to be successful, and the position/location of that particular value is returned. If the value is not found, then the search will be said to be unsuccessful. There are two methods for searching: linear search and binary search. In this chapter, we will only discuss linear search in detail, and the binary search technique will be discussed in the upcoming chapters. Now, we will learn how linear search works.

Linear search is a very simple technique used to search for a particular value in an array. It is also called a *sequential search*, as it works by comparing the values to be searched with every element of the array in a sequence until a match is found.

For example, let us take an array of ten elements, which is declared as:

```
int array[10] = { 23, 15, 47, 9, 30, 33, 90, 77, 100, 11 }
```

Let's search for `90` in the array; then, every element of the array will be compared to `90` until `90` is found:

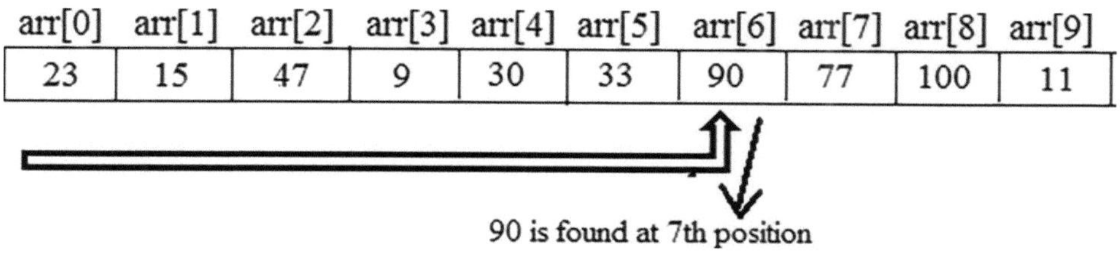

In this way, linear search is used to search for a particular value in the array. The following is the program for a linear search.

Write a program to search for an element in an array using the linear search technique:

```
# include <stdio.h>
intlinear_search( intarr[], int n, int value ) ;
void main()
{
intarr[10], n, i, r, value ;
clrscr() ;
printf("***LINEAR SEARCH***") ;
printf("\nEnter no of elements") ;
scanf("%d", &n) ;
printf("Enter the elements of array") ;
 for( i=0 ; i<n ; i++ )
 {
printf("\nEnter element %d", i+1) ;
scanf("%d", &arr[i]) ;
   }
printf("\nEnter value to search") ;
scanf("%d", &value) ;
  r = linear_search(arr, n, value) ;
  if(r == -1)
printf("value not found") ;
  else
```

```
printf("%d value found at %d", value, r+1) ;
getch() ;
  }
intlinear_search( intarr[], int n, int value )
   {
int i;
     for( i=0 ; i<n ; i++ )
     {
       if(arr[i] == value)
       return i ;
     }
  return (-1);
}
```

Here is the output:

```
***LINEAR SEARCH***
Enter no of elements  7

Enter elements of array
Enter element 1   15
Enter element 2   65
Enter element 3   87
Enter element 4   99
Enter element 5   29
Enter element 6     6
Enter element 7   33

Enter value to search   99
99 value found at 4th position
```

Merging of Two Arrays

The merging of two arrays means copying the elements of the first and second arrays into the third array. Here, we will take two sorted arrays, and the resultant merged array will also be sorted. The concept of merging is explained as follows.

Let us consider two sorted arrays, *Array* 1 and *Array* 2, and an *Array* 3 in which the elements will be placed after sorting:

Array 1

arr[0]	arr[1]	arr[2]	arr[3]	arr[4]	arr[5]	arr[6]	arr[7]	arr[8]	arr[9]
6	12	18	24	30					

Array 2

arr[0]	arr[1]	arr[2]	arr[3]	arr[4]	arr[5]	arr[6]	arr[7]	arr[8]	arr[9]
7	14	21	28	35					

Array 3

arr[0]	arr[1]	arr[2]	arr[3]	arr[4]	arr[5]	arr[6]	arr[7]	arr[8]	arr[9]
6	7	12	14	18	21	24	28	30	35

Array 3 shows how the merged array is formed using the sorted *Arrays* 1 and 2. Here, we compare the elements of the two arrays. First, the first element of *Array* 1 is compared with the first element of *Array* 2, and as 6 is less than 7 (6 < 7), `6` will be the first element in the merged array. Now, the second element of *Array* 1 is compared to the first element of *Array* 2, and as 7 is less than 12 (7 < 12), `7` will be the second element in the merged array. Now, the second element of *Array* 1 is compared with the second element of *Array* 2, and as 12 is less than 14 (12 < 14), `12` will be the third element in *Array* 3. This procedure is repeated until the elements of both *Arrays* 1 and 2 are placed in the right positions in the merged array, that is, *Array* 3.

Practical Application:

A real life example of merging would be if there are two different lines and both lines need to be merged according to the height of the people standing in that line, then merging would be done into a new line where the new line would consist of people from both lines in which people would be standing in order according to their heights.

Write a program to merge two sorted arrays:

```
#include<stdio.h>
#include<conio.h>
void main()
{
int arr1[5], arr2[5], arr3[10] ;
int m1, m2, s ;
inti=0, j=0 , k=0 ;
clrscr();
printf("Enter number of elements in array 1: ") ;
scanf("%d", &m1) ;
printf("Enter elements of array 1: ") ;
  for( s=0 ; s<m1 ; s++)
  {
printf("Enter element %d", s) ;
scanf("%d", &arr1[s]) ;
    }
printf("Enter number of elements in array 2: ") ;
scanf("%d", &m2) ;
printf("Enter elements of array 2: ") ;
  for( s=0 ; s<m2 ; s++)
  {
printf("Enter element %d", s) ;
```

```
scanf("%d", &arr2[s]) ;
    }
   while( i<m1 && j<m2)
   {
     if( arr1[i]<arr2[j] )
     {
       arr3[k]=arr1[i] ;
       k++ ;
       i++ ;
       }
     else if( arr2[j] < arr1[i] )
     {
       arr3[k] = arr2[j] ;
       j++ ;
       k++ ;
       }
     else
     {
      arr3[k] = arr2[i] ;
      k++ ;
      i++ ;
      arr3[k] = arr2[j] ;
      k++ ;
      j++ ;
  }
}// End of while loop
while( i<m1 )
{
  arr3[k] = arr1[i] ;
  k++ ;
  i++ ;
 }
while( j<m2 )
{
   arr3[k] = arr2[j] ;
   k++ ;
   j++ ;
   }
printf("After merging new array is: ") ;
 for( s=0 ; s<(m1 + m2) ; s++)
 {
printf("%d" , arr3[s]) ;
   }
getch() ;
}// End of main
```

Here is the output:

```
Enter number of elements in array 1:
 5
Enter number of elements in array 1:
arr[0] = 6
arr[1] = 12
```

```
arr[2] = 18
arr[3] = 24
arr[4] = 30

Enter number of elements in array 2:
 5
Enter number of elements in array 2:
arr[0] = 7
arr[1] = 14
arr[2] = 21
arr[3] = 28
arr[4] = 35

After merging new array is:
arr[0] = 6
arr[1] = 7
arr[2] = 12
arr[3] = 14
arr[4] = 18
arr[5] = 21
arr[6] = 24
arr[7] = 28
arr[8] = 30
arr[9] = 35
```

Sorting an Array

Sorting an array means arranging the data elements of a data structure in a specified order, either in ascending or descending order. Sorting refers to the process where, for example, in a class of 60 students who have received grades on their examination, the names of the students will be printed according to their grades, either in ascending or descending order.

For example, if we have an array of ten elements declared as:

```
int array[10] = { 78, 12, 47, 55, 61, 6, 99, 84, 32, 10 }
```

Then, after sorting, the new array will be:

```
array[10] = { 6, 10, 12, 32, 47, 55, 61, 78, 84, 99 }
```

There are various types of sorting techniques, which include selection sort, insertion sort, and merge sort; we will learn about selection sort in this chapter, and the other techniques will be discussed in the upcoming chapter.

Selection sort is a sorting technique that works by finding the smallest value in the array and placing it in the first position. After that, it then finds the second smallest value and places it in the second position. This process is repeated until the whole array is sorted. It is a very simple technique, and it is also easier to implement than any other sorting technique. Selection sort is generally used for sorting large records.

Let's look at the selection sort technique.

Consider an array with N elements:

- *Pass* 1 – Find the position POS of the smallest value in the array of N elements and interchange ARR[POS] with ARR[0]. Thus, A[0] is sorted.
- *Pass* 2 – Find the position POS of the smallest value in the array of N-1 elements and interchange ARR[POS] with A[1]. Thus, A[1] is sorted.
- And so on, until…
- *Pass* N–1 – Find the position POS of the smaller of the elements ARR[N-2] and ARR[N-1] and interchange ARR[POS] with ARR[N-2]. Thus, ARR[0], ARR[1], …, ARR[N-1] is sorted.

For example, sort the given array using selection sort:

arr[0]	arr[1]	arr[2]	arr[3]	arr[4]	arr[5]	arr[6]	arr[7]	arr[8]	arr[9]
16	82	48	34	75	9				

Pass	POS	Array[0]	Array[1]	Array[2]	Array[3]	Array[4]	Array[5]
1	5	9	82	48	34	75	16
2	5	9	16	48	34	75	82
3	3	9	16	34	48	75	82
4	3	9	16	34	48	75	82
5	4	9	16	34	48	75	82
6	5	9	16	34	48	75	82

In this way, the selection sort technique works. Below is the program given for selection sort.

Write a program to sort an array using the selection sort technique:

```
# include <stdio.h>
# include <conio.h>
void main()
{
int i, j, min, pos, arr[10], n, temp;
clrscr();
printf("enter no of elements in the array") ;
scanf("%d", &n) ;
printf("\nelements in the array are ") ;
 for( i=0 ; i<n ; i++ )
 {
printf("\nenter element %d", i+1) ;
scanf("%d", &arr[i]);
  }
printf("Selection Sort") ;
   for( i=1 ; i<n ; i++ )
```

```
    {
      min = arr[ i-1 ] ;
pos = i-1;
      for( j=I ; j<n ; j++ )
       {
         if( arr[j] < min )
         {
              min = arr[j] ;
pos = j ;
            }
         }
         if( pos != i-1 )
         {
           temp = arr[pos] ;
arr[pos] = arr[i-1] ;
arr[i-1] = temp ;
         }
       }
printf("after sorting new array is") ;
     for( i=0 ; i<n ; i++ )
     {
printf(" \t%d", arr[i]) ;
     }
getch() ;
}
```

Here is the output:

```
Enter number of elements    5

Enter elements of array
Enter element 1    80
Enter element 2    47
Enter element 3    51
Enter element 4    12
Enter element 5    67

Selection Sort
After sorting new array is
12    47    51    67    80
```

2D/TWO-DIMENSIONAL ARRAYS

We have already discussed 1D/one-dimensional arrays and their various types and operations. Now, we will learn about 2D/two-dimensional arrays. Unlike 1D arrays, 2D arrays are organized in the form of grids or tables. They are a collection of 1D arrays. 1D arrays are linearly organized in memory. A 2D array consists of two subscripts:

- First subscript, which denotes the row
- Second subscript, which denotes the column

A 2D array is represented as shown in Figure 3.5:

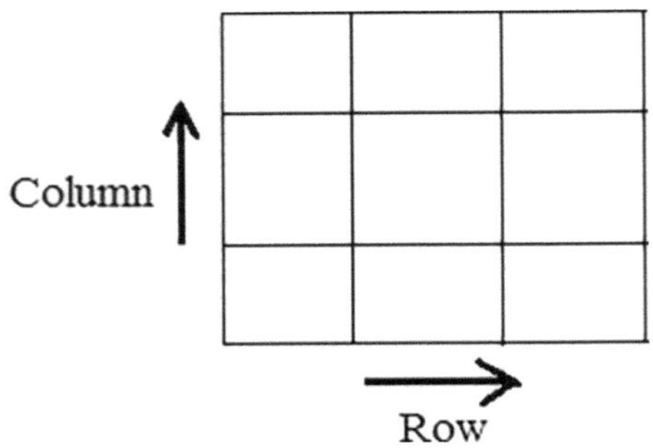

FIGURE 3.5 Representation of a 2D array.

DECLARATION OF TWO-DIMENSIONAL ARRAYS

As we declared 1D arrays, similarly, we can declare 2D arrays. For declaring 2D arrays, we must know the name of the array, the data type of each element, and the size of each dimension (size of rows and columns).

Here is the syntax: `data_type array_name [row_size][column_size] ;`

A 2D array is also called an $m \times n$ array, as it contains $m \times n$ elements, where each element in the array can be accessed by i and j, where $i<=m$ and $j<=n$, and where i, j, m, and n are defined as follows:

- i and j = Subscripts of array elements
- m = Number of rows
- n = Number of columns

For example, let us take an array of 3 x 3 elements. Therefore, the array is declared as:

```
int marks [3][3] ;
```

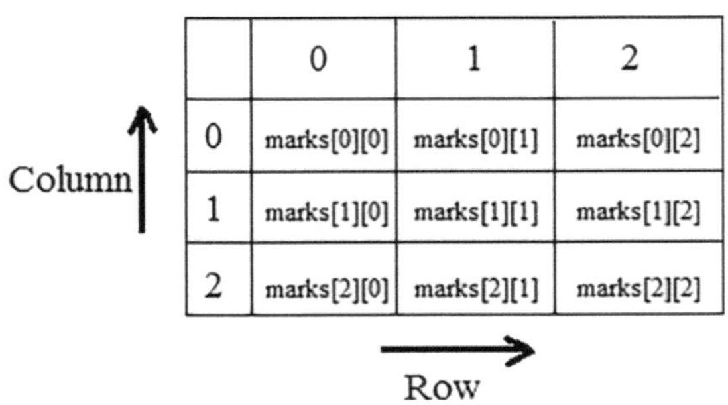

FIGURE 3.6 Array representation of 2 D array.

In the previous diagram, the array has three rows and three columns. The first element in the array is denoted by `marks[0][0]`. Similarly, the second element will be denoted by `marks[0][1]`, and so on. Data elements in an array can be stored in memory in two ways:

- *Row major order* – In row major order, the elements of the first row are stored before the elements of the second, third, and n rows. Here, the data elements are stored on a row-by-row basis:

00	01	02	10	11	12	20	21	22					

- *Column major order* – In column major order, the elements of the first column are stored before the elements of the second, third, and n columns. Here, the data elements are stored on a column-by-column basis:

00	10	20	01	11	21	02	12	22					

Now, we will calculate the base address of elements in a 2D array, as the computer does not store the address of each element. It just stores the address of the first element and calculates the addresses of other elements from the base address of the first element of the array. Hence, the addresses of other elements can be calculated in C from the given base address as follows:

- Elements in row major order:

$$Address(A[i][j]) = Base\ address(BA) + w(n(i–0) + (j–0)\ )$$

- Elements in column major order:

$$Address(A[i][j]) = Base\ address(BA) + w\ (m(j–0) + (i–0))$$

where w is the size in bytes to store one element.

Frequently Asked Questions

Q3. Consider a 25 x 5 two-dimensional array of students, which has a base address of 500, and the size of each element is 2, with starting row and column index 1. Now, calculate the address of the element student [15] [3], assuming that the elements are stored in

a) Row major order

b) Column major order

Answer.

a) Row major order:

Here we are given that $w = 2$, base address $= 500$, $n = 5$, $i = 15$, and $j = 3$:

$$Address(A[i][j]) = Base\ address(BA) + w\ (n(i–1) + (j–1))$$

$$Address(student[15][3]) = 500 + 2\ (5(15–1) + (3–1))$$

$= 500 + 2\,(5(14) + 2\,)$

$= 500 + 2\,(72)$

$= 500 + 144$

Address(student[15][3]) = 644

b) Column major order:

Here, we are given that $w = 2$, base address = 500, $m = 25$, $i = 15$, and $j = 3$:

Address(A[i][j]) = Base address(BA) + w (m(j–1) + (i–1))

Address(student[15][3]) = 500 + 2 (25(3–1) + (15–1))

= 500 + 2 (25(2) + 14)

= 500 +2 (64)

Address(student[15][3]) = 500 + 128 = 628

OPERATIONS ON 2D ARRAYS

There are various operations that are performed on two-dimensional arrays, which include:

- *Sum* – Let A_{ij} and B_{ij} be the two matrices that are to be added together, storing the result in the third matrix, C_{ij}. Two matrices will be added when they are compatible with each other; that is, they should have the same number of rows and columns:

$$C_{ij} = A_{ij} + B_{ij}$$

- *Difference* – Let A_{ij} and B_{ij} be the two matrices that are to be subtracted together, storing the result in the third matrix, C_{ij}. Two matrices will be subtracted when they are compatible with each other; that is, they should have the same number of rows and columns:

$$C_{ij} = A_{ij} - B_{ij}$$

- *Product* – Let A_{ij} and B_{ij} be the two matrices that are to be multiplied together, storing the result in the third matrix, C_{ij}. Two matrices will be multiplied with each other if the number of columns in the first matrix is equal to the number of rows in the second matrix. Therefore, an $m \times n$ matrix A can be multiplied with a $p \times q$ matrix B if $n=p$:

$$C_{ij} = A_{ik} \times B_{kj} \text{ for } k = 1 \text{ to } n$$

- *Transpose* – The transpose of an $m \times n$ matrix A is equal to an $n \times m$ matrix B:

$$B_{ij} = A_{ij}$$

Write a program to read and display a 3 × 3 array:

```
#include<stdio.h>
#include<conio.h>
void main()
{
int array[3][3], i, j ;
clrscr() ;
printf("Enter the elements of array ") ;
  for( i=0 ; i<3 ; i++ )
     {
        for( j=0 ; j<3 ; j++ )
        {
scanf("%d", &array[i][j]) ;
           }
         }
printf(" The array is : ) ;
  for( i=0 ; i<3 ; i++ )
  {
printf(" \n ") ;
    for( j=0 ; j<3 ; j++ )
     {
printf(" array[%d][%d] = %d", i, j, array[i][j])  ;
       }
    }
getch() ;
}
```

Here is the output:

```
Enter the elements of array
41
63
78
9
68
12
36
99
10

The array is:
array[0][0] = 41
array[0][1] = 63
array[0][2] = 78
array[1][0] =   9
array[1][1] = 68
array[1][2] = 12
array[2][1] = 36
array[2][2] = 99
array[2][3] = 10
```

Write a program to find the sum of two matrices:

```
#include<stdio.h>
#include<conio.h>
void main()
{
int i, j ;
int A[3][3], B[3][3], c[3][3] ;
clrscr() ;
printf("Enter the elements of A matrix: ") ;
  for( i=0 ; i<3 ; i++ )
  {
   for( j=0 ; j<3 ; j++ )
    {
scanf("%d", &A[i][j]) ;
    }
  }
printf("Enter the elements of B matrix: ") ;
  for( i=0 ; i<3 ; i++ )
  {
   for( j=0 ; j<3 ; j++ )
    {
scanf("%d", &B[i][j]) ;
    }
  }
for( i=0 ; i<3 ; i++ )
  {
   for( j=0 ; j<3 ; j++ )
    {
       C[i][j] = A[i][j] + B[i][j] ;
    }
  }
printf("Resultant matrix is : ") ;
for( i=0 ; i<3 ; i++ )
  {
   for( j=0 ; j<3 ; j++ )
    {
printf("%d", C[i][j]) ;
     }
   }
 } //End of main
```

Here is the output:

```
Enter the elements of A matrix
1  2  3  4  5  6  7  8  9
Enter the elements of B matrix
9  8  7  6  5  4  3  2  1

Resultant matrix is :
10  10  10  10  10  10  10  10  10
```

Write a program to find the transpose of a 3 × 3 matrix:

```
#include<stdio.h>
#include<conio.h>
void main()
{
int i, j ;
int matrix[3][3], transpose[3][3] ;
clrscr() ;
printf("Enter the elements of the matrix: ") ;
  for( i=0 ; i<3 ; i++ )
  {
   for( j=0 ; j<3 ; j++ )
    {
scanf("%d", &matrix[i][j]) ;
    }
  }
printf("The elements of matrix are ") ;
  for( i=0 ; i<3 ; i++ )
  {
printf("\n") ;
    for( j=0 ; j<3 ; j++ )
    {
printf("\t %d", matrix[i][j]) ;
    }
  }
for( i=0 ; i<3 ; i++ )
  {
   for( j=0 ; j<3 ; j++ )
    {
       transpose[i][j] = matrix[j][i] ;
    }
  }
printf("Elements of transposed matrix are ") ;
for( i=0 ; i<3 ; i++ )
  {
printf("\n") ;
    for( j=0 ; j<3 ; j++ )
    {
printf("\t %d", transpose[i][j]) ;
     }
   }
 } //End of main
```

Here is the output:

```
Enter the elements of the matrix
10  30  50  70  90  11  13  15  17
The elements of the matrix are
10   30   50
70   90   11
13   15   17
```

```
The elements of transpose matrix are
10   70   13
30   90   11
13   15   17
```

MULTIDIMENSIONAL/N-DIMENSIONAL ARRAYS

A *multidimensional array* is also known as an *n-dimensional* array. It is an array of arrays. It has n indices in it, which also justifies its name of n-dimensional array. An n-dimensional array is an $m_1 \times m_2 \times m_3 \times \ldots \times m_n$ array as it contains $m_1 \times m_2 \times m_3 \times \ldots \times m_n$ elements. Multidimensional arrays are declared and initialized in the same way as one-dimensional and two-dimensional arrays.

CALCULATING THE ADDRESS OF 3D ARRAYS

Just like 2D arrays, we can store 3D arrays in two ways, row major order and column major order:

- Elements in row major order:

$$Address\ ([i][j][k]) = Base\ Address\ (BA) + w\ (L_3(L_2\ (E_1) + E_2) + E_3)$$

- Elements in column major order:

$$Address\ ([i][j][k]) = Base\ Address\ (BA) + w\ ((E_3 L_2 + E_2)L_1 + E_1)$$

Where L is the length of the index, $L = upper\ bound - lower\ bound + 1$, and E is the effective address, $E = i - lower\ bound$.

Frequently Asked Questions

Q4. Let us take a 3D array A(4:12, −2:1, 8:14) and calculate the address of A(5, 4, 9) using row major order and column major order, where the base address is 500 and w = 4.

Answer.

Length of the three dimensions of A:

$L1 = 12 - 4 + 1 = 9$

$L2 = 1 - (-2) + 1 = 4$

$L3 = 14 - 8 = 6$

Therefore, A contains 9 x 4 x 6 = 216 elements.

Now, $E1 = 5 - 4 = 1$

$E2 = 4 - (-2) = 8$

$E3 = 9 - 8 = 1$

a) *Row major order:*

$$Address\,(5, 4, 9 = 500 + 4\,(6\,(4(1) + 8) + 1)$$

$$= 500 + 4\,(6\,(12) + 1)$$

$$= 500 + 4\,(73)$$

$$Address\,(5, 4, 9) = 500 + 292 = 792$$

b) *Column major order:*

$$Address\,(5, 4, 9) = 500 + 4\,((1.4 + 8)\,9 + 1)$$

$$= 500 + ((12)9 + 1)$$

$$Address\,(5, 4, 9) = 500 + 145 = 645$$

Write a program to read and display a 2 × 2 × 2 array:

```
#include<stdio.h>
#include<conio.h>
void main()
{
int array[2][2][2], i, j, k ;
clrscr() ;
printf("Enter the elements of array ") ;
 for( i=0 ; i<2 ; i++ )
     {
for( j=0 ; j<2 ; j++ )
        {
          for( k=0 ; k<2 ; k++ )
           {
scanf("%d", &array[i][j][k]) ;
}
         }
       }
printf(" The array is : ) ;
  for( i=0 ; i<2 ; i++ )
  {
printf(" \n ") ;
    for( j=0 ; j<2 ; j++ )
     {
printf(" \n ") ;
       for( k=0 ; k<2 ; k++ )
       {
```

```
printf(" array[%d][%d][%d] = %d", i, j, k, array[i][j][k] ) ;
        }
      }
    }
getch() ;
}
```

Here is the output:

```
Enter the elements of array
4
23
9
8
6
2
1
7
The array is :
array[0][0][0] = 4
array[0][0][1] = 23
array[0][1][0] = 9
array[0][1][1] = 8
array[1][0][0] = 6
array[1][0][1] = 2
array[1][1][0] = 1
array[1][1][1] = 7
```

ARRAYS AND POINTERS

A *pointer* is a special type of variable that is used to store addresses. Pointers can be used to access and manipulate data stored in memory. Pointers are very frequently used in arrays because pointers are more efficient in handling arrays and data tables. A pointer can be referred to as the address of a person's home, which can help us easily reach the destination. The concept of arrays is very much bound to the concept of pointers. Now, let us take an example. Let's say we have an array of ten elements declared as

```
int array[10] = { 10, 20 ,32, 7, 19, 89, 60, 54, 12, 77 }
```

The previous array will be stored in memory as shown in Figure 3.7:

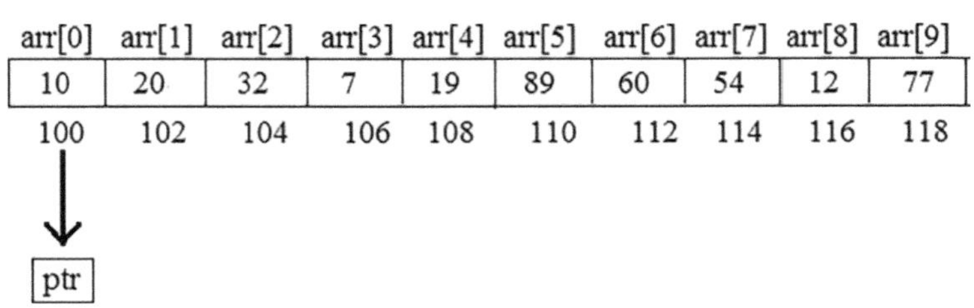

FIGURE 3.7 Memory representation of array.

The starting address of the array (`100`) is the base address of the array. The base address is the address of the first element in the array. Now, we will use a pointer variable to store the address of the first element; in other words, `ptr` will point to the first element of the array:

```
int * ptr ;
ptr = &array[0] ;
```

We can also store the addresses of other elements. If the pointer variable holds the address of the first element in the array, then the address of the successive elements can be calculated by `ptr++`:

```
int * ptr ;
ptr = &array[0] ;
ptr++ ;
printf(" The value of second element is %d", *ptr) ;
```

A pointer variable can be prefixed or postfixed with increment or decrement operators. Increment or decrement will depend upon the data type, for example, `int` – 2 bytes, `float` – 4 bytes, `char` – 1 byte, and so on.

ARRAY OF POINTERS

An array of pointers is declared as `int * ptr[10]`.

Therefore, from the previous statement, an array of ten pointers is declared where each of the pointers points to a variable. The code for an array of pointers is given as follows:

```
#include<stdio.h>
void main()
{
int i, j, k, l, m ;
int * arr[4] ;
  i = 25;
  j = 50;
  k = 75;
  l = 100;
arr[0] = &i
arr[1] = &j
arr[2] = &k
arr[3] = &l
  for( m=0 ; m<4 ; m++ )
  {
printf("%u\n", arr[m]) ;
  }
}
```

Here is the output:

```
32012
35260
31010
30058
```

Note: Here, the addresses are random in memory.

ARRAYS AND THEIR APPLICATIONS

Arrays are very frequently used in C as they have various applications that are very useful. These applications include:

- Sorting elements in ascending or descending order
- Implementing various other data structures, such as stacks, queues, and hash tables
- Implementing matrices, vectors, and various other kinds of rectangular tables
- Various other operations can be performed on the arrays, which include searching, merging, and sorting

Frequently Asked Questions

Q5. List some of the applications of arrays.

Answer.

1. Arrays are very useful in storing data in contiguous memory locations.
2. Arrays are used for implementing various other data structures, such as stacks, queues, etc.
3. Arrays are very useful as we can perform various operations on them.

SPARSE MATRICES

A *sparse matrix* is a matrix with a relatively high proportion of zero entries. A sparse matrix is used as it utilizes the memory space efficiently. Storing null elements in the matrix is a waste of memory, so we adopt a technique to store only non-null elements in the sparse matrices.

An example is shown in Figure 3.8:

$$\begin{bmatrix} 0 & 0 & 6 & 0 & 0 \\ 1 & 0 & 0 & 0 & 0 \\ 0 & 0 & 0 & 0 & 2 \\ 0 & 5 & 0 & 0 & 0 \end{bmatrix}$$

FIGURE 3.8 Representation of a sparse matrix.

TYPES OF SPARSE MATRICES

There are three types of sparse matrices, which are:

1. *Lower-triangular matrix* – In this type of sparse matrix, all the elements above the main diagonal must have a zero value, or in other words, we can say that elements below the main diagonal may contain non-zero elements. This type of matrix is called a lower-triangular matrix and is shown in Figure 3.9.

$$\begin{bmatrix} 5 & 0 & 0 & 0 \\ 4 & 6 & 0 & 0 \\ -3 & 9 & -5 & 0 \\ 2 & 1 & 7 & 3 \end{bmatrix}$$

FIGURE 3.9 Lower-triangular matrix.

2. *Upper-triangular matrix* – In this type of sparse matrix, all the elements above the main diagonal may contain non-zero elements, or in other words, we can say that all the elements below the main diagonal should have a zero value. This type of matrix is called an upper-triangular matrix, and is shown in Figure 3.10.

$$\begin{bmatrix} 1 & 2 & 3 & 4 \\ 0 & 6 & -1 & 5 \\ 0 & 0 & -7 & 8 \\ 0 & 0 & 0 & 9 \end{bmatrix}$$

FIGURE 3.10 Upper-triangular matrix.

3. *Tri-diagonal matrix* – In this type, elements with a non-zero value can appear only on the diagonal or adjacent to the diagonal. This type of matrix is called a tri-diagonal matrix, and it is shown in Figure 3.11.

$$\begin{bmatrix} 6 & 2 & 0 & 0 \\ 8 & 9 & -2 & 0 \\ 0 & 5 & -7 & 3 \\ 0 & 0 & 1 & 4 \end{bmatrix}$$

FIGURE 3.11 Tri-diagonal matrix.

REPRESENTATION OF SPARSE MATRICES

There are two ways in which the sparse matrices can be represented, which are:

1. *Array representation/three-tuple representation* – This representation contains three rows in which the first row represents the number of rows, columns, and non-zero entries/values in the sparse matrix. Elements in the other rows give information about the location and value of non-zero elements.

   For example, let us consider a sparse matrix:

$$\begin{bmatrix} 0 & 0 & 0 & 0 & 1 \\ 0 & 0 & 0 & 0 & 0 \\ 0 & 0 & 3 & 0 & 0 \\ 0 & 5 & 0 & 0 & 0 \end{bmatrix}$$

FIGURE 3.12 Sample sparse matrix.

An array representation of the previous sparse matrix will be:

Row	Column	Non-Zero Value
0	4	1
2	2	3
3	1	5

2. *Linked representation* – A sparse matrix can also be represented in a linked way. In this representation, we store the number of rows, columns, and non-zero entries in a single node, and there is a pointer that points to the next location. Let us consider an example to understand more clearly.

   Let us consider a sparse matrix:

$$\begin{bmatrix} 0 & 6 & 0 & 0 & 0 \\ 0 & 0 & 0 & 0 & 0 \\ 0 & 0 & 0 & 8 & 3 \end{bmatrix}$$

FIGURE 3.13 Sample sparse matrix.

The linked representation of the sparse matrix will be as follows:

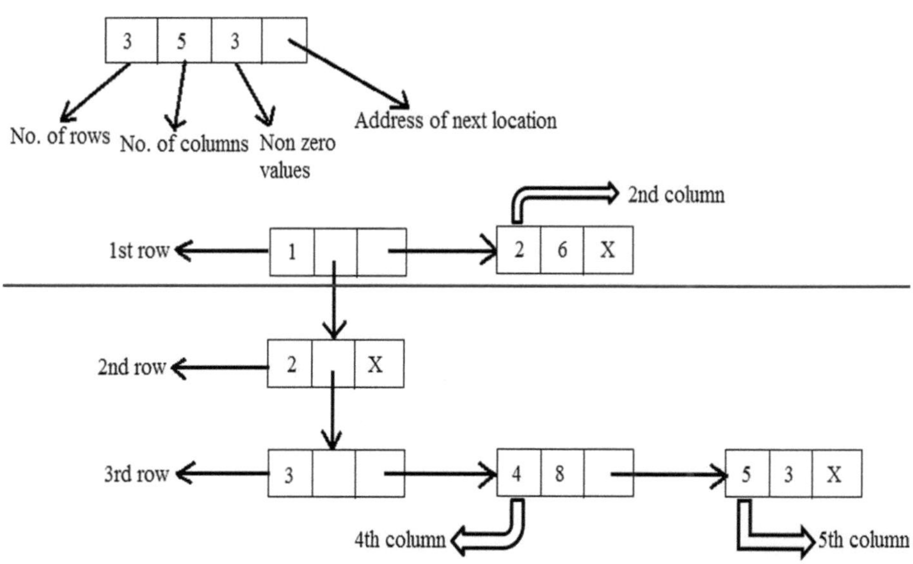

FIGURE 3.14 Linked representation of a sparse matrix.

Frequently Asked Questions

Q6. Explain the sparse matrix.

Answer.

A matrix in which the number of zero entries is much higher than the number of non-zero entries is called a sparse matrix. The natural method of representing matrices in memory as two-dimensional arrays may not be suitable for sparse matrices. One may save space by storing only non-zero entries. We can represent a sparse matrix by using a three-tuple method of storage:

- Row major method
- Column major method

SUMMARY

- An array is a collection of homogeneous (similar) types of data elements in contiguous memory. An array is a linear data structure because all elements of an array are stored in linear order.
- An array must be declared before it is used.
- The initialization of the elements of an array at compile time is done in the same way as when we initialize the normal or ordinary variables at the time of their declaration.
- Initialization of elements of an array at runtime refers to the method of inputting the values from the keyboard.
- The address of the elements in a 1D array can be calculated very easily, as an array stores all its data elements in contiguous memory locations, storing the base address.
- Traversing an array means accessing each and every element in an array exactly once so that it can be processed.
- Insertion of an element in an array refers to the operation of adding an element to the array. It can be done in three ways.
- Deleting an element from an array refers to the operation of the removal of an element from an array. Deletion is also done in three ways.
- Searching for an element in an array means finding whether a particular value exists in an array or not. If that particular value is found, then the search is said to be successful, and the position/location of that particular value is returned. If the value is not found, then the search will be said to be unsuccessful.

- A linear search is a very simple technique used to search for a particular value in an array.
- The merging of two arrays means copying the elements of the first and second arrays into the third array.
- Sorting an array means arranging the data elements of a data structure in a specified order, either in ascending or descending order.
- Selection sort is a sorting technique that works by finding the smallest value in the array and placing it in the first position. After that, it then finds the second smallest value and places it in the second position. This process is repeated until the whole array is sorted.
- Unlike 1D arrays, 2D arrays are organized in the form of grids or tables. They are a collection of 1D arrays.
- For declaring 2D arrays, we must know the name of the array, the data type of each element, and the size of each dimension (size of row and column).
- A multidimensional array is also known as an n-dimensional array. It is an array of arrays. It has n indices in it, which also justifies its name of an n-dimensional array.
- A pointer is a special type of variable that is used to store addresses. Pointers can be used to access and manipulate data stored in memory.
- A sparse matrix is a matrix with a relatively high proportion of zero entries. A sparse matrix is used because it utilizes the memory space efficiently.

EXERCISES

Theory Questions

1. What do we mean by an array, and how is it represented in memory?
2. What are the various operations that can be performed on arrays? Discuss in detail.
3. Explain the concept of two-dimensional arrays.
4. Briefly explain how arrays are related to pointers.
5. In how many ways can arrays be initialized? Explain in detail.
6. What do you understand by multidimensional arrays?
7. Consider a one-dimensional array declared as `intarr[10]`, and calculate the address of `arr[7]` if the base address is `200` and the size of each element is `2`.
8. What do we mean by sorting an array? Explain.
9. Write an algorithm to perform the selection sort technique.
10. Explain the process of merging two arrays along with the algorithm.
11. Consider a two-dimensional array declared as `intarray[10][10]`, and calculate the address of the element `array[5][6]` if the base address is `10000` and the size of each element is `2`, assuming the elements are to be stored in column-major order.
12. List some of the applications of arrays.

13. What do you understand by a linear search? Provide the algorithm.
14. What are the advantages of using the selection sort technique for sorting the elements in an array?
15. What is a sparse matrix? Explain its types.
16. Consider a three-dimensional array `A(2:6, -1:7, 9:10)`, and calculate the address of `A(9, 6, 8)` using row major order and column major order, where the base address is `2000` and `w = 4`.
17. Explain the linked representation of sparse matrices in detail.
18. Write the formulae for calculating the addresses of elements in row-major and column-major order in 2D and 3D arrays.

Programming Questions

1. Write a program to traverse an entire array.
2. Write a program to perform insertion at a specified position in a one-dimensional array.
3. Write a program to multiply two matrices.
4. Write a program that reads a matrix and displays:
   a) Sum of its rows' elements
   b) Sum of its columns' elements
   c) Sum of its diagonal elements
5. Write a program to show the concept of an array of pointers.
6. Write a program to perform the deletion of an element from the beginning.
7. Write a menu-driven program to perform various insertions and deletions in an array using the switch case.
8. Write a program to read and display a square matrix.
9. Write a program that reads an array of 50 integers. Display all the pairs of elements whose sum is `25`.
10. Write a program to convert a normal matrix into a sparse matrix.
11. Write a program to read an array of ten integers and then find the smallest and largest numbers in the array.
12. Write a program to add two sparse matrices.

Multiple Choice Questions

1. If an array is declared as `intarray[20][20]`, how many elements can it store?
   A. 20
   B. 40
   C. 400
   D. None of these
2. The elements of an array are always stored in ________ memory locations.
   A. Random
   B. Sequential
   C. Both
   D. None of these

3. What will be the output of the given program?

```
#include<stdio.h>
int main()
{
int arr[5] = {5, 1, 15, 20, 25} ;
int a, b, c ;
a = ++arr[1] ;
b = arr[1]++ ;
c = arr[a++] ;
printf("%d, %d, %d", a, b, c) ;
return 0 ;
}
```

A. `1, 2, 5`
B. `2, 1, 15`
C. `3, 2, 15`
D. `2, 3, 20`

4. What will be the output of the following program after execution?

```
void main()
{
int array[5] = {2, 4, 6, 8, 10} ;
printf("%d", array[6]) ;
 }
```

A. `0`
B. `10`
C. Garbage value
D. None of these

5. If an array is declared as `int array[]`, then the nth element can be accessed by:

A. `array[n]`
B. `*(array + n)`
C. `*(n + array)`
D. None of these
E. All of these

6. *Array*[5] = 19 initializes the ________element of the array with the value 19.

A. 4th
B. 5th
C. 6th
D. 7th

7. By default, the first subscript of the array is ________.

A. 2
B. 1
C. –1
D. 0

8. A multi-dimensional array, in simple terms, is an ________.

A. Array of arrays
B. Array of pointers
C. Array of addresses

9. A loop is used to access all the elements of an array.

   A. False
   B. True

10. Declaring an array means specifying the ________, ________, and ________.

   A. Data type, name, size
   B. Data elements, name, data type
   C. Name, size, address
   D. All of the above

CHAPTER 4

LINKED LISTS

INTRODUCTION

We have already learned that an array is a collection of data elements stored in contiguous memory locations. Also, we studied how arrays were static in nature; that is, the size of the array must be specified when declaring an array, which limits the number of elements to be stored in the array. For example, if we have an array declared as `intarray[15]`, then the array can contain a maximum of 15 elements and not more than that. This method of allocating memory is good when the exact number of elements is known, but if we are not sure of the number of elements, then there will be a problem, as, in data structures, our aim is to make programs efficient by consuming less memory space along with minimal time. To overcome this problem, we will use linked lists.

DEFINITION OF A LINKED LIST

A *linked list* is a linear collection of data elements. These data elements are called *nodes*, and they point to the next node by means of *pointers*. A linked list is a data structure that can be used to implement other data structures such as stacks, queues, trees, and so on. It is a sequence of nodes in which each node contains one or more data fields and a pointer that points to the next node. Linked lists are dynamic in nature; that is, memory is allocated as and when required. There is no need to know the exact size or the exact number of elements, as in the case of arrays. Figure 4.1 shows an example of a simple linked list that contains five nodes:

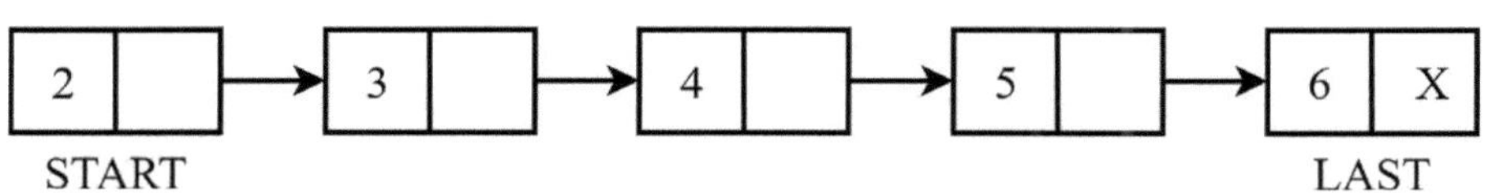

FIGURE 4.1 A linked list.

In the previous figure, we have made a linked list in which each node is divided into two parts:

1. The first part contains the information/data.
2. The second part contains the address of the next node.

The last node will not have another node connected to it, so it will store a special value called NULL represented by X symbol. Usually, `NULL` is defined by `-1`. Therefore, the `NULL` pointer represents the end of the linked list. There is another special pointer called `START` that stores the address of the first node of the linked list. Therefore, the `START` pointer represents the beginning of the linked list. If `start = NULL`, then it means that the linked list is empty. Since each node points to another node of the same type, a linked list is known as a *self-referential data type* or a *self-referential structure*.

The self-referential structure in a linked list is as follows:

```
struct node
{
  int info ;
  struct node * next ;
 } ;
```

Practical Application:

A simple real-life example is how each coach on a train is connected to the previous and next coach (except the first and last). In terms of programming, consider the coach body as a node and the connectors as links to the previous and next nodes.

The brain is also a good example of a linked list. In the initial stages of learning something by heart, the natural process is to link one item to another. It's a subconscious act. Also, when we forget something and try to remember, our brain follows associations and tries to link one memory with another, and so on until we finally recall the lost memory.

Frequently Asked Questions

Q1. Define a linked list.

Answer.

A linked list is a linear collection of data elements, called nodes, where the linear order is given by means of pointers. It is a dynamic data structure. For every data item in a linked list, there is an associated pointer that gives the memory location of the next data item in the linked list. The data items in the linked list are not in consecutive memory locations.

Frequently Asked Questions

Q2. List the advantages and disadvantages of a linked list.

Answer.

Advantages of linked lists:

1. Linked lists are dynamic data structures; that is, they can grow or shrink during the execution of the program.
2. They have efficient memory utilization. Memory is allocated whenever it is required, and it is deallocated whenever it is no longer needed.
3. Insertion and deletion are easier and more efficient.
4. Many complex applications can be easily carried out with linked lists.

Disadvantages of linked lists:

1. They consume more space because every node requires an additional pointer to store the address of the next node.
2. Searching for a particular element in the list is difficult and time-consuming.

MEMORY ALLOCATION IN A LINKED LIST

The process or concept of linked lists supports dynamic memory allocation. Now, what is meant by dynamic memory allocation? The answer to this simple question is that the process of allocating memory during the execution of the program, or the process of allocating memory to the variables at runtime, is called *dynamic memory allocation*. Until now, we have studied arrays in which we used to declare the size of the array initially, such as `array[50]`. This statement, after execution, allocates the memory for 50 integers. There can be a problem, however, if we use only 30% of the memory, as the rest of the allocated memory is either wasted or not used. To overcome this problem of wastage of memory space (in other words, to utilize the memory efficiently), dynamic memory allocation is used, which allows us to allocate/reserve the memory that is actually required. This will overcome the problem of wastage of memory space, as in the case of arrays. Dynamic memory allocation is best when we are not aware of the memory requirements in advance. The C language provides some functions that are used to dynamically allocate memory, as shown in the following table:

Function	Use of the Function	Syntax
`malloc()`	Allocates the requested memory space and returns a pointer to the first byte of allocated memory space	`ptr = (cast-type *) malloc(byte-size) ;`
`calloc()`	Allocates memory space for elements of an array, initializes to zero, and returns a pointer to the memory	`ptr = (cast-type *) calloc(n, elem-size) ;`
`free()`	Deallocates the previously allocated memory space	`free(ptr) ;`
`realloc()`	Alters the size of memory allocated previously	`ptr = realloc(ptr, newsize) ;`

TYPES OF LINKED LISTS

There are different types of linked lists that we will discuss in this section. These include:

1. Singly linked lists
2. Circular linked lists
3. Doubly linked lists
4. Header linked lists

Now, we will discuss all of them in detail.

Singly Linked Lists

A *singly linked list* is the simplest type of linked list in which each node contains some information/data and only one pointer that points to the next node in the linked list. The traversal of data elements in a singly linked list can be done only in one way, as seen in Figure 4.2.

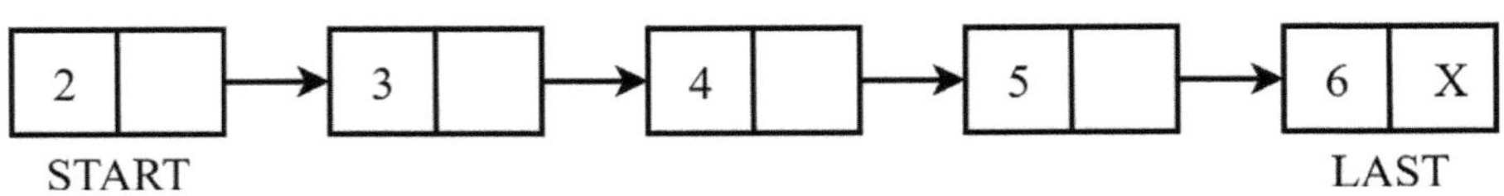

FIGURE 4.2 Singly linked list.

Operations on a Singly Linked List

Various operations can be performed on a singly linked list, which include:

- Traversing a linked list
- Searching for a given value in a linked list
- Inserting a new node in a linked list
- Deleting a node from a linked list
- Concatenating two linked lists
- Sorting a linked list
- Reversing a linked list

Let us now discuss all these operations in detail.

Traversing a Linked List

Traversing a linked list means accessing all the nodes of the linked list exactly once. A linked list will always contain a START pointer, which stores the address of the first node of the linked list and which also represents the beginning of the linked list, and a NULL pointer, which represents the end of the linked list. For traversing a linked list, we will use another pointer variable, PTR, which will point to the node that is currently being accessed. The algorithm for traversing a linked list is shown as follows:

```
Step 1: Set PTR = START
Step 2: Repeat Steps 3 & 4 while PTR != NULL
Step 3: Print PTR -> INFO
Step 4: Set PTR = PTR -> NEXT
            [End of Loop]
Step 5: Exit
```

Searching for a Given Value in a Linked List

Searching for a value in a linked list means finding a particular element/value in the linked list. As we discussed earlier, a node in a linked list contains two parts: one part is the information part, and the other is the address part. Hence, *searching* refers to the process of finding whether or not the given value exists in the information part of any node. If the value is present, then the address of that particular value is returned and the search is said to be successful; otherwise, the search is unsuccessful. A linked list will always contain a START pointer, which stores the address of the first node of the linked list and also represents the beginning of the linked list, and a NULL pointer, which represents the end of the linked list. There is another variable, PTR, which will point to the current node being accessed. SEARCH_VAL is the value to be searched in the linked list, and POS is the position/address of the node at which the value is found. The algorithm for searching a value in a linked list is given as follows:

```
Step 1: Set PTR = START
Step 2: Repeat Step 3 while PTR != NULL
Step 3: IF SEARCH_VAL = PTR -> INFO
              Set POS = PTR
              Print Successful Search!!
              Go to Step 5
              [End of If]
              ELSE
              Set PTR = PTR -> NEXT
              [End of Loop]
Step 4: Print Unsuccessful Search!!
Step 5: Exit
```

For example, if we have a linked list and we are searching for 8 in the list, then the steps are as shown in Figure 4.3:

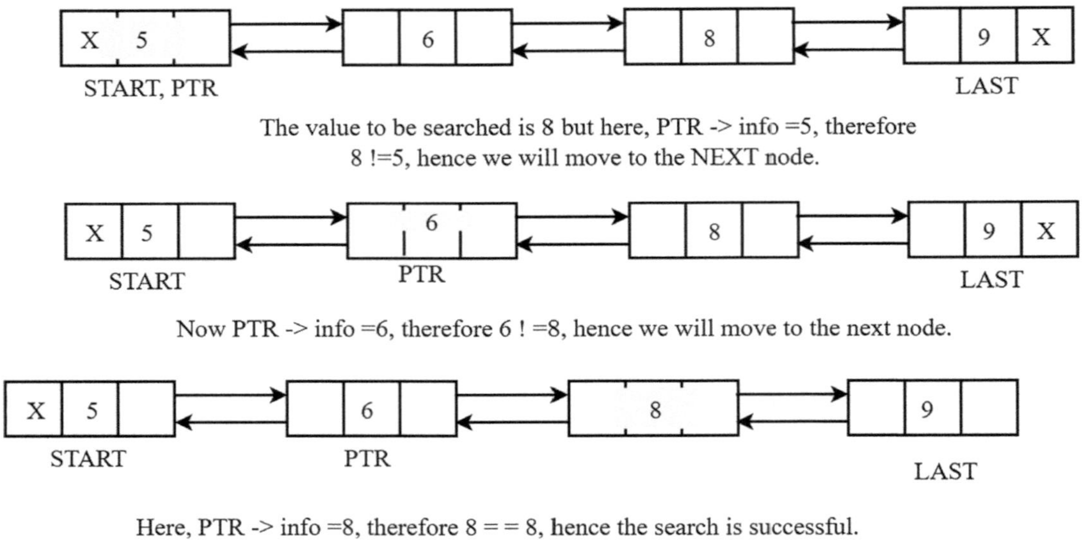

FIGURE 4.3 An example of searching a linked list.

Inserting a New Node in a Linked List

Here, we will learn how a new node is inserted into an existing linked list. We will discuss three cases in the insertion process:

1. A new node is inserted at the beginning of the linked list.
2. A new node is inserted at the end of the linked list.
3. A new node is inserted after the given node in a linked list.

Let us now discuss all of these cases in detail.

Inserting a New Node at the Beginning of a Linked List

In the case of inserting a new node at the beginning of a linked list, we will first check the overflow condition, which is whether the memory is available for a new node. If the memory is not available, then an overflow message is displayed; otherwise, the memory is allocated for the new node. Now, we will initialize the node with its information part, and its address part will contain the address of the first node of the list, which is the `START` pointer. Hence, the new node is added as the first node in the list, and the `START` pointer will point to the first node of the list. Now, to understand better, let us take an example. Consider a linked list as shown in Figure 4.4 with five nodes; a new node will be inserted at the beginning of the linked list.

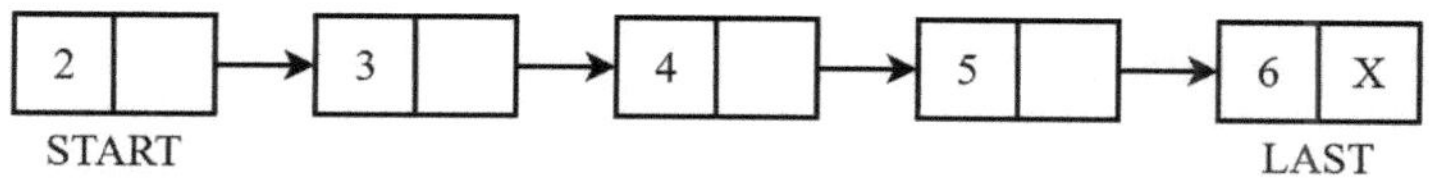

Now we will allocate memory for the new node and initialize with its info part

As the new node is to be inserted at the beginning of the linked list, now the address part (next part) of the NEW NODE will contain the address of the START pointer

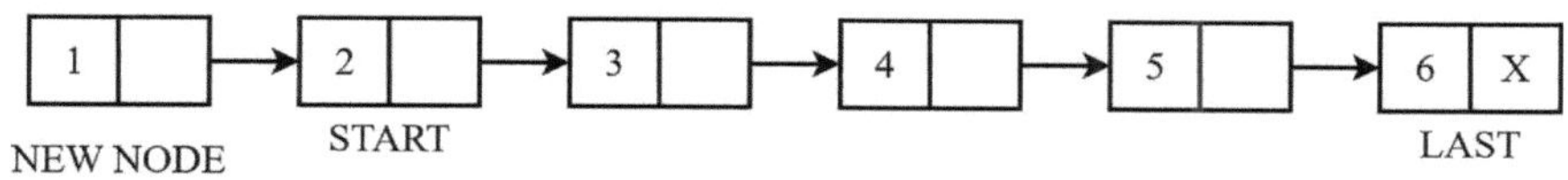

Now, the START pointer will store the address of and point to NEW NODE, which is now the first node in the list

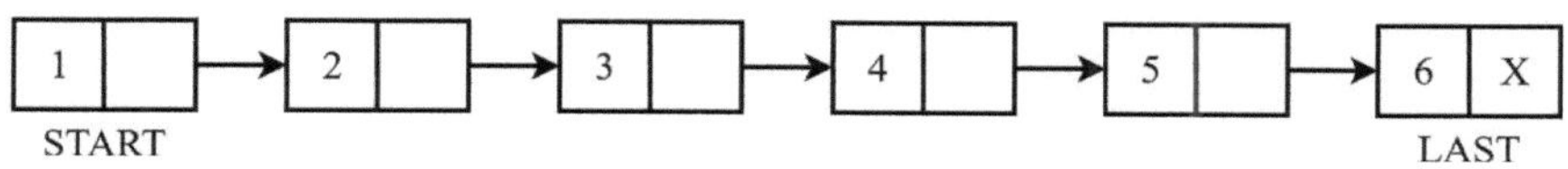

FIGURE 4.4 Inserting a new node at the beginning of a linked list.

From the previous example, it is clear how a new node will be inserted into an already existing linked list. Let us now understand its algorithm:

```
Step 1: START
Step 2: IF PTR = NULL
            Print OVERFLOW
            Go to Step 8
            [End of If]
Step 3: Set NEW NODE = PTR
Step 4: Set PTR = PTR -> NEXT
Step 5: Set NEW NODE -> INFO = VALUE
Step 6: Set NEW NODE -> NEXT = START
Step 7: Set START = NEW NODE
Step 8: EXIT
```

Inserting a New Node at the End of a Linked List

To insert the new node at the end of a linked list, we will first check the overflow condition, which is whether the memory is available for a new node. If the memory is not available, then an overflow message is displayed; otherwise, the memory is allocated for the new node. Then, a `PTR` variable is made, which will initially point to START and will be used to traverse the linked list until it reaches the last node. When it reaches the last node, the NEXT part of the last node will store the address of the new node, and the NEXT part of NEW NODE will contain NULL, which will denote the end of the linked list. Let us understand this with the help of an algorithm:

```
Step 1: START
Step 2: IF PTR = NULL
            Print OVERFLOW
            Go to Step 10
            [End of If]
Step 3: Set NEW NODE = PTR
Step 4: Set PTR = PTR -> NEXT
Step 5: Set NEW NODE -> INFO = VALUE
Step 6: Set NEW NODE -> NEXT = NULL
Step 7: Set PTR = START
Step 8: Repeat Step 8 while PTR -> NEXT != NULL
            Set PTR = PTR -> NEXT
            [End of Loop]
Step 9: Set PTR -> NEXT = NEW NODE
Step 10: EXIT
```

From the previous algorithm, we understand how to insert a new node at the end of an already existing linked list. Now, we will study further with the help of an example. Consider a linked list with four nodes, as shown in Figure 4.5. A new node will be inserted at the end of the linked list:

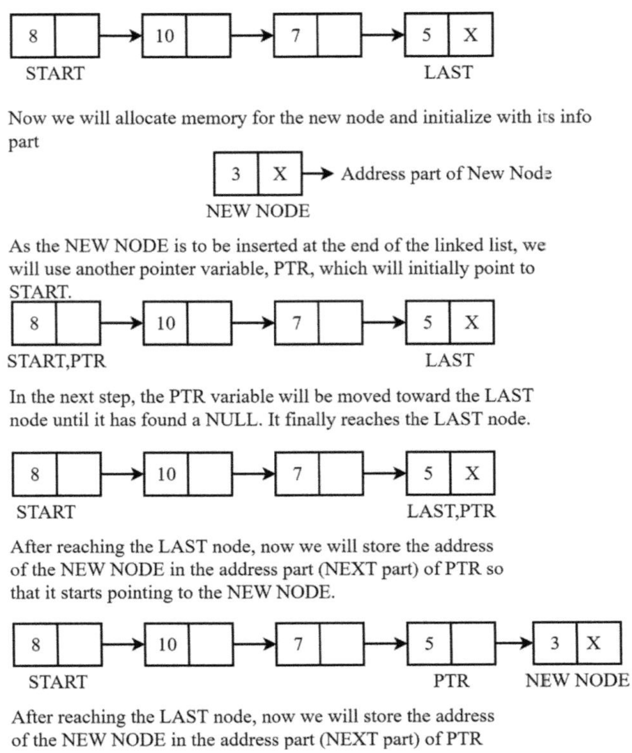

FIGURE 4.5 Inserting a new node at the end of a linked list.

Inserting a new node after a node in a linked list

In this case, a new node is inserted after a given node in a linked list. As in the other cases, we will again check the overflow condition. If the memory for the new node is available, it will be allocated; otherwise, an overflow message is printed. Then, a `PTR` variable is made, which will initially point to `START`, and the `PTR` variable is used to traverse the linked list until it reaches the value/node after which the new node is to be inserted. When it reaches that node/value, then the `NEXT` part of that node will store the address of the new node, and the `NEXT` part of the `NEW NODE` will store the address of its next node in the linked list. Let us understand this with the help of an example. Consider a linked list with four nodes, and a new node is to be inserted after the given node, as shown in Figure 4.6:

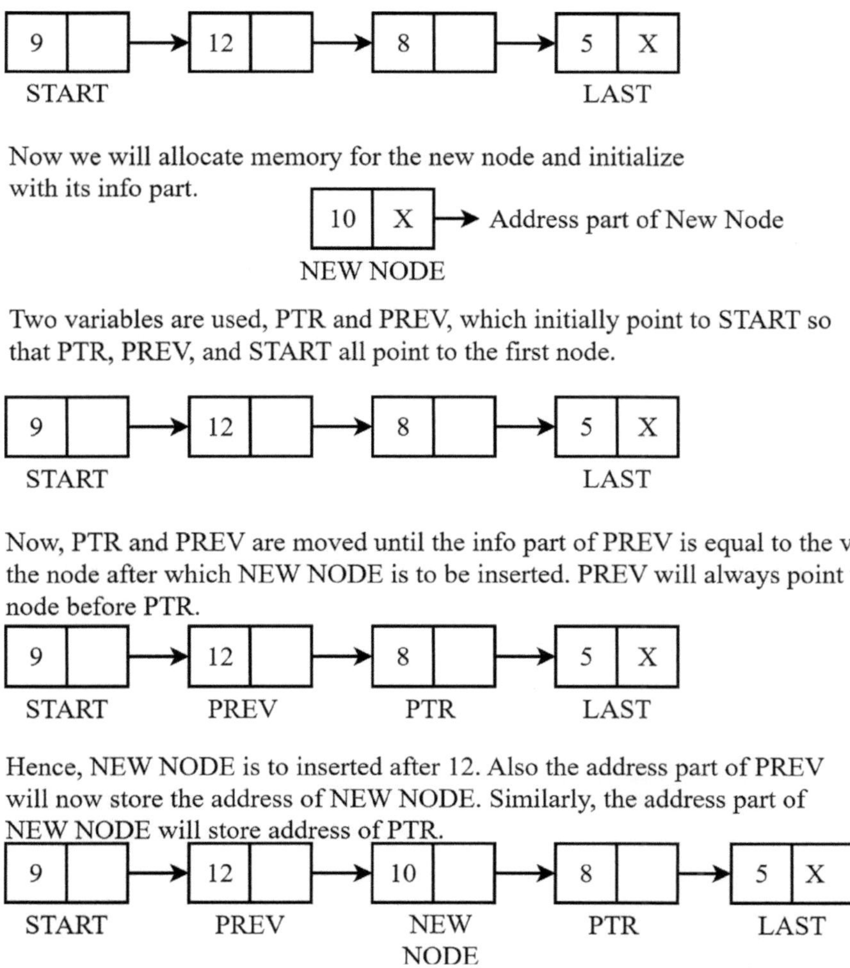

FIGURE 4.6 Inserting a new node after a given node in a linked list.

From the previous example, we have learned how a node can be inserted after a given node. Let's understand this with the help of an algorithm:

```
Step 1: START
Step 2: IF PTR = NULL
            Print OVERFLOW
            Go to Step 10
            [End of If]
Step 3: Set NEW NODE = PTR
Step 4: Set PTR = PTR -> NEXT
Step 5: Set NEW NODE -> INFO = VALUE
Step 6: Set PTR = START
Step 7: Set PREV = PTR
Step 8: Repeat Step 8 while PREV -> INFO != GIVEN_VAL
            Set PREV = PTR
            Set PTR = PTR -> NEXT
            [End of Loop]
Step 9: Set PREV -> NEXT = NEW NODE
Step 10: Set NEW NODE -> NEXT = PTR
Step 11: EXIT
```

Deleting a Node from a Linked List

In this section, we will learn how a node is deleted from an already existing linked list. We will discuss three cases in the deletion process:

1. The node is deleted from the beginning of the linked list.
2. The node is deleted from the end of the linked list.
3. The node is deleted after a given node from the linked list.

Let us now discuss all of these cases in detail.

Deleting a Node from the Beginning of the Linked List

In the case of deleting a node from the beginning of a linked list, we will first check the underflow condition, which occurs when we try to delete a node from a linked list that is empty. This situation exists when the `START` pointer is equal to `NULL`. If the condition is true, then the underflow message is printed on the screen; otherwise, the node is deleted from the linked list. Consider a linked list with five nodes, as shown in Figure 4.7; the node will be deleted from the beginning of the linked list.

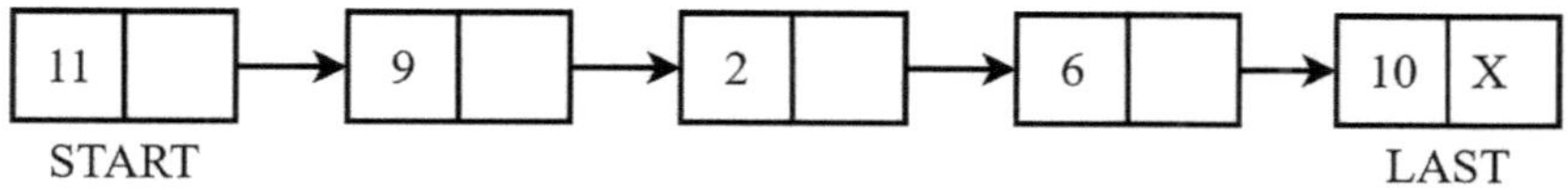

We will make another pointer variable, **PTR,** which will initially point to the START

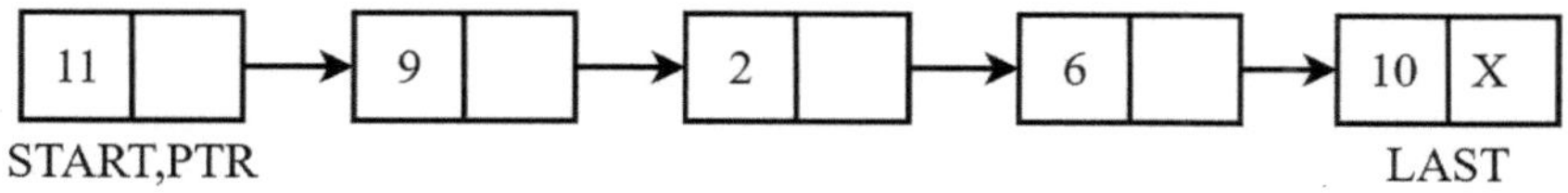

Now, as the first node is to be deleted, we will move the START pointer to the next node, which is 9.

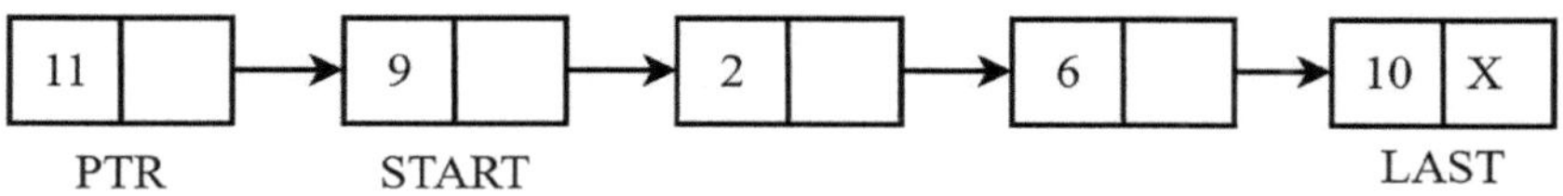

Finally, we will free the first node of the linked list. Hence, the new linked list will be:

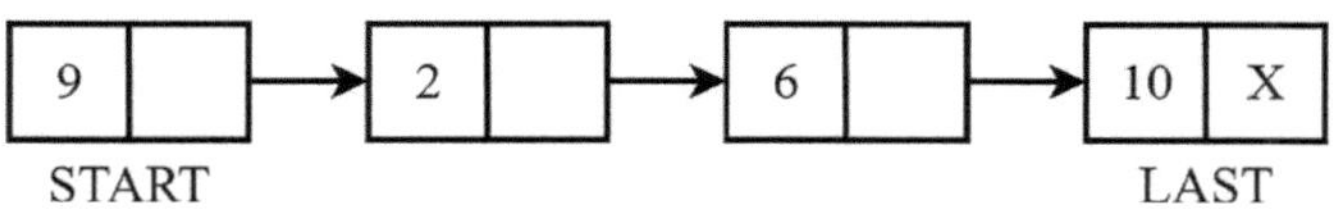

FIGURE 4.7 Deleting a node from the beginning of a linked list.

From the previous example, it is clear how a node is deleted from an already existing linked list. Let us now understand its algorithm:

```
Step 1: START
Step 2: IF START = NULL
             Print UNDERFLOW
             [End Of If]
Step 3: Set PTR = START
Step 4: Set START = START -> NEXT
Step 5: FREE PTR
Step 6: EXIT
```

In the previous algorithm, first, we check for the underflow condition, that is, whether there are any nodes present in the linked list or not. If there are no nodes, then an underflow message will be printed; otherwise, we move to Step 3, where we initialize `PTR` to `START`, that is, `PTR` will now store the address of the first node. In the next step, `START` is moved to the second node, so now `START` will store the address of the second node. Hence, the first node is deleted, and the memory that was occupied by `PTR` (initially the first node of the list) is free.

Deleting a Node from the End of the Linked List

In the case of deleting a node from the end of the linked list, we will first check the underflow condition. This situation exists when the `START` pointer is equal to `NULL`. Hence, if the condition is true, then the underflow message is printed on the screen; otherwise, the node is deleted from the linked list. Consider a linked list with five nodes, as shown in Figure 4.8; the node will be deleted from the end of the linked list.

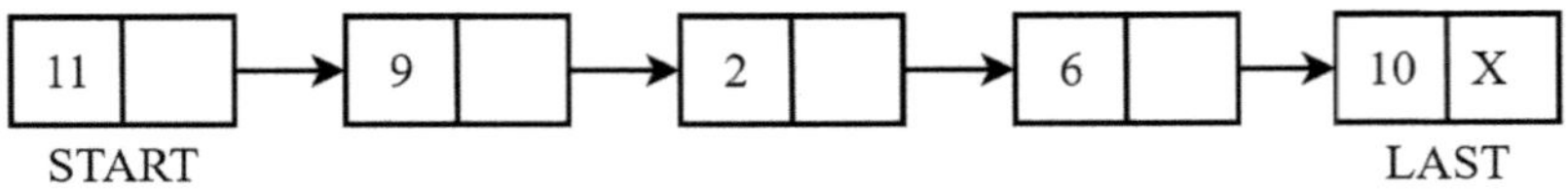

We will make another pointer variable, PTR, which will initially point to the START.

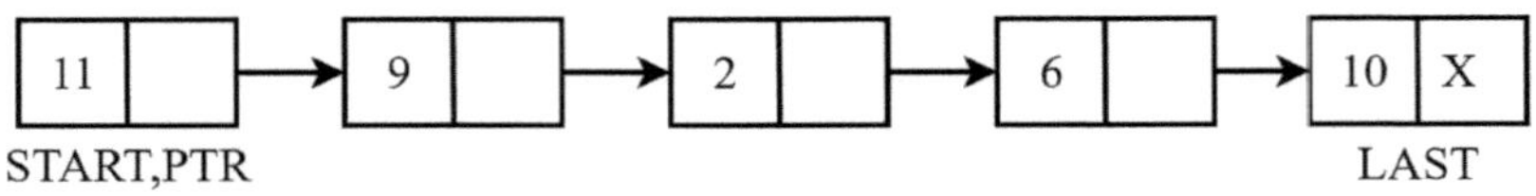

We will make another pointer variable, PREV, which will always point to the node just before the node pointed to by PTR. Move PTR till the next part of PTR = NULL.

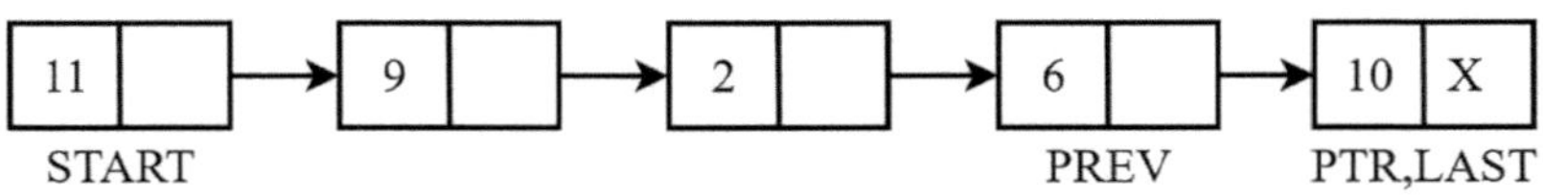

Finally, set the next part of PREV so it is equal to NULL. LAST will now point to PREV and the last node is deleted.

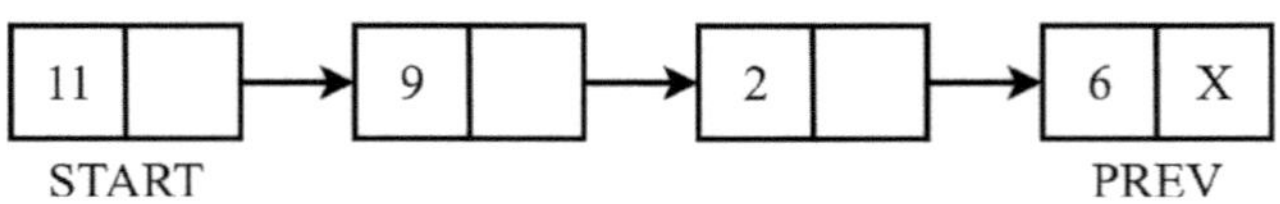

FIGURE 4.8 Deleting a node from the end of a linked list.

Let us now understand the algorithm for deleting a node from the end of a linked list:

```
Step 1: START
Step 2: IF START = NULL
             Print UNDERFLOW
             [End Of If]
Step 3: Set PTR = START
Step 4: Repeat while PTR -> NEXT != NULL
             Set PREV = PTR
             Set PTR = PTR -> NEXT
             [End of Loop]
Step 5: Set PREV -> NEXT = NULL
Step 6: FREE PTR
Step 7: EXIT
```

In the previous algorithm, we again check for the underflow condition. If the condition is true, then the underflow message is printed; otherwise, `PTR` is initialized to the `START` pointer, that is, `PTR` points to the first node of the list. In the loop, we have taken another pointer variable, `PREV`, which will always point to one node before the `PTR` node. After reaching the second-to-last and last node of the list, we will set the next part of `PREV` to `NULL`. Therefore, the last node is deleted, and the memory that was occupied by the `PTR` node is now free.

Deleting a Node after a Given Node from the Linked List

In the case of deleting a node after a given node from the linked list, we will again check the underflow condition as we checked in both of the other cases. This situation exists when the `START` pointer is equal to `NULL`. Hence, if the condition is true, then the underflow message is printed; otherwise, the node is deleted from the linked list. Consider a linked list with five nodes initially, as shown in Figure 4.9; the node will be deleted after a given node from the linked list.

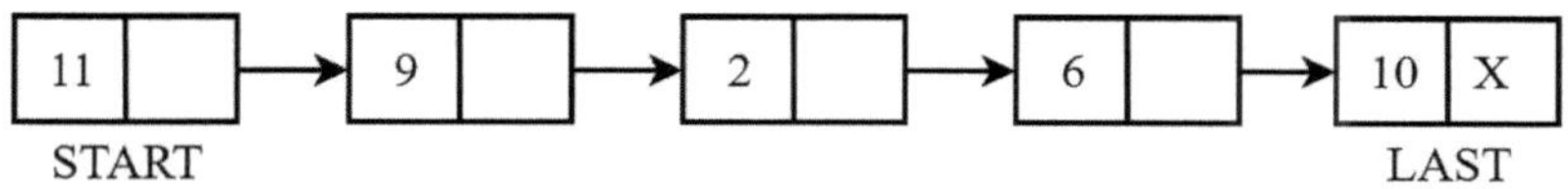

We will make another pointer variable, PTR, which will initially point to the START.

Two variables are used, PTR and PREV, which will initially point to START so that PTR, PREV, and START all point to the first node.

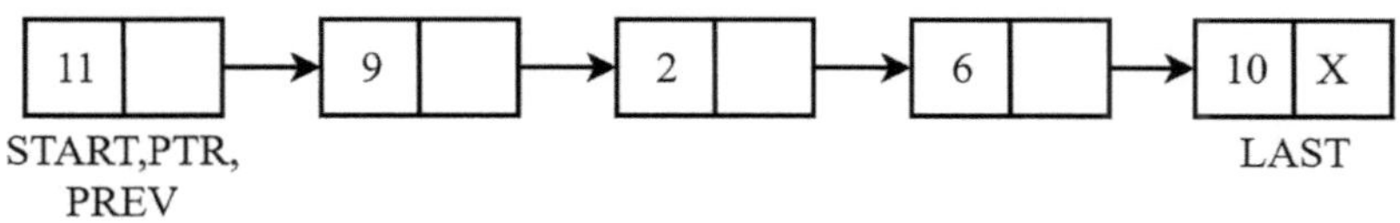

Now PTR and PREV are moved such that PREV points to the node containing the value after that the node is to be deleted and PTR points to the node that is to be deleted.

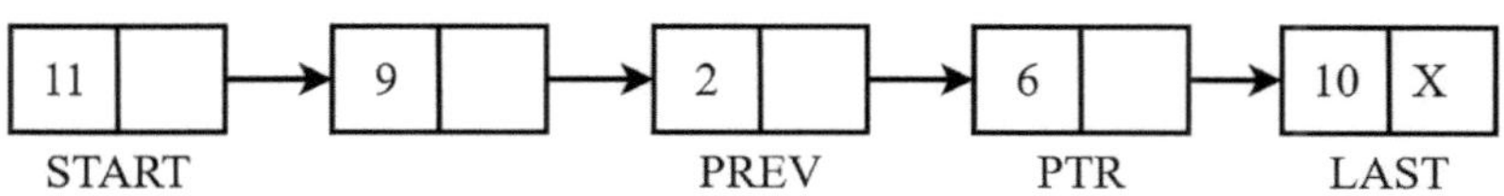

Now we know the node containing 6 is to be deleted from the linked list. So, we set the NEXT part of PREV to the NEXT part of PTR such that PREV will now point to the LAST node.

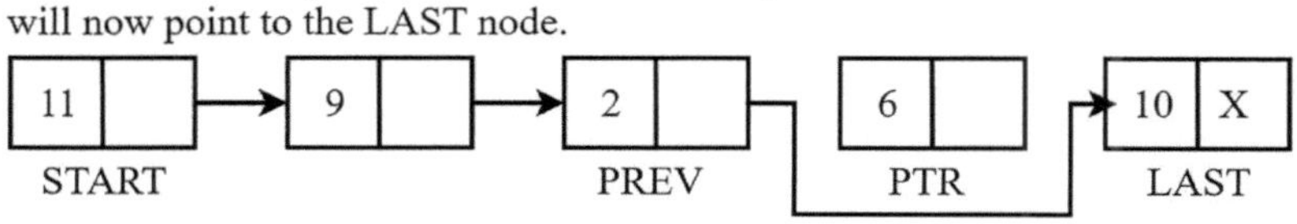

Now PTR is deleted and the new list after deletion is given below:

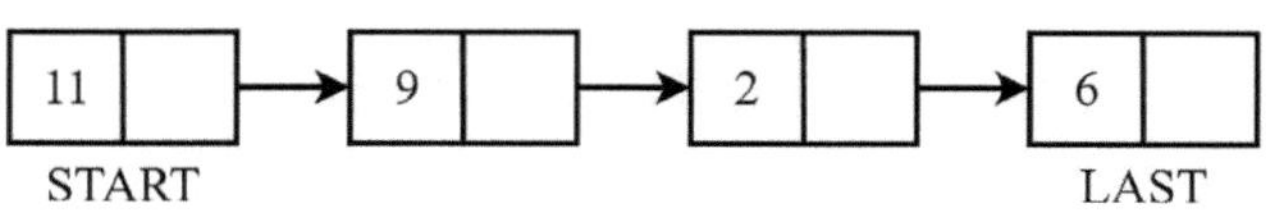

FIGURE 4.9 Deleting a node after a given node from the linked list.

Now, let us understand the previous case with the help of an algorithm:

```
Step 1: START
Step 2: IF START = NULL
             Print UNDERFLOW
             [End Of If]
Step 3: Set PTR = START
Step 4: Set PREV = START
Step 5: Repeat while PREV  -> INFO != GIVEN_VAL
             Set PREV = PTR
             Set PTR = PTR -> NEXT
             [End of Loop]
Step 6: Set PREV -> NEXT = PTR -> NEXT
Step 7: FREE PTR
Step 8: EXIT
```

In the previous algorithm, we first check for the underflow condition. If the condition is true, then the underflow message is printed; otherwise, PTR is initialized to the START pointer, that is, PTR points to the first node of the list. In the loop, we have taken another pointer variable, PREV, which will always point to one node before the PTR node. After reaching the node containing the given value that is to be deleted, we will set the next pointer of the node containing the given value to the address contained in the next part of the succeeding node. Therefore, the node is deleted, and the memory that was being occupied by the PTR is now free.

Concatenating Two Linked Lists

A concatenated linked list is created by the process of concatenating two different-sized linked lists into one linked list. Let us understand the concept of concatenation with the help of a C function:

```
void concatenate(struct node *head1, struct node *head2)
{
struct node *ptr;
ptr = head1;
 while(ptr->next!= NULL)
 {
   ptr = ptr->next;
 }
ptr->next = head2 ;
}
```

Sorting a Linked List

Sorting is the process of arranging the data elements in a sequence, either in ascending or descending order. In this, we are arranging the information of the linked list in a sequence. Let us understand it with the help of a C function:

```
void sorting()
{
int new;
struct node *ptr, *temp;
ptr = start;
 while(ptr->next!= NULL)
 {
  temp = ptr->next;
  while(temp!=NULL)
  {
   if(ptr->info > temp->info)
   {
    temp = ptr->info;
ptr->info = temp->info;
    temp->info = new;
   }
  temp = temp->next;
 }
ptr = ptr->next;
 }
}
```

Reversing a Linked List

In the process of reversing a linear linked list, we will take three pointer variables, PREV, PTR, and NEW, which will hold the addresses of the previous node, current node, and the next node, respectively, in the linked list. We will begin with the address of the first node, which is held in another pointer variable, START, which is assigned to PTR, and PREV is assigned to NULL. Now, let us understand it with the help of a C function:

```
void reverse_list()
{
ptr = start;
prev = NULL;
  while(ptr != NULL)
  {
    new = ptr->next;
    ptr->next = prev;
    prev = ptr;
    ptr = new;
  }
}
```

Write a menu-driven program for singly linked lists, performing insertion and deletion in all cases:

```
# include <stdio.h>
# include <conio.h>
# include <alloc.h>
struct node                          // Self Referential Structure
{
int data ;
struct node * next ;
  }* start ;
void add_at_beg();
void add_after(int item);
void add_before(int item);
void add_at_end();
void delete_at_beg();
void delete_at_mid();
void delete_at_end();
void main()
{
int choice, item;
 start= NULL:
clrscr();
 do
{
  printf(" SINGLY LINKED LISTS");
  printf("\n\t\t MENU");
printf(" \n 1 : addition at beginning");
printf(" \n 2 : addition after given node");
printf(" \n 3 : addition before given node");
printf(" \n 4 : addition at end");
printf(" \n 5 : deletion at beginning");
printf(" \n 6 : deletion at middle");
printf(" \n 7 : deletion at end");
printf(" \n 8 :Exit");
printf(" \n Enter your choice: ");
scanf("%d", &choice);
 switch(choice)
{
  case 1:      // Case 1 is for addition at beginning
add_at_beg();
  break;

  case 2: // Case 2 is for addition after a given node
printf(" \n enter item");
scanf(" %d", &item);
add_after(item);
  break;

  case 3: // Case 3 is for addition before a given node
printf(" \n enter item");
scanf(" %d", &item);
add_before(item);
```

```
  break;

  case 4:
add_at_end();                  // Case 4 is for addition at end
  break;

  case 5:
delete_at_beg();       // Case 5 is for deletion at beginning
  break;

  case 6:
printf(" \n enter item\n");
scanf(" %d",&item);      // Case 6 is for deletion at middle
delete_at_mid(item);
  break:

  case7:
delete_at_end();               // Case 7 is for deletion at end

  case 8:
  exit(0);

  default:
printf("wrong choice");
  break;
  }
 }
 while(ch!=  0);
getch();
}

void_add_at_beg()
{
struct node *ptr = (struct node *) malloc (size of( struct node));
//memory is allocated for ptr
printf(" enter value of node");
scanf("%d", &ptr->data);
ptr->next = start;
start= ptr;
printf("Successful Insertion!!");
getch();
return;
}

void add_after(int item)
{
struct node *ptr = (struct node *) malloc (size of( structnode));
struct node *loc ;
loc = start;
printf(" enter new node");
scanf("%d", &ptr->ata);
 if(start==NULL)
 {
printf(" list is empty , create a new list");
```

```
ptr->next = start;
   start= ptr;
   return;
  }
 while(loc != NULL)
 {
  if(loc->data==item)
  {
ptr->next = loc->next;
loc->next = ptr;
printf(" data is entered");
   return;
  }
 else
 {
loc = loc->next;
  }
 }
}

void add_before(int item)
{
struct node *ptr = (struct node *) malloc ( size of (struct node));
struct node *loc, *old;
 old = start;
loc= start->next;
printf(" enter new node value ");
scanf("%d", &ptr->data);
 if(start==NULL)
 {
printf(" list is empty ,create 1st  node");
ptr->next =start;
   start= ptr;
  }
  while(loc!= NULL)
  {
   if(loc->data==item)
   {
ptr->next = loc;
    old->next = ptr;
printf(" data is entered before item");
    return;
   }
  else
  {
   old=loc;
loc = loc->next;
  }
 }
}

void add_at_end()
{
struct node*ptr = (struct node *) malloc (size of( struct node));
```

```
struct node*loc ;
loc = start;
printf(" enter new node");
scanf("%d", &ptr->data);
  if(start==NULL)
  {
printf(" list is empty , create new node");
ptr->next = start;
    start= ptr;
  }
  while(loc->next!= NULL)
  {
loc = loc->next;
  }
ptr->next = NULL;
loc->next = ptr;
printf(" data is entered at the end");
 return;
}

void delete_at_beg()
{
struct node *ptr;
 if(start == NULL)
 {
printf(" list is empty , create new");
ptr->next = start;
  start = ptr;
  return;
 }
 else
 {
ptr = start;
  start = start->next;
 }
printf(" %d data is deleted",p->data);
  free(ptr);
}

void delete_at_mid(int item)
{
struct node*ptr ,*loc *old;
  if(start==NULL)
  {
printf(" list is empty , create 1st  node ");
ptr->next = start;
   start= ptr;
  }
  else
 {
  old = start;
loc = start->next;
  while(loc!=NULL)
```

```
 {
  if(loc->data == item)
  {
ptr = loc;
   old->next=loc->next;
   free(ptr);
   return;
  }
  else
  {
    old = loc;
loc = loc->next;
 }
printf("item not found");
  }
 }
}

void delete_at_end()
{
struct node *ptr,*loc;
  if(start == NULL)
  {
printf(" list is empty , create 1st node ");
ptr->next = start;
    start = ptr;
   }
  else
  {
loc = start;
ptr = start->next;
  while(ptr->next!=NULL)
 {
loc = ptr;
ptr = ptr->next;
  }
loc->next= NULL;
  free(ptr);
 }
}
```

Here is the output:

```
SINGLY LINKED LIST
 MENU
1: Addition at beginning
2: Addition after given node
3: Addition before given node
4: Addition at end
5: Deletion at beginning
6: Deletion at middle
7: Deletion at end
8: Exit
```

```
Enter your choice: 1
Enter value for node
15
Successful Insertion!!
```

After discussing the singly linked list, we will now learn about another type of linked list, the circular linked list.

Circular Linked Lists

Circular linked lists are a type of singly linked list in which the address part of the last node stores the address of the first node, unlike singly linked lists, in which the address part of the last node stores a unique value, `NULL`. While traversing a circular linked list, we can begin from any node, and we traverse the list in any direction, because a circular linked list does not have a first or last node. The memory declarations for representing a circular linked list are the same as for a linear linked list, as shown in Figure 4.10.

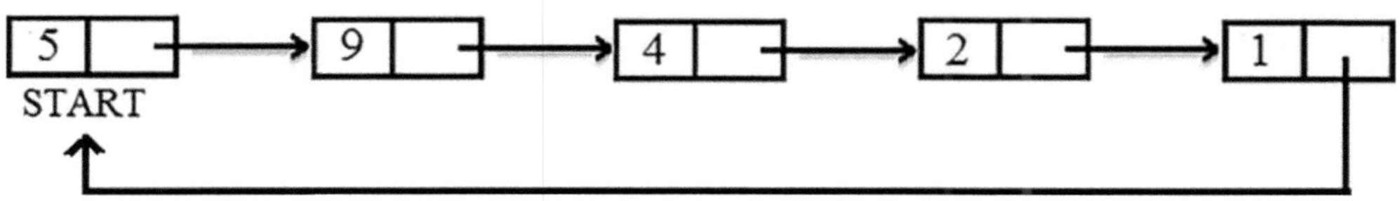

FIGURE 4.10 Circular linked list.

Operations on a Circular Linked List

Various operations can be performed on a circular linked list, which include:

a) Inserting a new node in a circular linked list
b) Deleting a node from a circular linked list

Let us now discuss both these cases in detail.

Inserting a New Node in a Circular Linked List

Here, we will learn how a new node is inserted into an existing linked list. We will discuss two cases in the insertion process:

1. A new node is inserted at the beginning of a circular linked list.
2. A new node is inserted at the end of a circular linked list.

A new node can also be inserted after a given node (same as that for a singly linked list).

Inserting a New Node at the Beginning of a Circular Linked List

In the case of inserting a new node at the beginning of a circular linked list, we will first check the overflow condition, that is, whether the memory is available for a new node. If the memory is not available, then an overflow message is printed; otherwise, the memory is allocated for the new node. Now, we will initialize the node with its info part, and its address part will contain the address of the first node of the list, which is the START pointer. Hence, the new node is added as the first node in the list, and the START pointer will point to the first node of the list. Now, let us take an example. Consider a linked list (as shown in Figure 4.11) with four nodes; a new node is to be inserted at the beginning of the circular linked list.

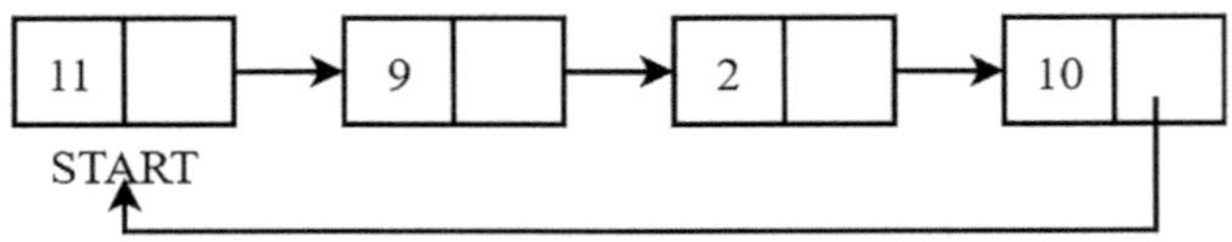

Now memory is allocated for the new node and we initialize its info part with 6.

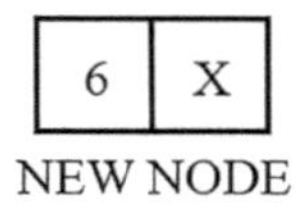

We will take another pointer variable, PTR, which will initially point to the START.

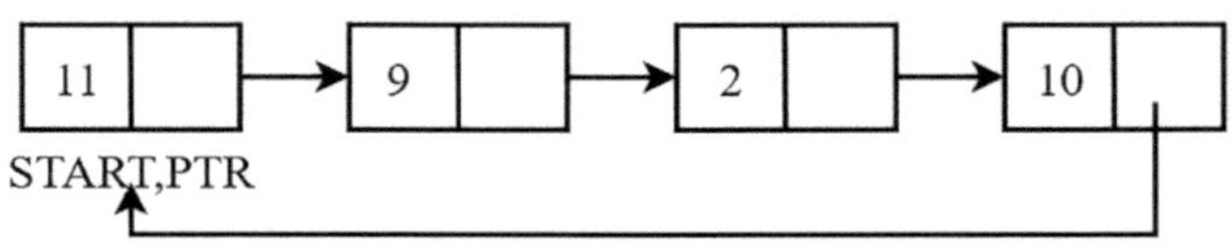

Now PTR is moved so that it points to the last node.

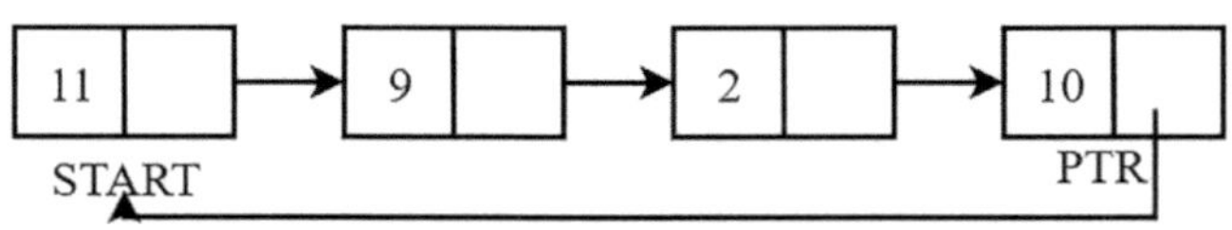

As the new node is to be inserted at the beginning of the list, PTR will now store the address of the new node, and the new node will store the address of START.

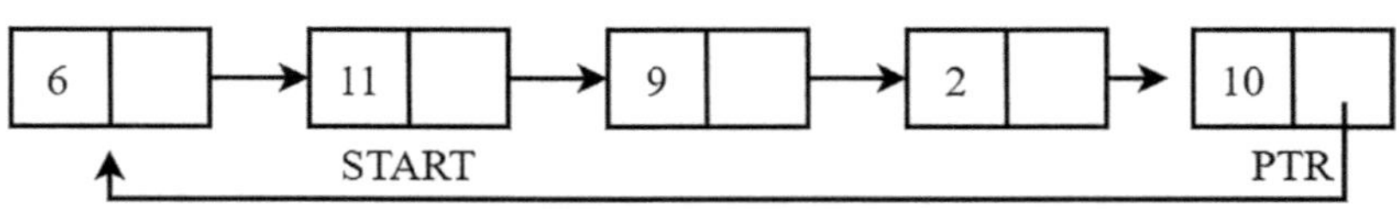

Now the new node is inserted at the beginning so START will point to the first node of the list, i.e., new node.

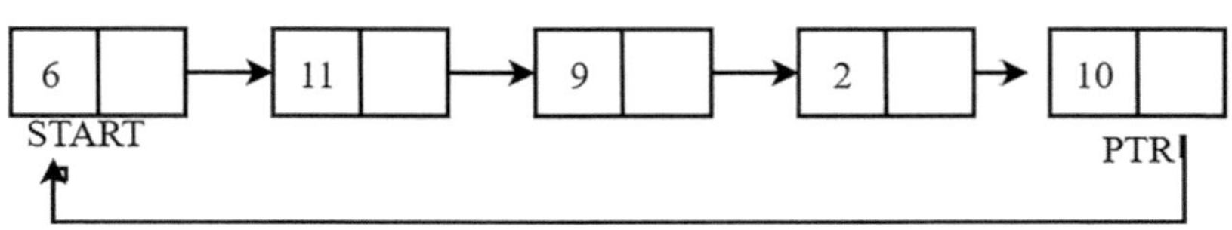

FIGURE 4.11 Inserting a new node at the beginning of a circular linked list.

Now, let us understand the previous case with the help of an algorithm:

```
Step 1: START
Step 2: IF TEMP = NULL
            Print OVERFLOW
            [End Of If]
Step 3: Set NEW NODE = TEMP
Step 4: Set NEW NODE -> INFO = VAL
Step 5: Set PTR = START
Step 6: Repeat while PTR  -> NEXT != START
            Set PTR = PTR -> NEXT
            [End of Loop]
Step 7: Set NEW NODE -> NEXT = START
Step 8: Set PTR -> NEXT = NEW NODE
Step 9: Set START = NEW NODE
Step 10: EXIT
```

Inserting a New Node at the End of a Circular Linked List

In this case, we will first check the overflow condition, that is, whether the memory is available for a new node. If the memory is not available, then an overflow message is printed; otherwise, the memory is allocated for the new node. Then, a `PTR` variable is made, which will initially point to `START`, and will be used to traverse the linked list until it reaches the last node. When it reaches the last node, the `NEXT` part of the last node will store the address of the new node, and the `NEXT` part of `NEW NODE` will contain the address of the first node of the linked list, which is denoted by `START`. Let us understand it with the help of an algorithm:

```
Step 1: START
Step 2: IF TEMP = NULL
            Print OVERFLOW
            [End Of If]
Step 3: Set NEW NODE = TEMP
Step 4: Set NEW NODE -> INFO = VAL
Step 5: Set NEW NODE -> NEXT = START
Step 6: Set PTR = START
Step 7: Repeat while PTR  -> NEXT != START
            Set PTR = PTR -> NEXT
            [End of Loop]
Step 8: Set PTR -> NEXT = NEW NODE
Step 9: EXIT
```

Let us take an example to understand it. Consider a linked list with four nodes, as shown in Figure 4.12; a new node is to be inserted at the end of the circular linked list.

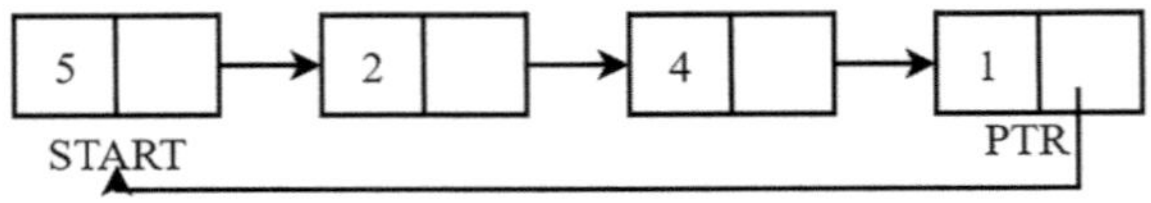

Now memory is allocated for the new node and we initialize its info part with 7

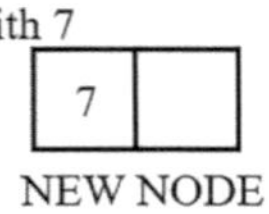

We will take another pointer variable, PTR, which will initially point to the START.

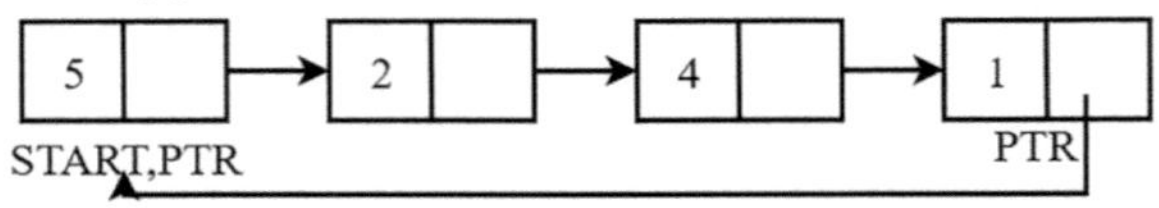

Now PTR is moved so that it points to the last node.

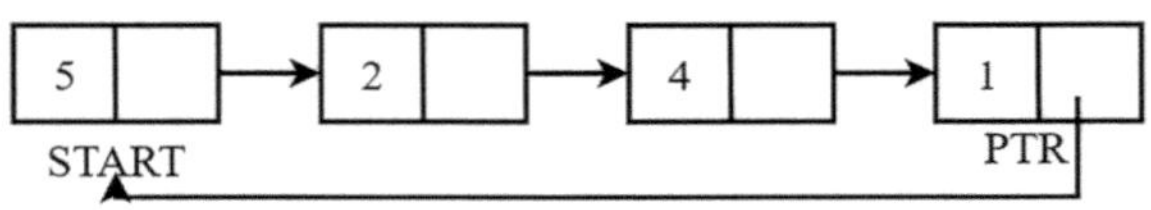

As the New Node is to be inserted at the end of the list, PTR will store the address of New Node and New Node will store the address of START.

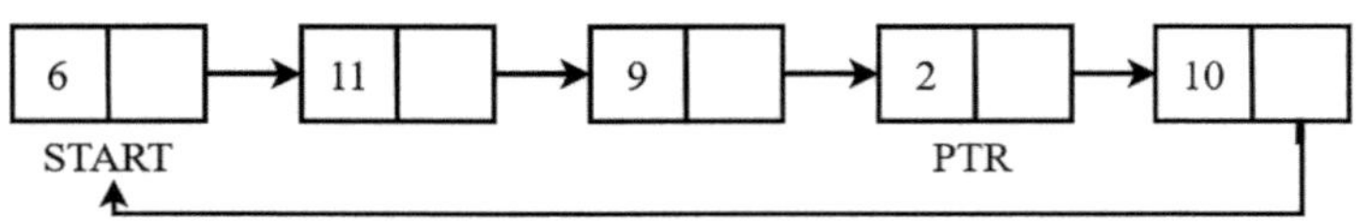

So New Node is inserted at the end of the circular linked list and is now the last node of the circular linked list.

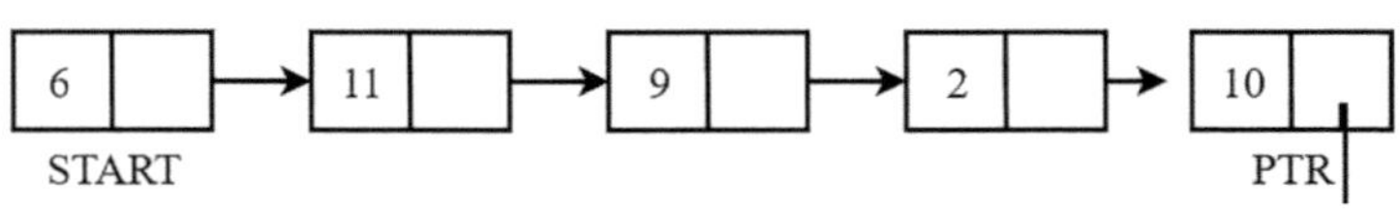

FIGURE 4.12 Inserting a new node at the end of a circular linked list.

Deleting a Node from a Circular Linked List

In this section, we will learn how a node is deleted from an already existing circular linked list. We will discuss two cases in the deletion process:

1. The node is deleted from the beginning of the circular linked list.
2. The node is deleted from the end of the circular linked list.

A node can also be deleted after a given node in the same way as for a singly linked list.

Deleting a Node from the Beginning of a Circular Linked List

In the case of deleting a node from the beginning of a linked list, we will first check the underflow condition, which occurs when we try to delete a node from a linked list that is empty. This situation exists when the `START` pointer is equal to `NULL`. Hence, if the condition is true, then an underflow message is displayed; otherwise, the node is deleted from the linked list. Consider a linked list with four nodes, as shown in Figure 4.13; the first node will be deleted from the linked list.

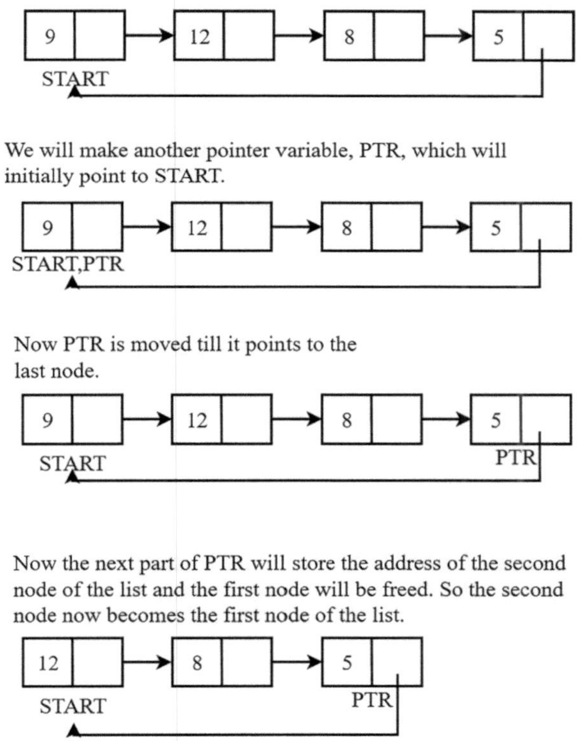

FIGURE 4.13 Deleting a node from the beginning of a circular linked list.

From the previous example, it is clear how a node will be deleted from an already existing linked list. Let us now understand its algorithm:

```
Step 1: START
Step 2: IF START = NULL
            Print UNDERFLOW
            [End Of If]
Step 3: Set PTR = START
Step 4: Repeat while PTR -> NEXT != START
Set PTR = PTR -> NEXT
            [End of Loop]
```

```
Step 5: Set PTR -> NEXT = START -> NEXT
Step 6: FREE START
Step 7: Set START = START -> NEXT
Step 8: EXIT
```

The previous algorithm shows how a node is deleted from the beginning of the linked list. First, we check for the underflow condition. Now, a pointer variable, `PTR`, is used, which will traverse the entire list until it reaches the last node of the list. Now, we change the next part of `PTR` to store the address of the second node of the list. Hence, the memory that occupied the first node is freed. Finally, the second node now becomes the first node of the linked list.

Deleting a Node from the End of a Circular Linked List

In this case, we will first check the underflow condition, which is when we try to delete a node from a linked list that is empty. This situation occurs when the `START` pointer is equal to `NULL`. Hence, if the condition is true, then an underflow message is printed; otherwise, the node is deleted from the linked list. Consider a linked list with four nodes, as shown in Figure 4.14; the last node will be deleted from the linked list.

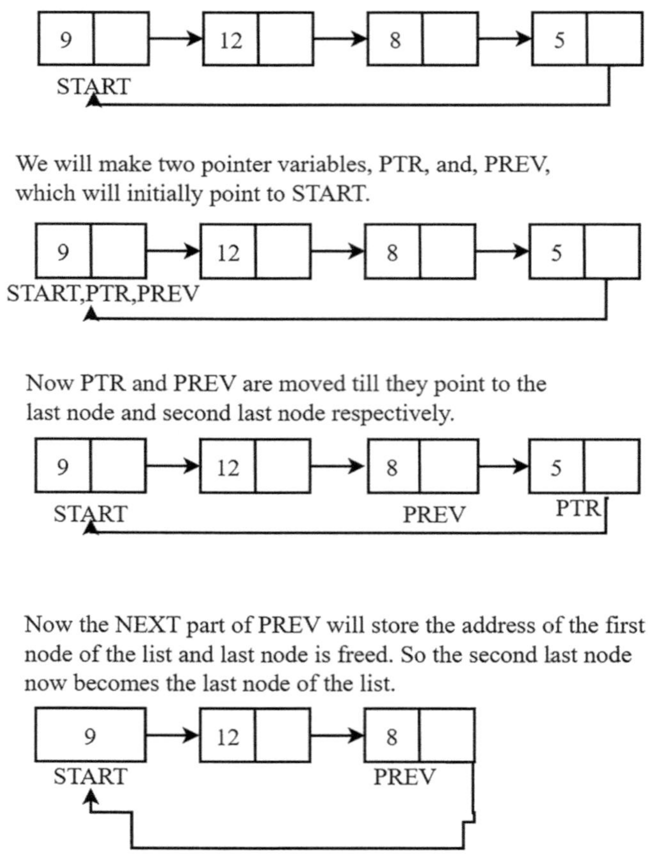

FIGURE 4.14 Deleting a node from the end of a circular linked list.

Let us now understand its algorithm:

```
Step 1: START
Step 2: IF START = NULL
             Print UNDERFLOW
             [End Of If]
Step 3: Set PTR = START
Step 4: Repeat while PTR -> NEXT != START
Set PREV = PTR
Set PTR = PTR -> NEXT
             [End of Loop]
Step 5: Set PREV -> NEXT = START
Step 6: FREE PTR
Step 7: EXIT
```

The previous algorithm shows how a node is deleted from the end of the linked list. First, we are checking for the underflow condition. Now, a pointer variable, `PTR`, is used to traverse the entire list until it reaches the last node of the list. In the `while` loop, we will use another pointer variable, `PREV`, which will always point to the node preceding `PTR`. When we reach the last node and its preceding node, that is, the second-to-last node, we will now change the next part of `PREV` to store the address of `START`. Hence, the memory occupied by the last node is freed. Finally, the second-to-last node now becomes the last node of the linked list. In this way, the deletion of a node from the end is done in a circular linked list. The practical implementation of insertion and deletion operations in a circular linked list is discussed with the help of code as follows.

Write a menu-driven program for circular linked lists, performing insertion and deletion in all cases:

```
# include <stdio.h>
# include <conio.h>
# include <alloc.h>
struct node        //Self Referential Structure
{
int data ;
struct node * next ;
  }* start ;
void add_at_beg();
void add_after(int item);
void add_before(int item);
void add_at_end();
void delete_at_beg();
void delete_at_mid();
void delete_at_end();
void main()
{
int choice, item;
 start = NULL:
```

```
clrscr();
 do
{
printf(" CIRCULAR LINKED LISTS");
printf("\n\t MENU");
printf(" \n 1 : addition at beginning");
printf(" \n 2 : addition after given node");
printf(" \n 3 : addition before given node");
printf(" \n 4 : addition at end");
printf(" \n 5 : deletion at beginning");
printf(" \n 6 : deletion at middle");
printf(" \n 7 : deletion at end");
printf(" \n 8 : Exit");
scanf("%d", &choice);
 switch(choice)
{
  case 1:                //Case 1 is for addition at beginning
add_at_beg();
  break;

  case 2:          //Case 2 is for addition after given node
printf(" \n enter item");
scanf(" %d", &item);
add_after(item);
  break;

  case 3: //Case 3 is for addition before given node
printf(" \n enter item");
scanf(" %d", &item);
add_before(item);
  break;

  case 4:
add_at_end();                 //Case 4 is for addition at end
  break;

  case 5:
delete_at_beg();        //Case 5 is for deletion at beginning
  break;

  case 6:
printf(" \n enter item\n");
scanf(" %d", &item);      //Case 6 is for deletion at middle
delete_at_mid(item);
  break:

  case7:
delete_at_end();              //Case 7 is for deletion at end

  case 8:
  exit(0);

  default:
```

```
printf("wrong choice");
  break;
  }
 }
 while(ch!=  0);
getch();
}

void_add_at_beg()
{
struct node *temp;
struct node *ptr = (struct node *) malloc (size of( struct node));
printf(" enter value of node");
scanf("%d", &ptr->data);
temp = start;
while(temp->next != start)
{
  temp = temp->next;
}
ptr->next = start;
start= ptr;
temp->next = start;
printf("Successful Insertion!!");
getch();
return;
}

void add_after(int item)
{
struct node *ptr = (struct node *) malloc (size of( struct node));
struct node *loc ;
loc = start;
printf(" enter new node");
scanf("%d", &ptr->data);
 if(start==NULL)
 {
printf(" list is empty , create a new list");
ptr->next = start;
   start= ptr;
   return;
  }
 while(loc != NULL)
 {
  if(loc->data==item)
  {
ptr->next = loc->next;
loc->next = ptr;
printf(" data is entered");
   return;
  }
```

```
 else
 {
loc = loc->next;
  }
 }
}

void add_before(int item)
{
struct node *ptr = (struct node *) malloc ( size of (struct node));
struct node *loc, *old;
 old = start;
loc= start->next;
printf(" enter new node value ");
scanf("%d", &ptr->data);
 if(start==NULL)
 {
printf(" list is empty ,create 1st  node");
ptr->next =start;
   start= ptr;
  }
  while(loc!= NULL)
  {
   if(loc->data==item)
   {
ptr->next = loc;
    old->next = ptr;
printf(" data is entered before item");
    return;
   }
  else
  {
   old=loc;
loc = loc->next;
  }
 }
}

void add_at_end()
{
struct node *ptr = (struct node *) malloc (size of( struct node));
struct node *temp ;
  temp = start;
printf(" enter new node");
scanf("%d", &ptr->data);
  if(start == NULL)
  {
printf(" list is empty , create new node");
ptr->next = start;
    start= ptr;
  }
  while(temp->next != start)
  {
    temp = temp->next;
```

```
    temp->next = ptr;
ptr->next = start;
printf(" data is entered at the end");
    return;
}

void delete_at_beg()
{
struct node *ptr, *temp;
 if(start == NULL)
 {
printf(" list is empty , create new");
ptr->next = start;
  start = ptr;
  return;
 }
  temp = start;
  while(temp->next != start);
  temp = temp->next;
ptr = start;
  start = ptr->next;
  temp->next = start;
printf(" %d data is deleted",ptr->data);
  free(ptr);
}

void delete_at_mid(int item)
{
struct node*ptr ,*loc *old;
  if(start==NULL)
  {
printf(" list is empty , create 1st  node ");
ptr->next = start;
   start= ptr;
  }
  else
 {
  old = start;
loc = start->next;
  while(loc!=NULL)
 {
  if(loc->data == item)
  {
ptr = loc;
   old->next=loc->next;
   free(ptr);
   return;
  }
  else
  {
    old = loc;
loc = loc->next;
 }
```

```
printf("item not found");
  }
 }
}

void delete_at_end()
{
struct node *ptr, *temp;
  if(start == NULL)
  {
printf(" list is empty , create 1st node ");
ptr->next = start;
    start = ptr;
   }
   temp = start;
ptr = start->nex t;
  while(temp->next != start)
 {
ptr = temp;
   temp = temp->next;
  }
ptr->next =start;
  free(temp);
 }
```

Here is the output:

```
CIRCULAR LINKED LIST
             MENU
1: Addition at beginning
2: Addition after given node
3: Addition before given node
4: Addition at end
5: Deletion at beginning
6: Deletion at middle
7: Deletion at end
8: Exit

Enter your choice: 1
Enter value for node
34
Successful Insertion!!
```

Doubly Linked Lists

A doubly linked list is also called a two-way linked list; it is a special type of linked list that can point to the next node as well as the previous node in the sequence. In a doubly linked list, each node is divided into three parts, as shown in Figure 4.15:

1. The first part is called the previous pointer, which contains the address of the previous node in the list.
2. The second part is called the information part, which contains the information about the node.
3. The third part is called the next pointer, which contains the address of the succeeding node in the list.

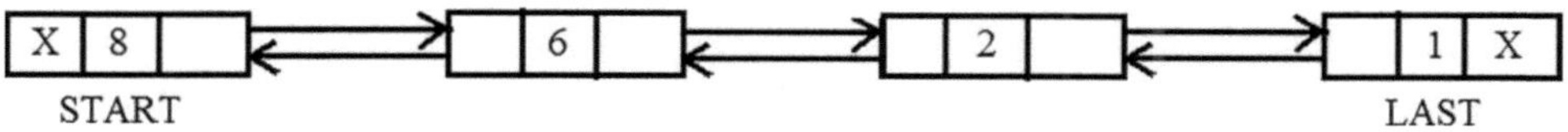

FIGURE 4.15 Doubly linked list.

The structure of a doubly linked list is given as follows:

```
struct node
{
struct node * prev ;
int info ;
struct node * next ;
 } ;
```

The first node of the linked list will contain a `NULL` value in the previous pointer to indicate that there is no element preceding it in the list, and, similarly, the last node will also contain a `NULL` value in the next pointer field to indicate that there is no element succeeding it in the list. Doubly linked lists can be traversed in both directions.

Operations on a Doubly Linked List

Various operations can be performed on a doubly linked list, which include:

1. Inserting a new node in a doubly linked list
2. Deleting a node from a doubly linked list

Let us now discuss both these operations in detail.

Inserting a New Node in a Doubly Linked List

In this section, we will learn how a new node is inserted into an already existing doubly linked list. We will consider four cases for the insertion process in a doubly linked list:

1. A new node is inserted at the beginning.
2. A new node is inserted at the end.
3. A new node is inserted after a given node.
4. A new node is inserted before a given node.

Inserting a New Node at the Beginning of a Doubly Linked List

In the case of inserting a new node at the beginning of a doubly linked list, we will first check for the overflow condition, that is, whether the memory is available for a new node. If the memory is not available, then an overflow message is displayed; otherwise, the memory is allocated for the new node. Now, we will initialize the node with its info part, and its address part will contain the address of the first node of the list, which is the `START` pointer. Hence, the new node is added as the first node in the list, and the `START` pointer will point to the first node of the list. Now, to understand better, let us take an example. Consider a linked list with four nodes, as shown in Figure 4.16; a new node will be inserted at the beginning of the linked list.

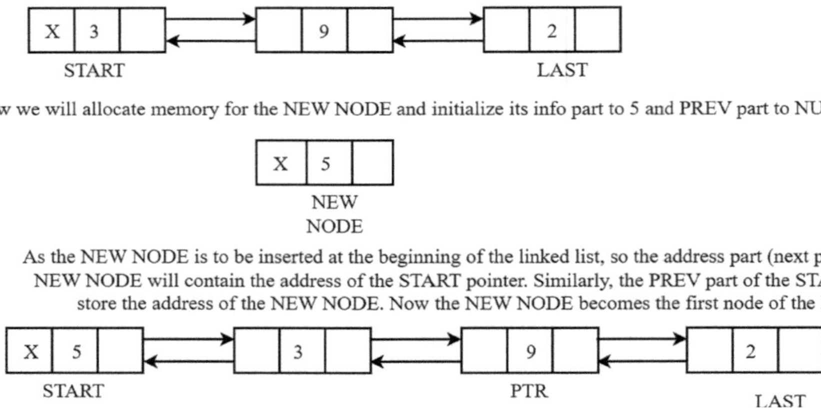

FIGURE 4.16 Inserting a new node at the beginning of a doubly linked list.

From the above example, it is clear how a new node will be inserted into an already existing doubly linked list. Let us now understand its algorithm:

```
Step 1: START
Step 2: IF PTR = NULL
            Print OVERFLOW
            Go to Step 9
            [End of If]
Step 3: Set NEW NODE = PTR
Step 4: Set NEW NODE -> INFO = VALUE
Step 5: Set NEW NODE -> PREV = NULL
Step 6: Set NEW NODE -> NEXT = START
Step 7: Set START -> PREV = NEW NODE
Step 8: Set START = NEW NODE
Step 9: EXIT
```

Inserting a New Node at the End of a Doubly Linked List

In the case of inserting the new node at the end of the linked list, we will first check the overflow condition, which is whether the memory is available for a new node. If the memory is not available, then an overflow message is printed; otherwise, the memory is allocated for the new node. Then, a PTR variable is made, which will initially point to START, and a PTR variable will be used to traverse the list until it reaches the last node. When it reaches the last node, the NEXT part of the last node will store the address of the new node, and the NEXT part of NEW NODE will contain NULL, which will denote the end of the linked list. The PREV part of NEW NODE will store the address of the node pointed to by PTR. Let us understand it with the help of an algorithm:

```
Step 1: START
Step 2: IF PTR = NULL
             Print OVERFLOW
             [End of If]
Step 3: Set NEW NODE = PTR
Step 4: Set PTR = PTR -> NEXT
Step 5: Set NEW NODE -> INFO = VALUE
Step 6: Set NEW NODE -> NEXT = NULL
Step 7: Set PTR = START
Step 8: Repeat while PTR -> NEXT != NULL
             Set PTR = PTR -> NEXT
             [End of Loop]
Step 9: Set PTR -> NEXT = NEW NODE
Step 10: Set NEW NODE -> PREV = PTR
Step 11: EXIT
```

Hence, from the previous algorithm, we understand how to insert a new node at the end of a doubly linked list. Now, we will study this further with the help of an example. Consider a linked list with four nodes, as shown in Figure 4.17; a new node will be inserted at the end of the doubly linked list:

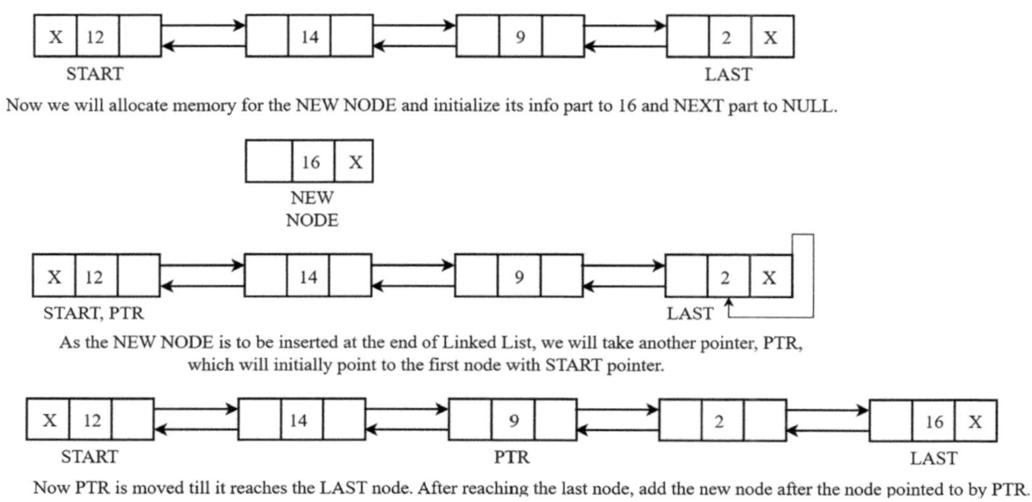

FIGURE 4.17 Inserting a new node at the end of a doubly linked list.

Inserting a New Node After a Given Node in a Doubly Linked List

In this case, a new node is inserted after a given node in a doubly linked list. As in the other cases, we will again check the overflow condition in it. If the memory for the new node is available, then it will be allocated; otherwise, an overflow message is displayed. Then, a `PTR` variable is made, which will initially point to `START`, and the `PTR` variable is used to traverse the linked list until its value becomes equal to the value after which the new node is to be inserted. When it reaches that node/value, then the `NEXT` part of that node will store the address of the new node, and the `PREV` part of `NEW NODE` will store the address of the preceding node. Let us understand it with the help of the following algorithm:

```
Step 1: START
Step 2: IF PTR = NULL
            Print OVERFLOW
            Go to Step 10
            [End of If]
Step 3: Set NEW NODE = PTR
Step 4: Set NEW NODE -> INFO = VALUE
Step 5: Set PTR = START
Step 6: Repeat while PTR -> INFO != GIVEN_VAL
            Set PTR = PTR -> NEXT
            [End of Loop]
Step 7: Set NEW NODE -> NEXT = PTR -> NEXT
Step 8: Set NEW NODE -> PREV = PTR
Step 9: Set PTR -> NEXT = NEW NODE
Step 10: EXIT
```

Now, we will understand it more with the help of an example. Consider a doubly linked list with four nodes, as shown in Figure 4.18; a new node will be inserted after a given node in the linked list:

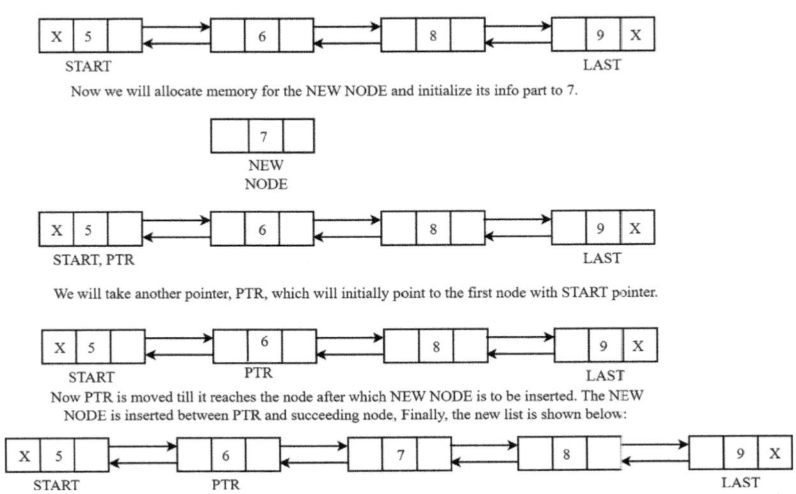

FIGURE 4.18 Inserting a new node after a given node in a doubly linked list.

Inserting a New Node Before a Given Node in a Doubly Linked List

In this case, a new node is inserted before a given node in a doubly linked list. As in the other cases, we will again check the overflow condition in it. If the memory for the new node is available, then it will be allocated; otherwise, an overflow message is displayed. Then, a `PTR` variable is made, which will initially point to `START`, and the `PTR` variable is used to traverse the linked list until its value becomes equal to the value before which the new node is to be inserted. When it reaches that node/value, then the `PREV` part of that node will store the address of `NEW NODE`, and the `NEXT` part of `NEW NODE` will store the address of the succeeding node. Now, to understand it better, let us take an example. Consider a linked list with four nodes, as shown in Figure 4.19; a new node will be inserted before a given node in the linked list.

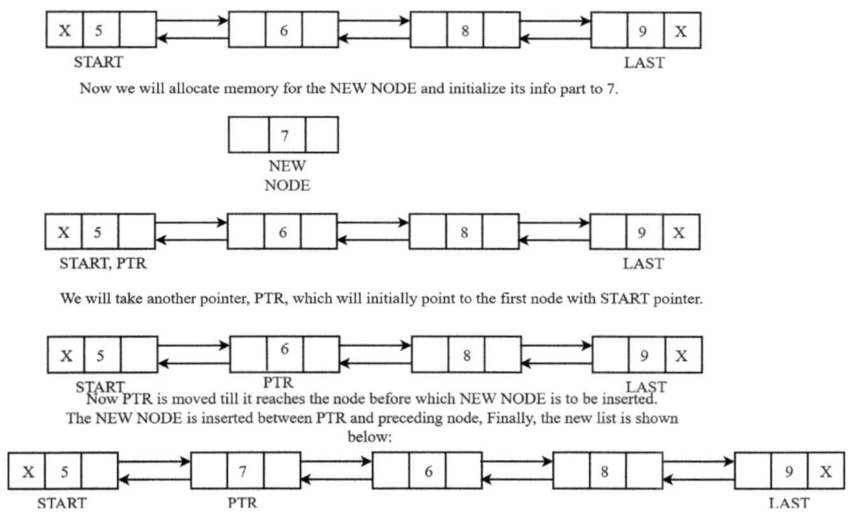

FIGURE 4.19 Inserting a new node before a given node in a doubly linked list.

From the previous example, it is clear how a new node will be inserted into an already existing doubly linked list. Let us now understand its algorithm:

```
Step 1: START
Step 2: IF PTR = NULL
             Print OVERFLOW
             Go to Step 10
             [End of If]
Step 3: Set NEW NODE = PTR
Step 4: Set NEW NODE -> INFO = VALUE
Step 5: Set PTR = START
Step 6: Repeat while PTR -> INFO != GIVEN_VAL
             Set PTR = PTR -> NEXT
             [End of Loop]
```

```
Step 7: Set NEW NODE -> NEXT = PTR
Step 8: Set NEW NODE -> PREV = PTR -> PREV
Step 9: Set PTR -> PREV = NEW NODE
Step 10: EXIT
```

Deleting a Node from a Doubly Linked List

In this section, we will learn how a node is deleted from an already existing doubly linked list. We will consider four cases for the deletion process in a doubly linked list:

1. A node is deleted from the beginning of the linked list.
2. A node is deleted from the end of the linked list.
3. A node is deleted after a given node from the linked list.
4. A node is deleted before a given node from the linked list.

Now, let us discuss all the previous cases in detail.

Deleting a Node from the Beginning of the Doubly Linked List

In the case of deleting a node from the beginning of the doubly linked list, we will first check the underflow condition, which occurs when we try to delete a node from a linked list that is empty. This situation exists when the `START` pointer is equal to `NULL`. Hence, if the condition is true, then the underflow message is displayed; otherwise, the node is deleted from the linked list. Consider a linked list with five nodes, as shown in Figure 4.20; the node will be deleted from the beginning of the linked list.

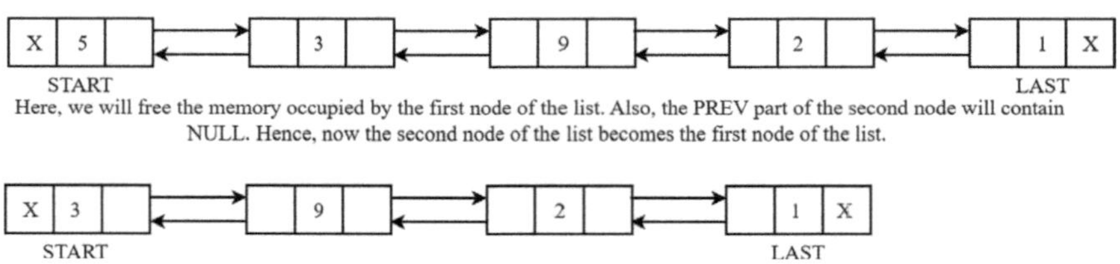

FIGURE 4.20 Deleting a node from the beginning of the doubly linked list.

Let us understand this with the help of an algorithm:

```
Step 1: START
Step 2: IF START = NULL
             Print UNDERFLOW
             [End Of If]
Step 3: Set PTR = START
Step 4: Set START = START -> NEXT
Step 5: Set START -> PREV = NULL
Step 6: FREE PTR
Step 7: EXIT
```

In the previous algorithm, first, we are checking for the underflow condition, which is whether there are any nodes present in the linked list or not. If there are no nodes, then an underflow message will be printed; otherwise, we move to `Step 3`, where we are initializing `PTR` to `START`, that is, `PTR` will now store the address of the first node. In the next step, `START` is moved to the second node, as now `START` will store the address of the second node. Also, the `PREV` part of the second node will now contain a value, `NULL`. Hence, the first node is deleted, and the memory that was being occupied by `PTR` is free (initially, the first node of the list).

Deleting a Node from the End of the Doubly Linked List

In the case of deleting a node from the end of the linked list, we will first check the underflow condition. This situation exists when the `START` pointer is equal to `NULL`. Hence, if the condition is true, then the underflow message is printed on the screen; otherwise, the node is deleted from the linked list. Consider a linked list with five nodes, as shown in Figure 4.21; the node will be deleted from the end of the linked list.

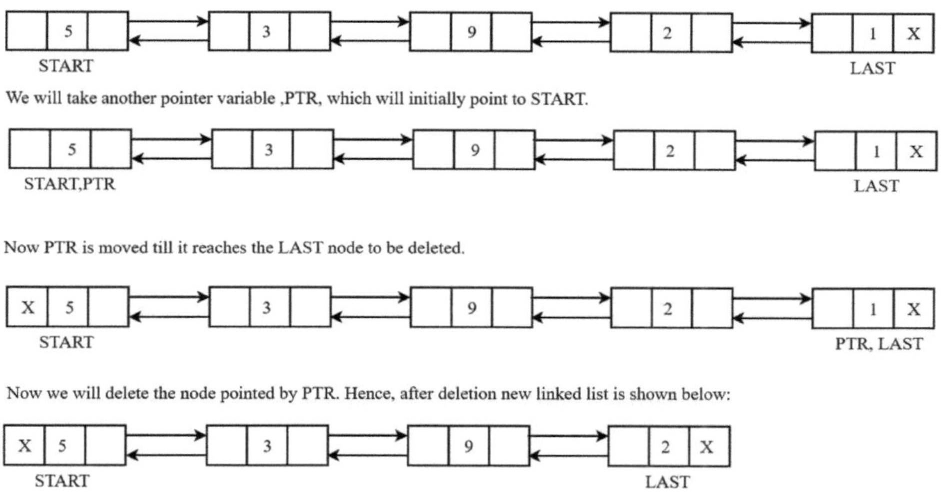

FIGURE 4.21 Deleting a node from the end of the doubly linked list.

From the previous example, it is clear how a node will be deleted from an already existing doubly linked list. Let us now understand its algorithm:

```
Step 1: START
Step 2: IF START = NULL
             Print UNDERFLOW
             [End Of If]
Step 3: Set PTR = START
Step 4: Repeat while PTR -> NEXT != NULL
             Set PTR = PTR -> NEXT
             [End of Loop]
Step 5: Set PTR -> PREV -> NEXT= NULL
Step 6: FREE PTR
Step 7: EXIT
```

In the previous algorithm, again, we are checking for the underflow condition. If the condition is true, then the underflow message is printed; otherwise, `PTR` is initialized to the `START` pointer, that is, `PTR` is pointing to the first node of the list. In the loop, `PTR` is traversed until it reaches the last node of the list. After reaching the last node of the list, we can also access the second-to-last node by taking the address from the `PREV` part of the last node. Therefore, the last node is deleted, and the memory is now free, which was occupied by the `PTR`.

Deleting a Node After a Given Node from the Doubly Linked List

In the case of deleting a node after a given node from the linked list, we will again check the underflow condition as we checked in both of the other cases. This situation exists when the `START` pointer is equal to `NULL`. Hence, if the condition is true, then the underflow message is displayed; otherwise, the node is deleted from the linked list. Consider a linked list with five nodes, as shown in Figure 4.22; the node will be deleted after a given node in the linked list.

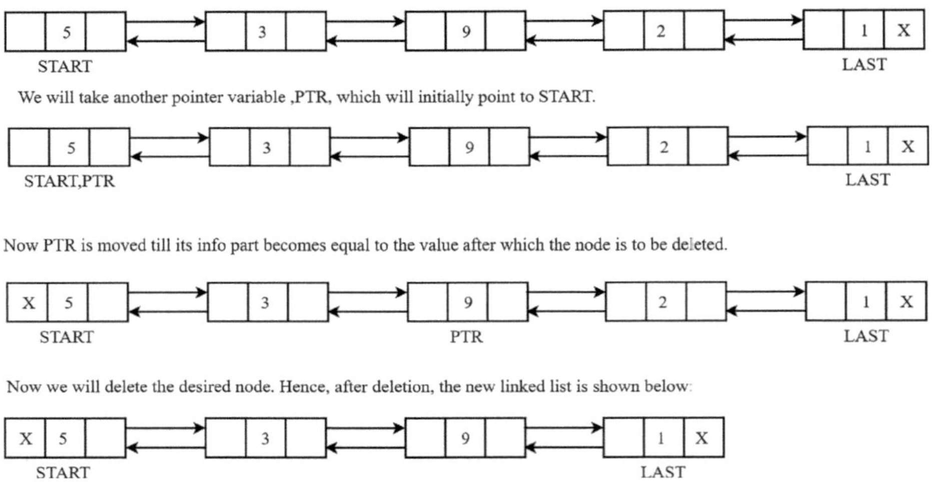

FIGURE 4.22 Deleting a node after a given node from the doubly linked list.

Now, let us understand the previous case with the help of an algorithm:

```
Step 1: START
Step 2: IF START = NULL
            Print UNDERFLOW
            [End Of If]
Step 3: Set PTR = START
Step 4: Repeat while PTR  -> INFO != GIVEN_VAL
            Set PTR = PTR -> NEXT
            [End of Loop]
Step 5: Set TEMP = PTR -> NEXT
Step 6: Set PTR -> NEXT = TEMP -> NEXT
Step 7: Set LAST -> PREV = PTR
Step 8: FREE TEMP
Step 9: EXIT
```

In the previous algorithm, first, we are checking for the underflow condition. If the condition is true, then the underflow message is printed; otherwise, `PTR` is initialized to the `START` pointer, that is, `PTR` is pointing to the first node of the list. In the loop, `PTR` is moved until its info part becomes equal to the node after which the node is to be deleted. After reaching that node of the list, we can also access the succeeding node by taking the address from the `NEXT` part of that node. Therefore, the node is deleted, and the memory that was being occupied by `TEMP` is now free.

Deleting a Node Before a Given Node from the Doubly Linked List

In the case of deleting a node before a given node from the linked list, we will again check the underflow condition as we checked in both the other cases. This situation occurs when the START pointer is equal to NULL. Hence, if the condition is true then the underflow message is printed; otherwise, the node is deleted from the linked list. Consider a linked list with five nodes, as shown in Figure 4.23; the node will be deleted before a given node from the linked list.

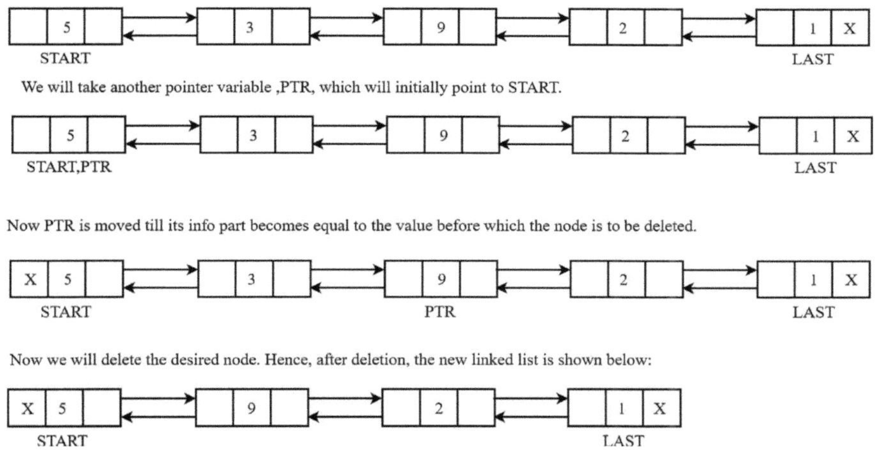

FIGURE 4.23 Deleting a node before a given node from the doubly linked list.

From the previous example, it is clear how a node will be deleted from an already existing doubly linked list. Let us now understand its algorithm:

```
Step 1: START
Step 2: IF START = NULL
              Print UNDERFLOW
              [End Of If]
Step 3: Set PTR = START
Step 4: Repeat while PTR  -> INFO != GIVEN_VAL
              Set PTR = PTR -> NEXT
              [End of Loop]
Step 5: Set TEMP = PTR -> PREV
Step 6: Set TEMP -> PREV -> NEXT = PTR
Step 7: Set PTR -> PREV = TEMP -> PREV
Step 8: FREE TEMP
Step 9: EXIT
```

In the above algorithm, first, we are checking for the underflow condition. If the condition is true, then the underflow message is printed; otherwise, PTR is initialized to the START pointer, that is, PTR is pointing to the first node of the list. In the loop, PTR is moved until its info part becomes equal to the node before which the node is to be deleted. After reaching that node of the list, we can also access the preceding node by taking the address from the PREV part of that node. Therefore, the node is deleted, and the memory that was being occupied by TEMP is now free. The practical implementation of insertion and deletion operations in a doubly linked list is discussed with the help of following code.

Write a menu-driven program for doubly linked lists, performing insertion and deletion in all cases:

```
# include <stdio.h>
# include <conio.h>
# include <alloc.h>
struct node //Self Referential Structure
{
struct node *prev ;
int data ;
struct node *next ;
 }* start ;
void add_at_beg() ;
void add_after(int item);
void add_before(int item);
void add_at_end();
void delete_at_beg();
void delete_at_mid();
void delete_at_end();
void main()
{
int choice, item;
  start= NULL:
clrscr();
  do
 {
printf(" DOUBLY LINKED LISTS");
printf("\n\t MENU");
printf(" \n 1 : addition at beginning");
printf(" \n 2 : addition after given node");
printf(" \n 3 : addition before given node");
printf(" \n 4 : addition at end");
printf(" \n 5 : deletion at beginning");
printf(" \n 6 : deletion at middle");
printf(" \n 7 : deletion at end");
printf(" \nEnter your choice: ");
scanf("%d", &choice);
   switch(choice)
  {
    case 1:          //Case 1 is for addition at beginning
add_at_beg();
    break;
```

```
    case 2: //Case 2 is for addition after given node
printf("\n enter item");
scanf(" %d", &item);
add_after(item);
    break;

    case 3: //Case 3 is for addition before given node
printf(" \n enter item");
scanf(" %d", &item);
add_before(item);
    break;
    case 4:
add_at_end();              //Case 4 is for addition at end
    break;

     case 5:
delete_at_beg();    //Case 5 is for deletion at beginning
     break;

     case 6: //Case 6 is for deletion at middle
printf(" \n enter item\n");
scanf(" %d", &item);
delete_at_mid(item);
     break:

     case7:
delete_at_end();          //Case 7 is for deletion at end

     case 8:
     exit(0);

     default:
printf("wrong choice");
     break;
    }
  }
 while(ch!=  0);
getch();
}

void_add_at_beg()
{
struct node *ptr = (struct node*) malloc (size of( struct node));
printf(" enter new node");
scanf("%d", &ptr->data);
  if(start == NULL)
  {
printf(" list is empty , create new");
ptr->next = NULL;
ptr->prev= NULL;
    start= ptr;
    return;
  }
```

```
  else
 {
ptr->next = start;
   start->prev = ptr;
ptr->prev = NULL;
   start = ptr;
   return;
  }
printf("data is entered");
}

void add_after(int item)
{
struct node *ptr = (struct node*) malloc (size of( struct node));
struct node*loc ;
loc = start;
printf(" enter new node");
scanf("%d", &ptr->data);
  if(start == NULL)
  {
printf(" list is empty , create new");
ptr->next = NULL;
ptr->prev= NULL;
   start = ptr;
   return;
  }
  while(loc!= NULL)
 {
   if(loc->data==item)
  {
loc->next->prev = ptr;
ptr->next = loc->next;
loc->next = ptr;
ptr->prev = loc;
    return;
  }
  else
 {
loc = loc->next;
  }
 }
printf(" data is entered");
}

void_add_before(int item)
{
struct node *ptr = (struct node *) malloc ( size of (struct node));
struct node *loc;
loc = start;
printf(" enter new node");
scanf("%d", &ptr->data);
  if(start == NULL)
  {
printf(" list is empty , create new");
```

```
ptr->next = NULL;
ptr->prev= NULL;
    start = ptr;
    return;
  }
  while(loc!= NULL)
  {
    if(loc->data==item)
   {
loc->prev->next = ptr;
ptr->prev= loc->prev;
loc->prev = ptr;
ptr->next = loc;
    return;
  }
  else
 {
loc = loc->next;
  }
 }
printf(" data is entered before item");
}

void add_at_end()
{
struct node *ptr = (struct node*) malloc (size of( struct node));
struct node *loc ;
loc = start;
printf(" enter new node");
scanf("%d", &ptr->data);
  if(start == NULL)
  {
printf(" list is empty , create new");
ptr->next = NULL;
ptr->prev = NULL;
   start = ptr;
   return;
  }
  while(loc!= NULL)
  {
   if(loc->data==item)
  {
loc->next = ptr;
ptr->prev = loc;
ptr->next= NULL;
   return;
  }
  else
  {
loc = loc->next;
   }
 }
printf(" data is entered at the end");
}
```

```
void delete_at_beg()
{
struct node *ptr;
struct node *loc ;
 if(start == NULL)
 {
printf(" list is empty , create new");
ptr->prev = NULL;
ptr->next = NULL;
   start = ptr;
   return;
  }
  else
 {
ptr = start;
   start = start->next;
   start->prev = NULL;
printf(" data is deleted");
   free(ptr);
  }
}

void delete_at_mid(int item)
{
struct node *ptr ,*loc ;
loc = start;
printf(" enter new node");
scanf("%d", &ptr->data);
  if(start==NULL)
 {
printf(" list is empty , create new");
ptr->next = NULL;
ptr->prev= NULL;
   start = ptr;
   return;
  }
  while(loc!= NULL)
  {
    if(loc->data==item)
  {
ptr = loc;
loc->prev->next= loc->next;
loc->next->prev= loc->prev;
printf("data is deeted");
    free(ptr);
    return;
  }
  else
  {
loc = loc->next;
  }
 }
}
```

```
void delete_at_end()
{
struct node *ptr ,*loc ;
loc = start;
printf(" enter new node");
scanf("%d", &ptr->data);
  if(start==NULL)
  {
printf(" list is empty , create new");
ptr->next = NULL;
ptr->prev= NULL;
   start = ptr;
   return;
  }
  while(loc->next != NULL)
  {
loc = loc->next;
  }
ptr = loc;
loc->prev->next=NULL;
printf("item is deleted");
 free(ptr);
}
```

Here is the output:

```
DOUBLY LINKED LIST
           MENU
1: Addition at beginning
2: Addition after given node
3: Addition before given node
4: Addition at end
5: Deletion at beginning
6: Deletion at middle
7: Deletion at end
8: Exit

Enter your choice: 1
Enter value for node
56
```

Now, let us discuss header linked lists.

Header Linked Lists

Header linked lists are a special type of linked list that always contains a special node, called the header node, at the beginning. This header node usually contains vital information about the linked list, like the total number of nodes in the list, whether the list is sorted or not, and so on. There are two types of header linked lists:

1. *Grounded header linked list* – This linked list stores a unique value of `NULL` in the address field (next part) of the last node of the list, as shown in Figure 4.24.

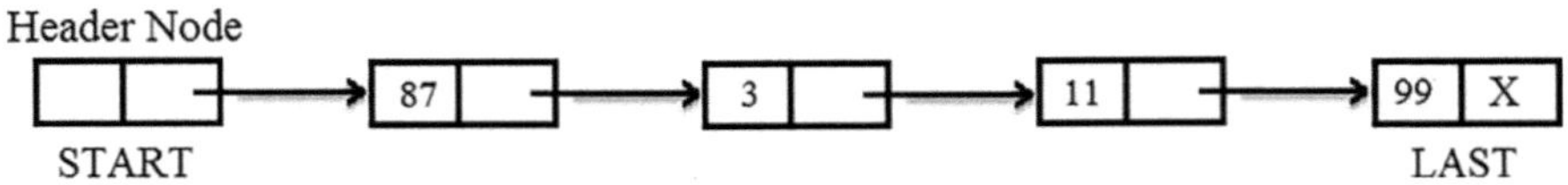

FIGURE 4.24 Grounded header linked list.

2. *Circular header linked list* – This linked list stores the address of the header node in the address field (next part) of the last node of the list, as shown in Figure 4.25. The practical implementation of header and circular header linkedlist with programming codes is discussed below.

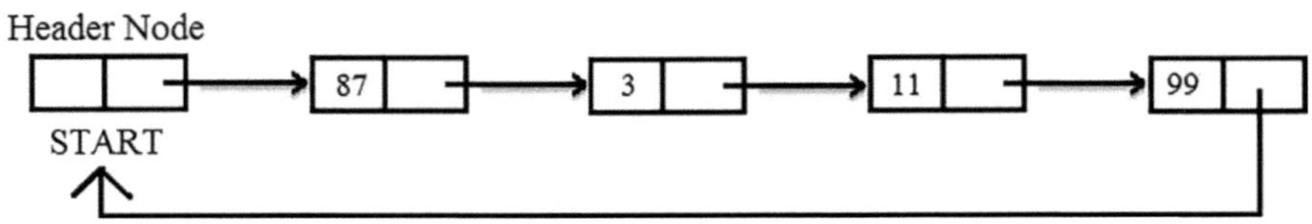

FIGURE 4.25 Circular header linked list.

Frequently Asked Questions

Q3. What are the uses of a header node in a linked list?

Answer.

The header node is a node of a linked list that may or may not have the same data structure as that of a typical node. The only common thing between a typical node and a header node is that they both have a pointer pointing to a typical node. Such a node can be used to store some extra pointers.

Write a program to implement a header linked list:

```
# include <stdio.h>
# include<conio.h>
# include <alloc.h>
struct node
{
int data;
struct node *next;
} *start = NULL;
struct node *create_header (struct node *);
struct node *display (struct node *);
```

```
int main()
{
int choice;
clrscr();
   while(choice != 3)
{
printf("HEADER LINKED LIST");
printf("\n\t\tMENU");
printf("\n 1 : Create list");
printf("\n 2 : Display ");
printf("\n 3 : Exit ");
printf("\n  enter your choice: ");
scanf("%d", & choice);
   switch(choice)
{
    Case 1 :
    start = create_header(start);
printf("\n HEADER LIST CREATED");
    break;

    Case 2:
    start = display (start);
    break;

    Case 3:
    exit(0);
  }
getch();
  return 0;
}

struct node *create_header (struct node * start)
{
struct node * new, *ptr;
int number;
printf("\n enter data: ");
scanf("%d", &number);
   while (number == 0)
  {
     new  = (struct node*) malloc (sizeof (struct node));
     new ->data =  number;
     new->next = NULL;
     if (start->next == NULL)
  {
    start = (struct node*) malloc (sizeof (struct node));
    start->next = new;
   }
  else
  {
ptr = start;
    while (ptr->next != NULL)
   {
ptr = ptr->next;
```

```
ptr->next = new;
  }
printf("\ n enter the data:");
scanf("%d" , & number);
  }
 return start;
}

struct node *display(struct node *start)
{
struct node *ptr;
ptr = start;
   while (ptr!= NULL)
printf(" %d", ptr->data);
ptr = ptr-> next;
  }
 return start;
}
```

Here is the output:

```
HEADER LINKED LIST
          MENU
1: Create list
2: Display
3: Exit
Enter your choice: 1
Enter data: 75
```

Write a program to implement a circular header linked list:

```
# include <stdio.h>
# include <conio.h>
# include <alloc.h>
structhnode//Self Referential Structure
{
int data;
hnode *next;
}* start;
void create()
{
hnode->start = new hnode();  //Creating a header linked list
   start->data = 0;
   start-> next = start;
   return 0;
 }
void begin();
void insert();
```

```
void display();
void delete(int);
void main()
{
int choice, item;
clrscr();
   create();
   do
  {
printf(" CIRCULAR HEADER LIST");
printf("\n MENU");
printf(" \n 1: INSERT ");
printf(" \n 2 : DELETE");
printf(" \n 3 : DISPLAY");
printf(" \n 4 : EXIT");
printf("\n Enter your choice:");
scanf("%d", &choice);
    switch (choice)
   {
      case 1:
      insert();           //Case 1 is for Insertion
      break;

      case 2:
      {
printf(" \n enter item you want to delete: ");
scanf(" %d", &item);       //Case 2 is for Deletion
         del (item);
         break;
         }

      case 3:
      display();                          //Case 3 is for Display
      break;

      case 4:
      exit(0);

      default :
printf ("wrong choice");
      break;
    }
  }
  while(choice !=  0);
getch();
}

void display()
{
hnode *ptr = new hnode;
   if(start->next = start)
  {
printf("\n list is empty");
```

```
        return;
    }
    else
    {
  printf (" \n current information is :", start->data);
      for (ptr = start->next ; ptr != start; ptr = ptr->next)
    {
  printf("data is:", ptr->data);
    }
    return;
  asdd

  void insert()
  {
  hnode *ptr = new hnode;
  hnode *old, *loc;
  printf("\nEnter Data");
  scanf("%d", &ptr->data);
     old = start;
  loc = start->next;
     while(loc!= start)
    {
       if (loc->data > = ptr->data)
    {
  ptr->next = old->next;
        old->next = ptr;
        (start->data)++;
        return;
     }
     else
     {
       old = loc ;
  loc = loc-> next;
     }
   }
  ptr->next = start;
    old->next = ptr;
    (start->next)++;
    return;
   }

  void delete()
  {
  hnode *ptr, *old, *loc;
     if (start->next ==NULL)
    {
  printf("\n list is empty");
       if (start->next== start)
     {
  printf(" \n cannot delete");
       return;
    }
     old = start;
```

```
loc = start->next;
   while ( loc!= start&& item>ptr->data)
  {
    if(loc->data==item)
  {
     old->next = loc->next;
printf("\n deleted item is:", loc->data);
     (start->data)--;
     return;
  }
 else
{
   old = loc;
loc = loc->next;
  }
 }
printf("\n item is  not in list\n");
 }
}

void display()
{
hnode *ptr = new hnode;
   if(start->next = start)
  {
printf("\n list is empty");
     return;
  }
  else
  {
printf (" \n current information is :", start->data);
    for (ptr = start->next ; ptr != start; ptr = ptr->next)
  {
printf("data is:", ptr->data);
  }
  return;
 }
}
```

Here is the output:

```
CIRCULAR HEADER LINKED LIST
                   MENU
1: INSERT
2: DELETE
3: DISPLAY
4: EXIT
Enter your choice: 1
Enter Data
78
```

APPLICATIONS OF LINKED LISTS

Linked lists have various applications, but one of the most important applications of linked lists is *polynomial representation*; linked lists can be used to represent polynomials, and there are different operations that can be performed on them. Now, let us see how polynomials can be represented in memory using linked lists.

Polynomial Representation

Consider a polynomial $10x^2 + 6x + 9$. In this polynomial, every individual term consists of two parts: first, a coefficient, and second, a power. Here, the coefficients of the expression are 10, 6, and 9, and 2, 1, and 0 are the respective powers of the coefficients. Now, every individual term can be represented using a node of the linked list. Figure 4.26 shows how a polynomial expression can be represented using a linked list:

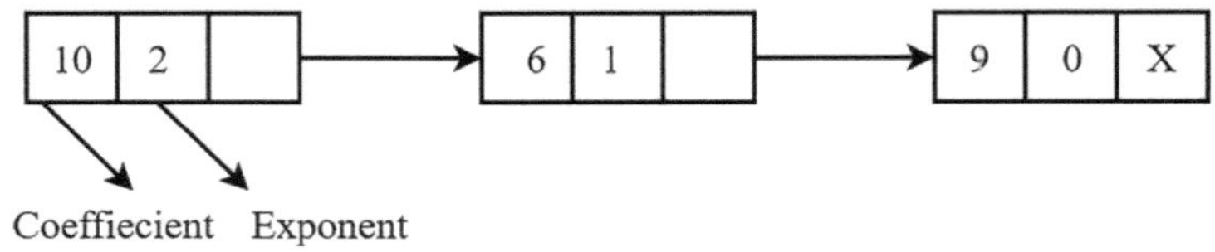

FIGURE 4.26 Linked representation of a polynomial.

SUMMARY

- A linked list is a sequence of nodes in which each node contains one or more data fields and a pointer that points to the next node.
- The process of allocating memory during the execution of the program, or the process of allocating memory to the variables at runtime, is called dynamic memory allocation.
- A singly linked list is the simplest type of linked list, in which each node contains some information/data and only one pointer that points to the next node in the linked list.
- Traversing a linked list means accessing all the nodes of the linked list exactly once.
- Searching for a value in a linked list means finding a particular element/value in the linked list.
- A circular linked list is also a type of singly linked list in which the address part of the last node stores the address of the first node.
- A doubly linked list is also called a two-way linked list; it is a special type of linked list that can point to the next node as well as the previous node in the sequence.
- A header linked list is a special type of linked list that always contains a special node, called the header node, at the beginning. This header node usually contains vital information about the linked list, such as the total number of nodes in the list, whether the list is sorted or not, and so forth.
- One of the most important applications of linked lists is polynomial representation, as linked lists can be used to represent polynomials, and there are different operations that can be performed on them.

EXERCISES

Theory Questions

1. What is a linked list? How is it different from an array?
2. How many types of linked lists are there? Explain in detail.
3. What is the difference between singly and doubly linked lists?
4. List the various advantages of linked lists over arrays.
5. What is a circular linked list? What are the advantages of a circular linked list over a linked list?
6. Define a header linked list and explain its utility.
7. Give the linked representation of the following polynomial: $10x^2y - 6x + 7$
8. Specify the use of a header node in a header linked list.
9. List the various operations that can be performed in linked lists.

Programming Questions

1. Write an algorithm/program to insert a node at a desired position in a circular linked list.
2. Write a program to insert and delete a node at the beginning of a doubly linked list.
3. Write an algorithm to reverse a singly linked list.
4. Write a program to delete a node from a header linked list.
5. Write an algorithm to concatenate two linked lists.
6. Write a program to implement a circular header linked list.
7. Write a program to count the non-zero values in a header linked list.
8. Write a program that inserts a node in the linked list before a given node.
9. Write an algorithm to search for an element in a given linear linked list.
10. Write a program that inserts a node in a doubly linked list after a given node.

Multiple Choice Questions

1. What are linked lists best suited for?
   - **A.** Applications where data and structure are hierarchical in nature
   - **B.** Applications where size of the structure and data are constantly changing
   - **C.** Applications where size of the structure and data are fixed
   - **D.** None of these
2. Each node in a linked list must contain at least ________ field(s).
   - **A.** Four
   - **B.** Three
   - **C.** One
   - **D.** Two

3. Which type of linked list stores the address of the header node in the address field of the last node?
   - A. Doubly linked list
   - B. Circular header linked list
   - C. Singly linked list
   - D. Header linked list
4. The situation in a linked list when `START = NULL` is known as what?
   - A. Overflow
   - B. Underflow
   - C. Both
   - D. None of these
5. Linked lists can be implemented in what type of data structures?
   - A. Queues
   - B. Trees
   - C. Stacks
   - D. All of these
6. Which type of linked list contains a pointer to the next as well as the previous nodes?
   - A. Doubly linked list
   - B. Singly linked list
   - C. Circular linked list
   - D. Header linked list
7. The first node in the linked list is called the ________.
   - A. End
   - B. Middle
   - C. Start
   - D. Begin
8. A linked list cannot grow and shrink during compile time. True or false?
   - A. False
   - B. It might grow
   - C. True
   - D. None of the above
9. Data elements in the linked list are known as ________.
   - A. Nodes
   - B. Pointers
   - C. Lists
   - D. All of the above
10. What does `NULL` represent in a linked list?
    - A. Start of list
    - B. End of list
    - C. None of the above

CHAPTER 5

QUEUES

INTRODUCTION

A *queue* is an important data structure that is widely used in many computer applications. A queue can be visualized with many examples from our day-to-day lives with which we are already familiar. A very simple example of a queue is a line of people standing outside waiting to enter a movie theater. The first person standing in the line will enter the movie theater first. Similarly, there are many daily life examples in which we can see a queue being implemented. Hence, we observe that whenever we talk about a queue, we see that the element at the first position will be served first. Thus, a queue can be described as a *first-in, first-out* (*FIFO*) data structure; that is, the element that is inserted first will be the first one to be taken out. Now, let us discuss queues in detail.

DEFINITION OF A QUEUE

A queue is a linear collection of data elements in which the element that is inserted first will be the element taken out first (i.e., a queue is a FIFO data structure). A queue is an abstract data structure, somewhat similar to a stack. Unlike a stack, a queue is open from both ends. A queue is a linear data structure in which the first element is inserted from one end, called the *rear* end (also called the *tail* end), and the deletion of the element takes place from the other end, called the *front* end (also called the *head*). One end is always used to insert data, and the other end is used to remove data.

Queues can be implemented by using arrays or linked lists. We will discuss the implementation of queues using arrays and linked lists in this section.

Practical Application:

A real-life example of a queue is people moving on an escalator. The people who got on the escalator first will be the first ones to step off it.

Another illustration of a queue is a line of people standing at the bus stop waiting for the bus. Therefore, the first person standing in the line will get on the bus first.

IMPLEMENTATION OF A QUEUE

Queues can be represented by two data structures:

- Arrays
- Linked lists

Now, let us discuss both of them in detail.

Implementation of Queues Using Arrays

Queues can be easily implemented using arrays. Initially, the front end (head) and the rear end (tail) of the queue point at the first position or location of the array. As we insert new elements into the queue, the rear keeps on incrementing, always pointing to the position where the next element will be inserted, while the front remains at the first position. The representation of a queue using an array is shown in Figure 5.1:

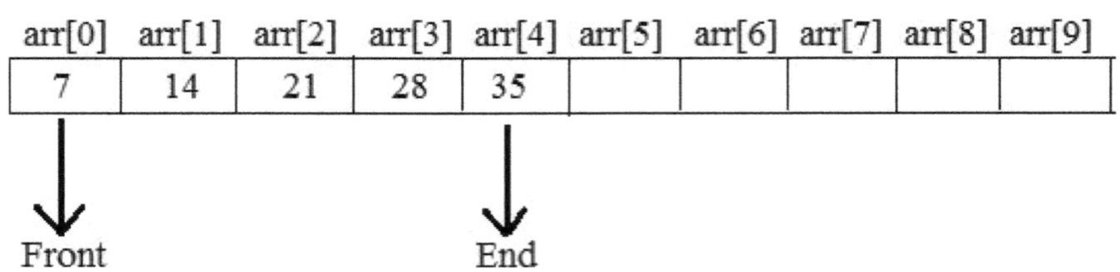

FIGURE 5.1 Array representation of a queue.

Implementation of Queues Using Linked Lists

We have already studied how a queue is implemented using an array. Now, let us discuss the same using linked lists. We already know that in linked lists, dynamic memory allocation takes place, that is, the memory is allocated at runtime. In the case of arrays, memory is allocated at the start of the program. We have already discussed this in Chapter 4, which was about linked lists. If we are aware of the maximum size of the queue in advance, then the implementation of a queue using arrays will be efficient. If the size is not known in advance, then we will use the concept of a linked list, in which dynamic memory allocation takes place. As we know, a linked list has two parts: the first part contains the information about the node, and the second part stores the address of the next element in the linked list. Similarly, we can also implement a linked queue, as shown in Figure 5.2. Now, the `START` pointer in the linked list will become the

FRONT pointer in a linked queue, and the end of the queue will be denoted by REAR. All insertion operations will be done at the rear end only. Similarly, all deletion operations will be done at the front end only.

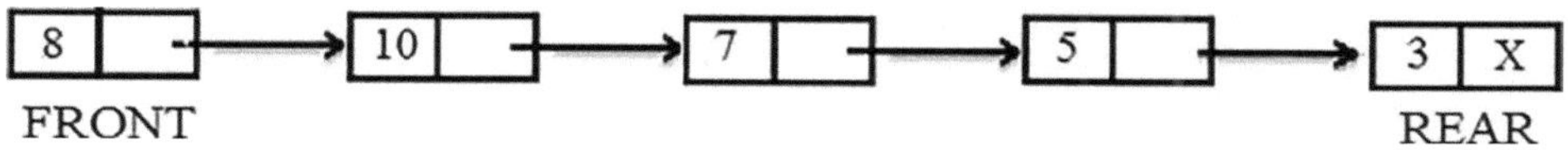

FIGURE 5.2 A linked queue.

Insertion in Linked Queues

Insertion is the process of adding new elements to an existing queue. The new elements in the queue will always be inserted from the rear. Initially, we will check whether FRONT = NULL. If the condition is true, then the queue is empty; otherwise, the new memory is allocated for the new node. We will understand it further with the help of an algorithm:

```
Step 1: START
Step 2: Set NEW NODE -> INFO = VAL
         IF FRONT = NULL
            Set FRONT = REAR = NEW NODE
            Set FRONT -> NEXT = REAR -> NEXT = NEW NODE
          ELSE
            Set REAR -> NEXT = NEW NODE
            Set NEW NODE -> NEXT = NULL
            Set REAR  = NEW NODE
        [End of If]
Step 3: EXIT
```

In the previous algorithm, first, we allocated the memory for the new node. Then, we are initializing it with the information to be stored in it. Next, we are checking whether the new node is the first node of the queue or not. If the new node is the first node of the queue, then we are storing NULL in the address part of the new node. In this case, the new node is tagged as FRONT as well as REAR. If the new node is not the first node of the queue, however, then it is inserted at the REAR end of the queue.

For example, consider a linked queue with five elements, as shown in Figure 5.3; a new element is to be inserted into the queue.

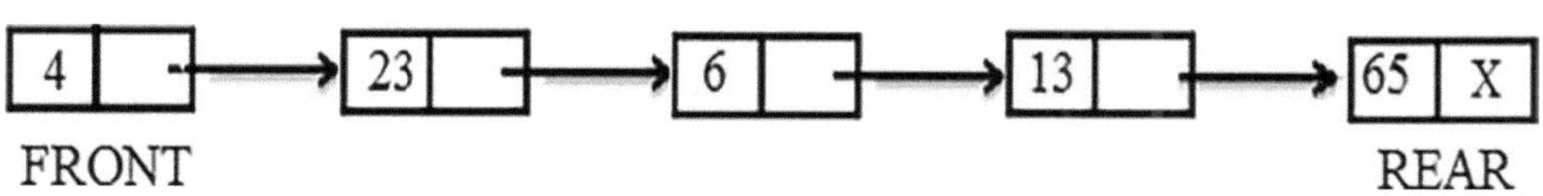

FIGURE 5.3 Linked queue before insertion.

After inserting the new element in the queue, the updated queue becomes as shown in Figure 5.4:

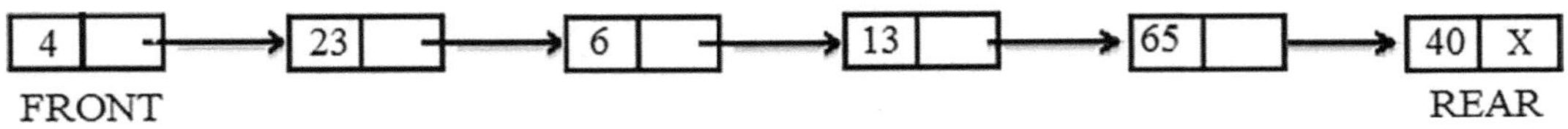

FIGURE 5.4 Linked queue after insertion.

Deletion in Linked Queues

Deletion is the process of removing elements from the already existing queue. The elements from the queue will always be deleted from the front end. Initially, we will check with the underflow condition, that is, whether `FRONT = NULL`. If the condition is true, then the queue is empty, which means we cannot delete any elements from it. Therefore, an underflow error message is displayed on the screen. We will understand it further with the help of an algorithm:

```
Step 1: START
Step 2: IF FRONT = NULL
            Print UNDERFLOW ERROR
        [End of If]
Step 3: Set TEMP = FRONT
Step 4: Set FRONT = FRONT -> NEXT
Step 5: FREE TEMP
Step 6: EXIT
```

In the previous algorithm, we first check for the underflow condition, that is, whether the queue is empty or not. If the condition is true, then an underflow error message will be displayed; otherwise, we will use a pointer variable, `TEMP`, which will point to `FRONT`. In the next step, `FRONT` is now pointing to the second node in the queue. Finally, the first node is deleted from the queue.

For example, consider a linked queue with five elements, as shown in Figure 5.5; an element is to be deleted from the queue.

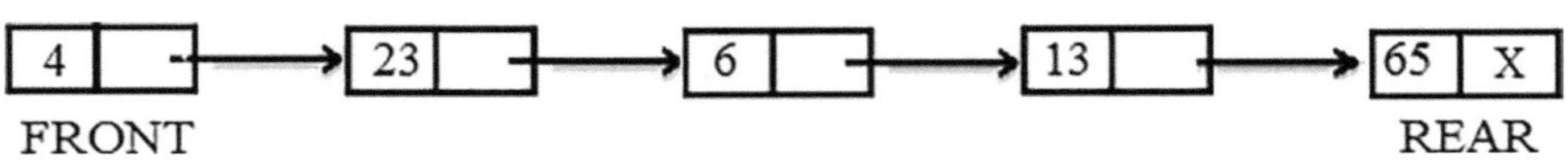

FIGURE 5.5 Linked queue before deletion.

After deleting an element from the queue, the updated queue becomes as shown in Figure 5.6:

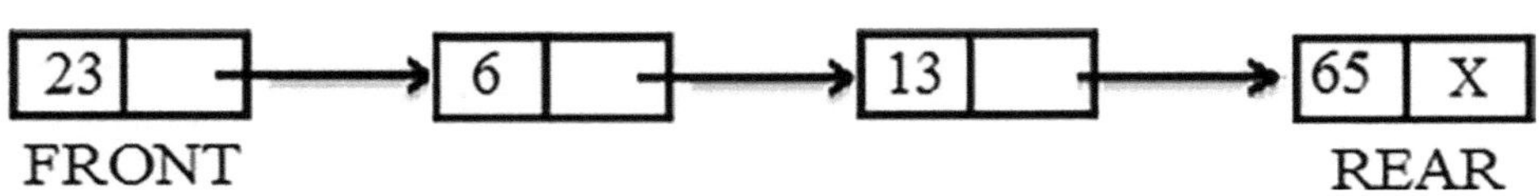

FIGURE 5.6 Linked queue after deletion.

Write a menu-driven program implementing a linked queue performing insertion and deletion operations:

```
# include<stdio.h>
# include<conio.h>
struct node
{
  int info ;
  struct node *next ;
 } *front = NULL, *rear = NULL ;
void insertion(int val) ;
void deletion() ;
void display() ;
void main()
{
 int val, choice ;
 while(1)
 {
  clrscr() ;
  printf(„\n***MENU***") ;
  printf(„\n1. INSERTION IN QUEUE") ;
  printf(„\n2. DELETION IN QUEUE") ;
  printf("\n3. DISPLAY") ;
  printf("\n4. EXIT") ;
  printf("\nenter your choice: ") ;
  scanf("%d", &choice) ;
  switch(choice)
  {
    case 1 :
    printf("enter value to insert: ") ;
    scanf("%d", &val) ;
    insertion(queue, val) ;
    break ;

    case 2 :
    deletion(queue, val) ;
    break ;

    case 3 :
    display(queue) ;
    break ;

    case 4 :
    printf("!!Exit!!") ;
    exit(0) ;

    default :
    printf("wrong choice") ;
   }
 }
}
```

```
void insertion(int val)
{
  struct node *Newnode ;
//Case 1 is for inserting an element in the linked queue
  Newnode = (struct node*)malloc(sizeof(struct node)) ;
  Newnode -> info = val ;
  Newnode -> next = NULL ;
  if(front == NULL)
  {
    front = rear = Newnode
  }
  else
 {
   rear -> next = Newnode ;
   rear = Newnode ;
  }
   printf("Success!!") ;
   getch() ;
 }
}

void deletion()
{
  struct node *temp ;
 //Case 2 is for deleting an element from the linked queue
  if(front == NULL)
   {
     printf("\nEmpty queue") ;
  }
else
{
 temp = front ;
 front = front -> next ;
 printf("deleted value is %d", n) ;
 free(temp) ;
 }
getch() ;
}

void display()
{
  struct node *ptr ;
 //Case 3 is for displaying the elements in the linked queue
  if(front == NULL)
  {
   printf("\nEmpty queue") ;
 }
 ptr = front ;
 while(ptr -> next != NULL)
 {
   printf("%d", ptr -> info) ;
   ptr = ptr -> next ;
    }
```

```
    printf("%d->NULL", ptr -> info) ;
  }
getch() ;
}
```

This is the output:

```
***MENU***
1: INSERTION IN QUEUE
2: DELETION IN QUEUE
3: DISPLAY
4: EXIT

Enter your choice: 1
Enter value to insert: 183
Success!!
```

Frequently Asked Questions

Q1. Define queues. In what ways can a queue be implemented?

Answer.

A queue is a linear data structure in which the first element is inserted from one end, called the `REAR` end (also called the tail end), and the deletion of the element takes place from the other end, called the `FRONT` end (also called the head). Each type of queue can be implemented in two ways:

- Array representation (static representation)
- Linked list representation (dynamic representation)

OPERATIONS ON QUEUES

The two basic operations that can be performed on queues are as follows.

Insertion

Insertion is the process of adding new elements to the queue. Before inserting any new element in the queue, however, we must always check for the overflow condition, which occurs when we try to insert an element in a queue that is already full. An overflow condition can be checked as follows: `REAR = MAX - 1`, where `MAX` is the size of the queue. Hence, if the overflow condition is true, then an overflow message is displayed on the screen; otherwise, the element is inserted into the queue. Insertion is always done at the rear end. Insertion is also known as enqueue.

For example, let us take a queue that has five elements in it. Suppose we want to insert another element, `50`, in it; then `REAR` will be incremented by 1. Thus, a new element is inserted at the position pointed to by `REAR`. Now, let us see how insertion is done in the queue in Figure 5.7:

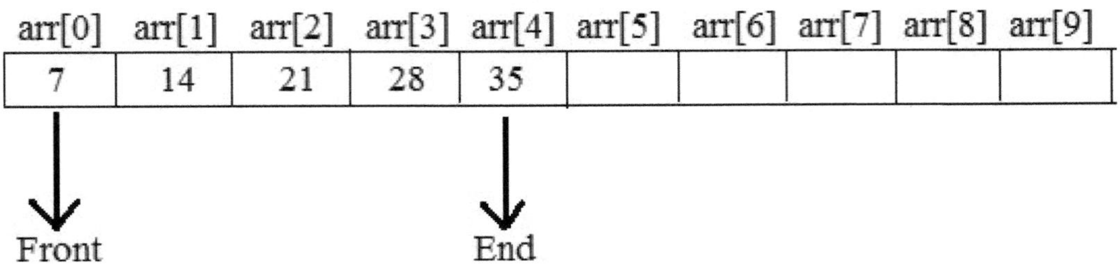

FIGURE 5.7 Queue before insertion.

After inserting `50` in it, the new queue will be as shown in Figure 5.8:

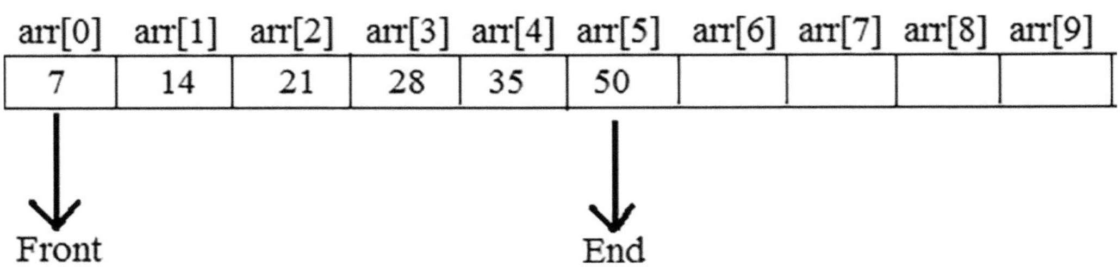

FIGURE 5.8 Queue after inserting a new element.

Here is the algorithm for inserting a new element in a queue:

```
Step 1: START
Step 2: IF REAR = MAX - 1
                Print OVERFLOW ERROR
                [End of If]
Step 3: IF FRONT = -1 && REAR = -1
                Set FRONT = 0
                Set REAR = 0
                ELSE
                REAR = REAR + 1
                [End of If]
Step 4: Set QUE[REAR] = ITEM
Step 5: EXIT
```

In the previous algorithm, first, we check for the overflow condition. In `Step 2`, we are checking to see whether the queue is empty or not. If the queue is empty, then both `FRONT` and `REAR` are set to zero; otherwise, `REAR` is incremented to the next position in the queue. Finally, the new element is stored in the queue at the position pointed to by `REAR`.

Deletion

Deletion is the process of removing elements from the queue. Before deleting any element from the queue, however, we must always check for the underflow condition, which occurs when we try to delete an element from a queue that is empty. An underflow condition can be checked as follows: `FRONT > REAR` or `FRONT = -1`. Hence, if the underflow condition is true, then an underflow message is displayed on the screen; otherwise, the element is deleted from the queue. Deletion is always done at the front end. Deletion is also known as dequeue.

For example, let us take a queue that has five elements in it. Suppose we want to delete an element, `7`, from the queue; then `FRONT` will be incremented by 1. Thus, the new element is deleted from the position pointed to by `FRONT`. Now, let us see how deletion is done in the queue in Figure 5.9:

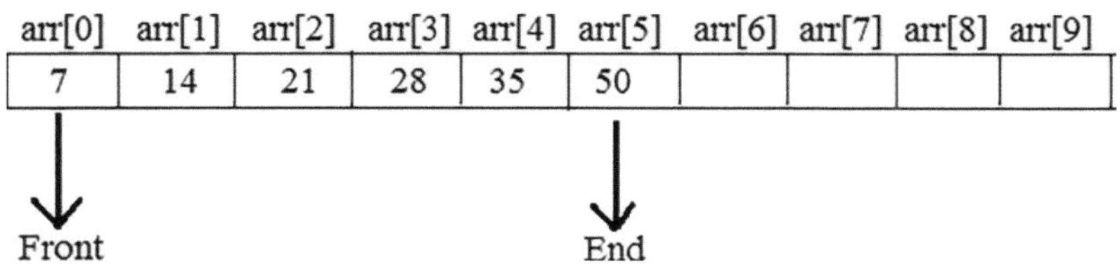

FIGURE 5.9 Queue before deletion.

After deleting `7` from it, the new queue will be as shown in Figure 5.10:

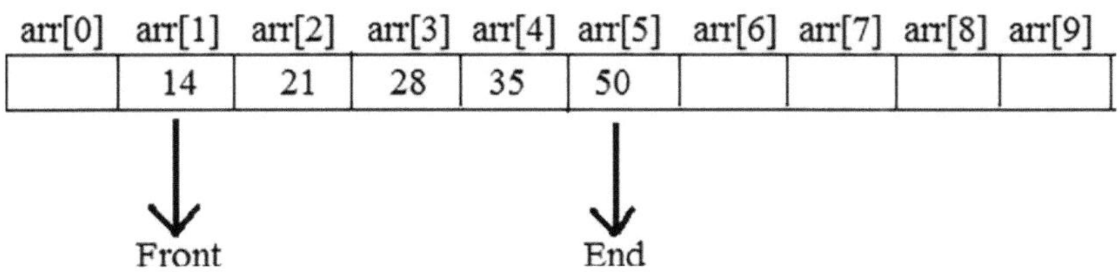

FIGURE 5.10 Queue after deleting an element.

Here is the algorithm for deleting an element from a queue:

```
Step 1: START
Step 2: IF FRONT > REAR or FRONT = -1
             Print UNDERFLOW ERROR
        [End of If]
Step 3: Set ITEM = QUE[FRONT]
Step 4: Set FRONT = FRONT + 1
Step 5: EXIT
```

In the previous algorithm, first, we check for the underflow condition, that is, whether the queue is empty or not. If the queue is empty, then no deletion takes place; otherwise, `FRONT` is incremented to the next position in the queue. Finally, the element is deleted from the queue. The practical implementation of insertion and deletion operations on a linear queue is demonstrated through following code.

Write a menu-driven program for a linear queue performing insertion and deletion operations:

```
# include<stdio.h>
# include<conio.h>
# define SIZE 10
void insertion(int queue[], int n) ;
void deletion(int queue[], int n) ;
void display(int queue[]) ;
int front = -1, rear = -1 ;
void main()
{
 int queue[SIZE], n, ch ;
 while(1)
 {
  clrscr() ;
  printf(„\n***MENU***") ;
  printf(„\n1. INSERTION IN QUEUE") ;
  printf(„\n2. DELETION IN QUEUE") ;
  printf("\n3. DISPLAY") ;
  printf("\n4. EXIT") ;
  printf("\nenter your choice: ") ;
  scanf("%d", &ch) ;
  switch(ch)
  {
    case 1 :
    printf("enter value to insert: ") ;
    scanf("%d", &n) ;
    insertion(queue, n) ;
    break ;

    case 2 :
    deletion(queue, n) ;
    break ;

    case 3 :
    display(queue) ;
    break ;

    case 4 :
    printf("!!Exit!!") ;
    exit(0) ;

    default :
    printf("wrong choice") ;
   }
 }
}

void insertion(int queue[], int n)
{
  //Case 1 is for inserting an element in the queue
    if(front == 0 || rear == SIZE-1)
```

```
    {
      printf("\noverflow error") ;
    }
    else
   {
     rear = rear + 1 ;
     queue[rear] = n ;
     printf("Success!!") ;
     getch() ;
   }
  }

  void deletion(int queue[], int n)
  {
     int i ;  //Case 2 is for deleting an element from the queue
     if(front == -1 && rear == -1)          {
       printf("\nunderflow error") ;
    }
  else
  {
   for(i = front ; i<rear ; i++)
   {
     queue[i] = queue[i+1] ;
    }
   rear -- ;
   printf("deleted value is %d", n) ;
   }
  getch();
  }

  void display(int queue[])
  {
     int i ;
  //Case 3 is for displaying the elements of the queue
     front = front + 1 ;
     for(i = rear ; i > front ; i--)
     {
       printf("\n%d", queue[i]) ;
   }
  getch() ;
  }
```

Here is the output:

```
***MENU***
1: INSERTION IN QUEUE
2: DELETION IN QUEUE
3: DISPLAY
4: EXIT

Enter your choice: 1
Enter value to insert: 56
Success!!
```

TYPES OF QUEUES

This section discusses the following types of queues:

- Circular queues
- Priority queues
- Dequeues (double-ended queues)

Let us discuss all of them one by one in detail.

Circular Queues

A *circular queue* is a special type of queue that is implemented in a circular fashion rather than in a straight line. A circular queue is a linear data structure in which the operations are performed based on the FIFO principle, and the last position is connected to the first position to make a circle. It is also called a *ring buffer*.

Limitation of Linear Queues

In linear queues, we studied how insertion and deletion take place. We discussed that while inserting a new element in the queue, it is only done at the rear end. Similarly, while deleting an element from the queue, it is only done at the front end. Now, let us consider a queue of ten elements given as shown in Figure 5.11:

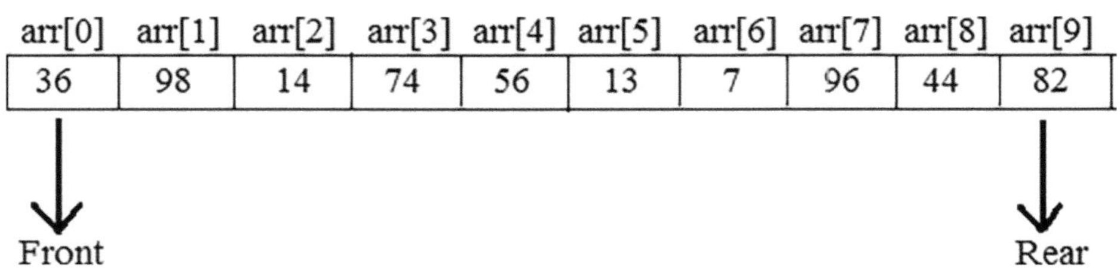

FIGURE 5.11 Queue before deletion.

The queue is now full, so we cannot insert any more elements into it. If we delete three elements from the queue, the queue will be as shown in Figure 5.12:

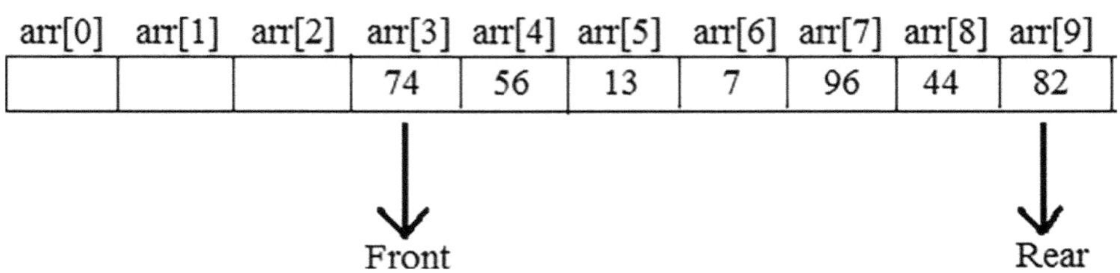

FIGURE 5.12 Queue after deletion.

Thus, we can see that even after the deletion of three elements from the queue, the queue is still full, as `REAR = MAX - 1`. We still cannot insert any new elements in it, as there is no space to store new elements. This is a major drawback of the linear queue.

To overcome this problem, we can shift all the elements to the left so that the new elements can be inserted from the `REAR` end, but shifting all the elements of the queue can be a very time-consuming procedure, as the practical queues are very large in size. Another solution to this problem is a circular queue. First of all, let us see how a circular queue looks, as shown in Figure 5.13:

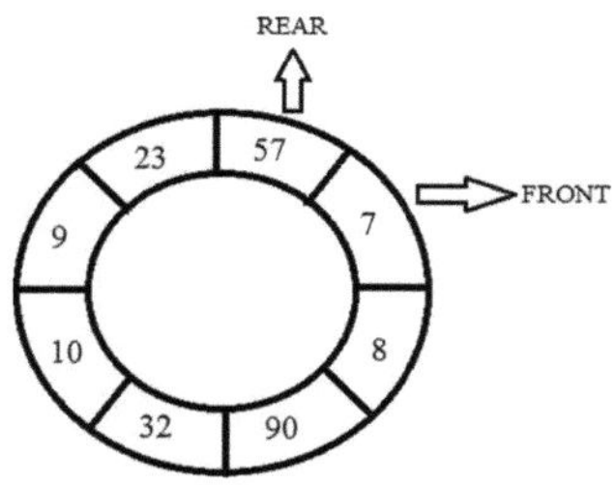

FIGURE 5.13 A circular queue.

In a circular queue, the elements are stored in a circular form such that the first element is next to the last element in the queue. A circular queue will be full when `FRONT = 0` and `REAR = MAX - 1` or `FRONT = REAR + 1`. In that case, an overflow error message will be displayed on the screen. Similarly, a circular queue will be empty when both `FRONT` and `REAR` are equal to zero. In that case, an underflow error message will be displayed on the screen. Now, let us study both insertion and deletion operations on the circular queue.

Practical Application:

A circular queue is used in operating systems for scheduling different processes.

Frequently Asked Questions

Q2. What is a circular queue? List the advantages of a circular queue over a simple queue.

Answer.

A circular queue is a particular kind of queue where new items are added to the rear end of the queue and items are read off from the front end of the queue, so there is a constant stream of data flowing in and out of the queue. A circular queue is also known as a circular buffer. It is a structure that allows data to be passed from one process to another, making the most efficient use of memory. The only difference between a linear queue and a circular queue is that in a linear queue, when the rear points to the last position in the array, we cannot insert data even if we have deleted some elements. In a circular queue, we can insert elements as long as there is free space available. The main advantage of a circular queue as compared to a linear queue is that it avoids the wastage of space.

Inserting an Element in a Circular Queue

While inserting a new element in an already existing queue, we will first check for the overflow condition, which occurs when we are trying to insert an element in a queue that is already full, as previously discussed. The position of the new element to be inserted can be calculated by using the following formula:

REAR = (*REAR* + *1*)% *MAX*), where *MAX* is equal to the size of the queue.

For example, let us consider a circular queue with three elements in it. Suppose we want to insert an element, `56`, into it. Let us see how insertion is done in the circular queue.

Initially, the queue contains three elements. `FRONT` denotes the beginning of the circular queue, and `REAR` denotes the end of it, as shown in Figure 5.14.

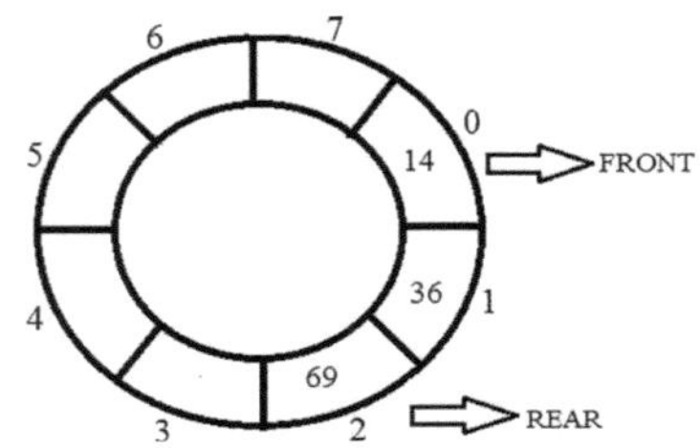

FIGURE 5.14 Initial circular queue without insertion.

Now, the new element is to be inserted into the queue. Hence, `REAR = REAR + 1`, that is, `REAR` will be incremented by 1 so that it points to the next location in the queue, as shown in Figure 5.15.

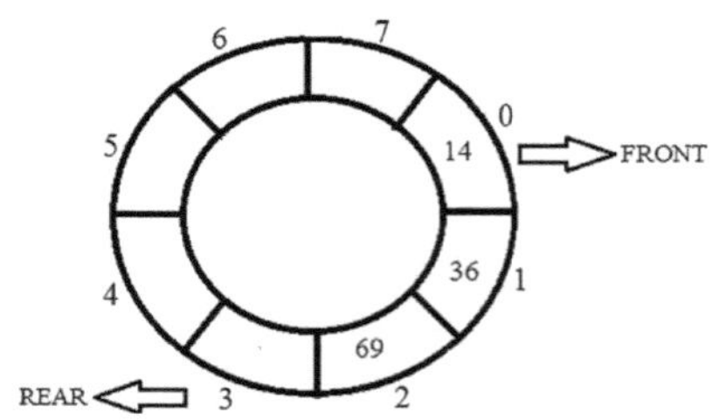

FIGURE 5.15 REAR is incremented by 1 so that it points to the next location.

Finally, in this step, the new element is inserted at the location pointed to by `REAR`. Hence, after insertion, the queue is as shown in Figure 5.16:

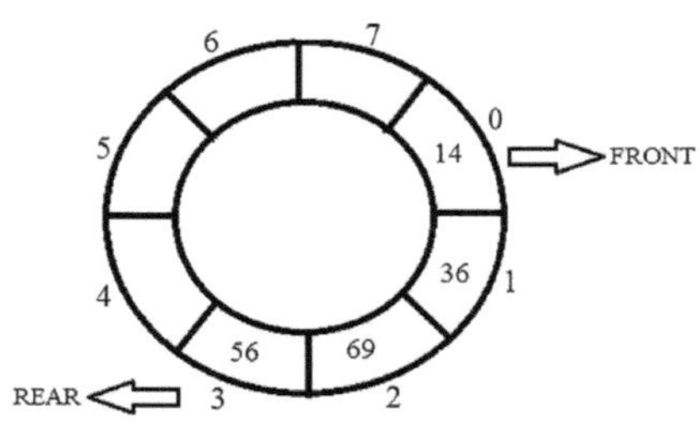

FIGURE 5.16 Final queue after inserting a new element.

Let's look at the algorithm for inserting an element in a circular queue. Here CQUEUE is an array with `N` elements. `FRONT` and `REAR` point to the front and rear elements of the queue. `ITEM` is the value to be inserted:

```
Step 1: START
Step 2: IF (FRONT = 0 && REAR = MAX - 1) OR (FRONT = REAR + 1)
             Print OVERFLOW ERROR
Step 3: ELSE
             IF (FRONT = -1)
             Set FRONT = 0
             Set REAR = 0
Step 4: ELSE
             IF (REAR = MAX - 1)
             Set REAR = 0
             ELSE
             REAR = REAR + 1
              [End of If]
             [End of If]
Step 5: Set CQUEUE[REAR] = ITEM
Step 6: EXIT
```

In the previous algorithm, first, we check for the overflow condition. Second, we check whether the queue is empty or not. If the queue is empty, then `FRONT` and `REAR` are set to zero. In `Step 4`, if `REAR` has reached its maximum capacity, then we set `REAR = 0`; otherwise, `REAR` is incremented by 1 so that it points to the next position where the new element is to be inserted. Finally, the new element is inserted into the queue.

Deleting an Element from a Circular Queue

While deleting an element from an already existing queue, we will first check for the underflow condition, which occurs when we are trying to delete an element from a queue that is empty. After deleting an element from the circular queue, the position of the `FRONT` end can be calculated by this formula:

FRONT = (*FRONT* +*1*) % *MAX*, where *MAX* is equal to the size of the queue.

For example, let us consider a circular queue with seven elements in it. Suppose we want to delete element `45` from it. Let us see how deletion is done in the circular queue.

Initially, the queue contains seven elements. `FRONT` denotes the beginning of the circular queue, and `REAR` denotes the end of it, as shown in Figure 5.17.

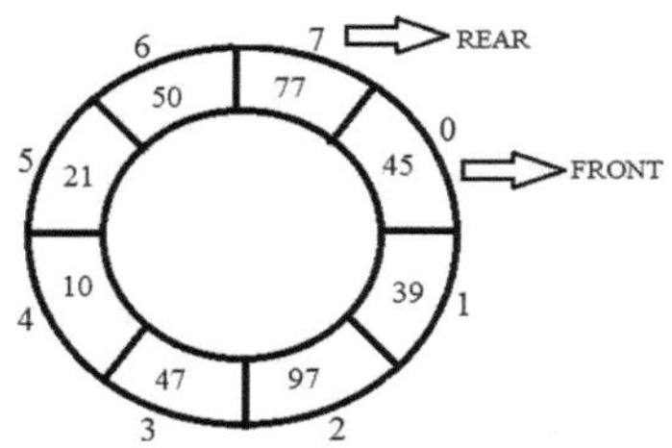

FIGURE 5.17 Initial circular queue without deletion.

Now, the element is to be deleted from the queue. Hence, `FRONT = FRONT + 1`, that is, `FRONT` will be incremented by 1 so that it points to the next location in the queue. Also, the value is deleted from the queue. Thus, the queue after deletion is shown in Figure 5.18:

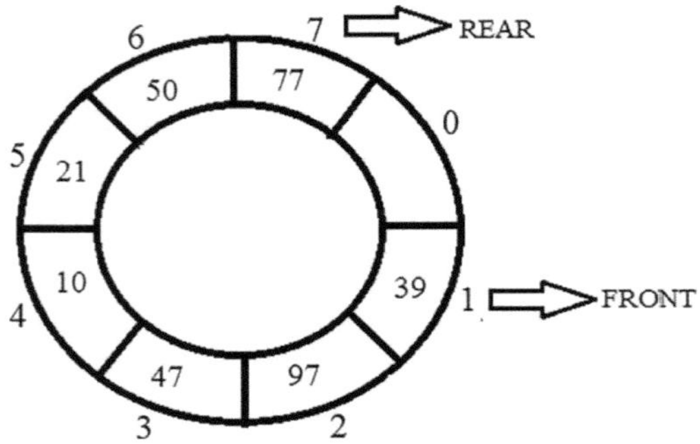

FIGURE 5.18 Final queue after deleting an element.

Let's look at the algorithm for deleting an element from a circular queue. Here, `CQUEUE` is an array with `N` elements. `FRONT` and `REAR` point to the front and rear elements of the queue. `ITEM` is the value to be deleted:

```
Step 1: START
Step 2: IF (FRONT = -1)
             Print UNDERFLOW ERROR
Step 3: ELSE
             Set ITEM = CQUEUE[FRONT]
Step 4: IF (FRONT = REAR)
             Set FRONT = -1
             Set REAR = -1
Step 5: ELSE IF (FRONT = MAX - 1)
             Set FRONT = 0
             ELSE
             FRONT = FRONT + 1
              [End of If]
             [End of If]
Step 6: EXIT
```

In the previous algorithm, we first check for the underflow condition. Second, we store the element to be deleted in `ITEM`. Third, we check to see whether the queue is empty or not after deletion. Also, if `FRONT` has reached its maximum capacity, then we set `FRONT = 0`; otherwise, `FRONT` is incremented by 1 so that it points to the next position. Finally, the element is deleted from the queue. **The practical implementation of insertion and deletion operations on a Circular Queue is discussed through following code.**

Write a menu-driven program for a circular queue performing insertion and deletion operations.

```
# include <stdio.h>
# define  MAX 10
int cqueue[MAX], front = -1, rear = -1 ;
int item ;
void main()
{
 int choice ;
 clrscr() ;
 while(1)
 {
 printf("\n MENU OF CIRCULAR QUEUE") ;
 printf("\n 1. Insertion in Circular Queue") ;
 printf("\n 2. Deletion in Circular Queue") ;
 printf("\n 3. Display") ;
 printf("\n 4. Exit");
 printf("\n enter your choice") ;
 scanf("%d", &choice);
 switch(choice)
 {
 case 1 :
 {
   printf(" \n enter item") ;
  //Case 1 is for inserting an element in the circular queue
   scanf("%d", &item) ;
   if((front == 0 && rear == MAX - 1) || (front == rear + 1))
   {
     printf("\ncircular queue overflow") ;
     getch() ;
     exit(0) ;
    }
   if((front == -1) && (rear == -1))
  {
    front = 0 ;
    rear = 0 ;
  }
  else
  {
  rear = (rear + 1) % MAX ;
  cqueue[rear] = item ;
  printf("success") ;
 }
 getch() ;
}
break ;

case 2 :
{
  if(front == -1 && rear == -1)
 //Case 2 is for deleting an element from the circular queue
 {
   printf("\n circular queue underflow") ;
```

```
      getch() ;
      exit(0) ;
     }
     else if (front == rear)
     {
       front = rear = -1 ;
       item = cqueue[front] ;
     }
     else
    {
       item = cqueue[front] ;
       front = (front + 1) % MAX ;
       printf("deleted value is %d", item) ;
     }
    getch() ;
   }
   break ;

   case 3 :
   {
     int i ; //Case 3 is for displaying the elements of the circular queue
     front = front + 1 ;
     for(i = front ; i <= rear; i++)
     {
       printf("%d\n", cqueue[i]) ;
     }
   }
   break ;

   case 4 :
   exit(0) ;

   default :
   {
    printf("wrong choice") ;
    exit(0) ;
      }
     }
    }
   }
```

Here is the output:

```
MENU OF CIRCULAR QUEUE
1: Insertion in Circular Queue
2: Deletion in Circular Queue
3: Display
4: Exit

Enter your choice
1
Enter item
18v
Success!!
```

Priority Queues

A *priority queue* is another variant of a queue in which elements are processed on the basis of assigned priority. Each element in a priority queue is assigned a special value called the *priority of the element*. The elements in a priority queue are processed based on the following rules:

1. The element with the highest priority is processed first, and then elements with a lower priority are processed.
2. If two elements have the same priority, then the elements are processed on a first-come, first-served basis.

The priority of the element is selected by its value, called the *implicit priority*, and the priority number given with each element is called the *explicit priority*.

A priority queue is like a modified queue or stack data structure, but where each element also has a priority associated with it. In a priority queue, insertion and deletion operations are also done according to the assigned priority. If we want to delete an element from the priority queue, then the element with the highest priority is processed first and is deleted. The case is the same with insertion. The priority given to the elements in the queue is based on several factors. Priority queues are commonly used in operating systems for executing higher priority processes first. The priority assigned to these processes may be based on the time taken by the CPU to execute these processes completely.

Practical Application:

In an operating system, if there are four processes to be executed where the first process needs 3 ns to complete, the second process needs 5 ns to complete, the third process needs 9 ns to complete, and the fourth needs 8 ns to complete, then the first process will be given the highest priority and will be the first to be executed among all the processes.

The priority queues are further divided into two types, as follows:

1. *Ascending priority queue* – In this type of priority queue, elements can be inserted in any order, but at the time of deletion of elements from the queue, the smallest element is searched and deleted first.
2. *Descending priority queue* – In this type of priority queue, elements can be inserted in any order. At the time of deletion of elements from the queue, the largest element is searched and deleted first.

For example, operating systems maintain a Priority Queue of various processes submitted to the system.

Frequently Asked Questions

Q3. Define a priority queue.

Answer.

A priority queue is a collection of elements in which each element has been assigned a priority, and the order in which elements are deleted and processed comes from the following rules:

- An element of higher priority is processed before any element of lower priority.
- Two elements with the same priority are processed according to the order in which they were added to the queue.

The array elements in a priority queue can have the following structure:

```
struct data
{
     int item ;
     int priority ;
     int order ;
} ;
```

Implementation of a Priority Queue

A priority queue can be implemented in two ways:

- Using arrays
- Using linked lists

Let us now discuss both of these implementations in detail.

Implementation of a Priority Queue Using Arrays

While implementing a priority queue using arrays, the following points must be considered:

- Maintain a separate queue for each level of priority or priority number.
- Each queue will appear in its own circular array and must have its own pairs of pointers, that is, `FRONT` and `REAR`.
- If each queue is allocated the same amount of memory, then a 2D array can be used instead of a linear array.

For example, `FRONT [K]` and `REAR [K]` are the pointers containing the front and rear values of row `K` of the queue, where `K` is the priority number. Suppose we want to insert an element with priority `K`, then we will add the element at the `REAR` end of row `K` (`K` is the row as well as the priority number of that element). If we add `F` with priority number `4`, then the queue will be given as shown in Figure 5.19:

FRONT	REAR
2	2
1	3
0	0
5	1
4	4

	1	2	3	4	5
1		A			
2	B	C	X		
3					
4	F			D	E
5				G	

FIGURE 5.19 Priority queue after inserting a new element.

Implementation of a Priority Queue Using Linked Lists

A priority queue can be implemented using a linked list. While implementing the priority queue using a linked list, every node will have three parts:

- Information part
- Priority number of the element
- Address of the next element

An element with higher priority will precede the element having lower priority. Also, priority number and priority are opposite to each other; that is, an element having a lower value of priority number means it has higher priority. For example, as shown in Figure 5.20, if there are two elements, *Q* and *S*, with priority numbers 2 and 5, respectively, then *Q* will be processed first because it has a higher priority.

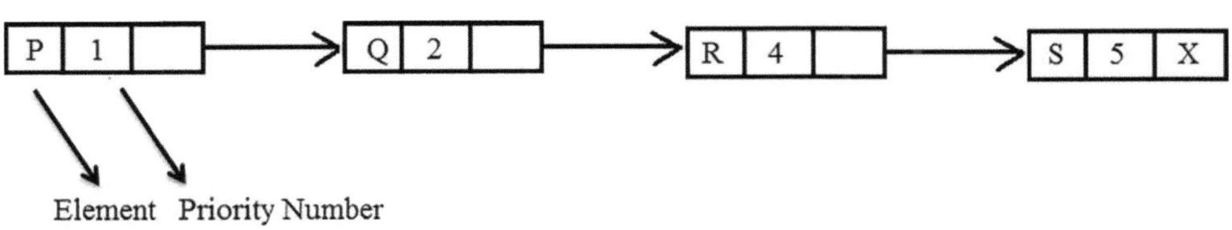

FIGURE 5.20 A linked priority queue.

Insertion in a Linked Priority Queue

While inserting a new element in a linked priority queue, first, we will traverse the entire queue until we find a node that has a lower priority than the new element. Thus, the new element is inserted before the element with the lower priority. Also, if there is an element in the queue that has the same priority as that of the new element, then the new element is inserted after that element.

For example, consider a priority queue with four elements, as shown in Figure 5.21:

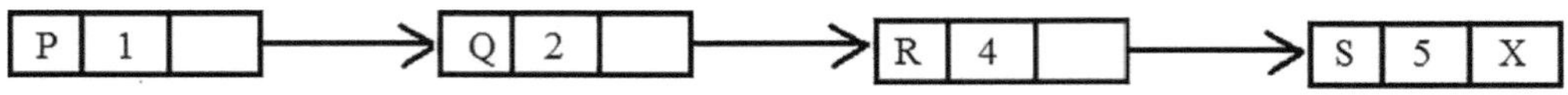

FIGURE 5.21 Linked priority queue before insertion.

Now, a new element with information *A* and the priority number *3* is to be inserted; hence, the element will be inserted before *R*, which has priority number *4*, which is lower than that of the new element. The priority queue after inserting a new element is shown in Figure 5.22:

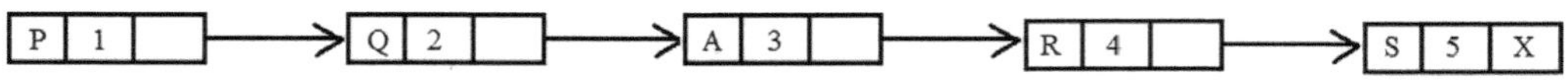

FIGURE 5.22 Linked priority queue after inserting a new element.

Deletion from a Linked Priority Queue

Deleting an element from a linked priority queue is a very simple process. In this case, the first node from the priority queue is deleted, and the information of that node is processed first. The practical implementation of insertion and deletion operations in a priority queue is discussed below:

For example, consider a priority queue with five elements, as shown in Figure 5.23:

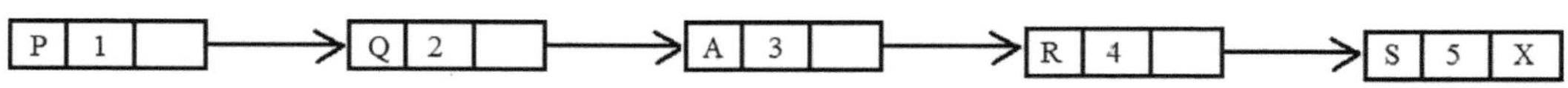

FIGURE 5.23 Linked priority queue before deletion.

Now, the first node from the queue is deleted. So, the priority queue after deletion is as shown in Figure 5.24:

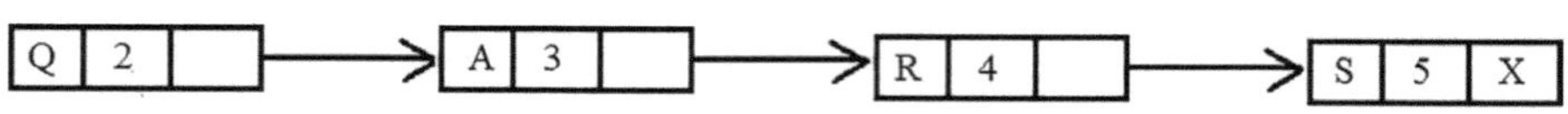

FIGURE 5.24 Linked priority queue after deleting the first node.

Write a menu-driven program for a priority queue performing insertion and deletion operations:

```
# include <stdio.h>
# include <conio.h>
# define MAX 5
void insert_by_priority(int) ;
void delete_by_priority(int) ;
void create() ;
void check(int) ;
void display_pqueue() ;
int p_queue[MAX] ;
```

```
int front, rear ;
void main()
{
    int n, choice ;
    printf("\n***MENU***) ;
    printf("\n1 - Insert an element into priority queue") ;
    printf("\n2 - Delete an element from priority queue") ;
    printf("\n3 - Display priority  queue") ;
    printf("\n4 - Exit");
    create() ;
    while (1)
   {
     printf("\nEnter your choice: ") ;
     scanf("%d", &choice) ;
     switch (choice)
    {
      case 1 :
      printf("\nEnter value to insert: ") ;
      scanf("%d", &n) ;
      insert_by_priority(n) ;
      break ;

      case 2 :
      printf("\nEnter value to delete : ") ;
      scanf("%d", &n) ;
      delete_by_priority(n) ;
      break ;

      case 3 :
      display_pqueue() ;
      break ;

      case 4 :
      exit(0) ;

      default :
      printf("\nChoice is incorrect ") ;
     }
   }
}

void create()
{
    front = rear = -1;
                  //Function to create an empty priority queue
}

 void insert_by_priority(int data)
{
    if (rear >= MAX - 1)
                //Function to insert value into priority queue
    {
        printf("\nQueue overflow") ;
        return ;
```

```
        }
        if ((front == -1) && (rear == -1))
        {
            front ++ ;
            rear ++ ;
            p_queue[rear] = data ;
            return ;
        }
        else
        check(data) ;
        rear ++ ;
    }

    void check(int data)
    {
        int i, j ;
                    //Function to check priority and place element

        for (i = 0 ; i <= rear ; i++)
        {
            if (data >= p_queue[i])
            {
                for (j = rear + 1 ; j > i ; j--)
                {
                    p_queue[j] = p_queue[j - 1] ;
                }
                p_queue[i] = data ;
                return ;
            }
        }
        p_queue[i] = data ;
    }

    void delete_by_priority(int data)
    {
        int i ;
    // Function to delete an element from queue
        if ((front==-1) && (rear==-1))
        {
            printf("\nQueue is empty") ;
            return ;
        }
      for (i = 0; i <= rear; i++)
       {
          if (data == p_queue[i])
            {
                for (; i < rear; i++)
                {
                    p_queue[i] = p_queue[i + 1] ;
                }
            p_queue[i] = -99 ;
            rear -- ;
```

```
        if (rear == -1)
            front = -1 ;
        return ;
        }
    }
    printf("\n%d not found in queue to delete", data) ;
}

void display_pqueue()
{
    if((front == -1)&&(rear == -1))
                        // Function to display queue elements
    {
        printf("\nQueue is empty") ;
        return ;
    }
  for (; front <= rear ; front++)
    {
        printf(" %d ", p_queue[front]) ;
      }
  front = 0 ;
}
```

This is the output:

```
  ***MENU***
1 - Insert an element into priority queue
2 - Delete an element from priority queue
3 - Display priority  queue
4 - Exit

Enter your choice 1
Enter value to insert
54
```

Dequeues (Double-Ended Queues)

A double-ended queue (or dequeue, pronounced *deck*) is a special type of data structure in which the insertion and deletion of elements is done at either end, that is, either at the front end or at the rear end of the queue. It is often called a *head-tail linked list* because elements are added or removed from either the head (front) end or tail (end). Dequeues are implemented using circular arrays in the computer's memory. The `LEFT` and `RIGHT` pointers are maintained in the dequeue, which point to either end of the queue. An example is shown in Figure 5.25 followed by discussion of practical implementation of a dequeue through a code

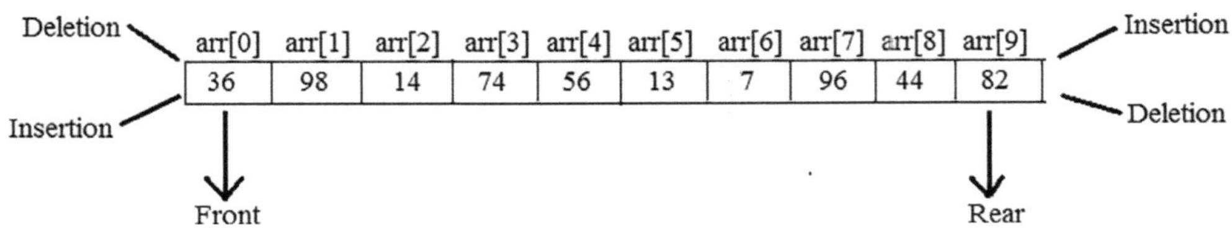

FIGURE 5.25 A double-ended queue.

Practical Application:

A real-life example of a dequeue is the way that, in a train station, the entry and exit of passengers can take place from both sides.

There are two types of double-ended queues:

1. *Input-restricted dequeue* – In this, the deletion operation can be performed at both ends (i.e., both the front and rear end) while the insertion operation can be performed only at one end (i.e., the rear end), as shown in Figure 5.26.

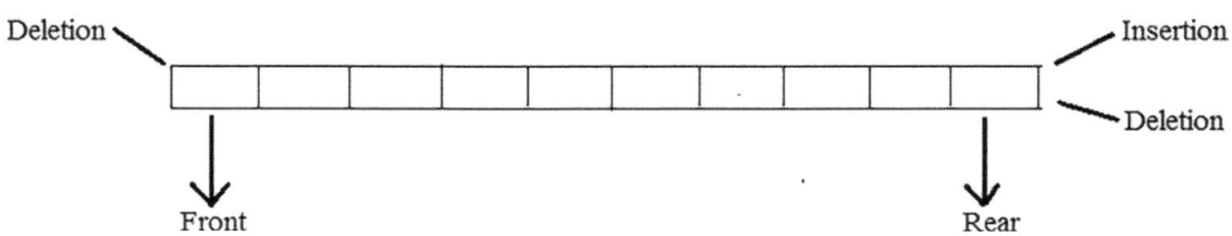

FIGURE 5.26 An input-restricted double-ended queue.

2. *Output-restricted dequeue* – In this, the insertion operation can be performed at both ends while the deletion operation can be performed only at one end (i.e., the front end), as shown in Figure 5.27.

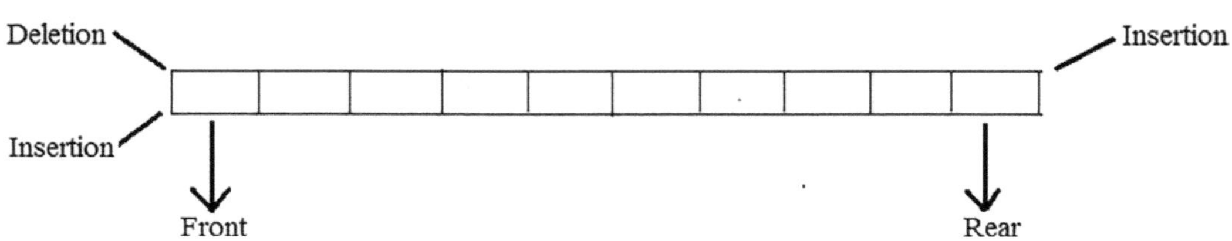

FIGURE 5.27 An output-restricted double-ended queue.

Write a menu-driven program for a double-ended queue performing insertion and deletion operations:

```
#include<stdio.h>
#include<conio.h>
#include<stdlib.h>
# define size[5]
int deque[size] ;
int front = -1, rear = -1 ;
void display() ;
void insert_front() ;
void insert_rear() ;
void delete_front() ;
void delete_rear() ;
int choice, item ;
void main()
{
  clrscr() ;
  while(1)
  {
   printf("\n Menu") ;
   printf("\n 1. Insert from Front") ;
   printf("\n 2. Insert from Rear") ;
   printf("\n 3. Delete from Front") ;
   printf("\n 4. Delete from Rear") ;
   printf("\n 5. Display") ;
   printf("\n 6. Exit") ;
   printf("\nEnter your choice: ") ;
   scanf("%d", &choice) ;
   switch(choice)
   {
     case 1:
     insert_front() ;
     getch() ;
     break ;

     case 2:
     insert_rear() ;
     getch() ;
     break ;

     case 3:
     delete_front() ;
     getch() ;
     break ;

     case 4:
     delete_rear() ;
     getch() ;
     break ;
```

```
        case 5:
        display() ;
        getch() ;
        break ;

        case 6:
        exit(0) ;

        default:
        printf("\n Invalid Choice") ;
        getch() ;
        break ;
       }
    }
}

void insert_front()
{
  if(front==0)
//Case 1 is for inserting an element in the queue from front end
  {
    printf("\n Queue is Full") ;
   }
  else
  front = front - 1 ;
  printf("\n Enter a no") ;
  scanf("%d", &item) ;
  deque[front] = item ;
}

void insert_rear()
{
 if(rear == max)
//Case 2 is for inserting an element in the queue from rear end
 {
  printf("\n Queue is Full") ;
  }
 else
 rear = rear + 1 ;
 printf("\n Enter a no") ;
 scanf("%d", &item) ;
 deque[rear] = item ;
}

void delete_front()
{
 if(front == max)
//Case 3 is for deleting an element in the queue from front end
 {
   printf("\n Queue is Empty") ;
  }
 else
 item = deque[front] ;
```

```
  front = front + 1 ;
  printf("\n No. deleted is %d", item) ;
 }

void delete_rear()
 {
  if(rear == 0)
//Case 4 is for deleting an element in the queue from rear end
  {
    printf("\n Queue is Empty") ;
   }
  else
  item = deque[rear] ;
  rear = rear - 1 ;
  printf("\n No. deleted is %d", item) ;
 }

void display()
 {
   int i ;
   printf("\n The Queue is:") ;
           //Case 5 is for displaying the elements of the queue
   for(i = front ; i <= rear ; i++)
   {
    printf("%d \n ", deque[i]) ;
   }
 }
```

This is the output:

```
      Menu
1. Insert from Front
2. Insert from Rear
3. Delete from Front
4. Delete from Rear
5. Display
6. Exit

Enter your choice:
1
Enter a no
25
```

APPLICATIONS OF QUEUES

In real life, call center phone systems use queues to hold people calling them in order until a service representative is free.

The handling of interruptions in real-time systems uses the concept of queues. The interruptions are handled in the same order as they arrive, that is, first-come, first-served.

The round-robin technique for processor scheduling is implemented using queues. Queues are used to manage order of printing various documents submitted to a printer.

SUMMARY

- A queue is a linear collection of data elements in which the element inserted first will be the element taken out first (i.e., a queue is a FIFO data structure).
- A queue is a linear data structure in which the first element is inserted from one end, called the `REAR` end, and the deletion of the element takes place from the other end, called the `FRONT` end.
- The implementation of queues can be done in two ways: through arrays and through linked lists.
- Insertion and deletion are the two basic operations that are performed on queues.
- A circular queue is a linear data structure in which the operations are performed based on a FIFO principle, and the first index comes after the last index.
- A priority queue is a queue in which elements are processed on the basis of assigned priority. Each element in a priority queue is assigned a special value called the priority of the element.
- When a priority queue is implemented using linked lists, then every node of the list will have three parts: a data part, the priority number of the element, and the address of the next element.
- A double-ended queue is a special type of data structure in which the insertion and deletion of elements are done at either end, that is, either at the front end or at the rear end of the queue.
- An input-restricted dequeue is a queue in which deletion can be done at both ends, but insertion is done only at the rear end.
- An output-restricted dequeue is a queue in which insertion can be done at both ends, but deletion is done only at the front end.
- There are various applications of queues.

EXERCISES

Theory Questions

1. What is a linear queue? Give a real-life example.
2. What is a circular queue, and how is it different from a linear queue?
3. Define priority queues.
4. Discuss the various operations that can be performed on the queues.
5. Define queues and in what ways a queue can be implemented. What do you understand by double-ended queues? Discuss the different types of dequeues in detail.
6. Give some of the applications of queues.
7. Why are queues known as first-in, first-out structures?
8. Explain the concept of a linked queue and also discuss how insertion and deletion take place in it.

Programming Questions

1. Write a program to create a linear queue containing nine elements.
2. Write an algorithm to implement a priority queue.
3. Write a code for insertion and deletion in a queue.
4. Give an algorithm for the insertion of an element in a circular queue. Write a program to implement a queue that allows insertion and deletion at both ends.
5. Write an algorithm that reverses the elements of a queue.
6. Write an algorithm for insertion and deletion in a queue using pointers. Write the functions for insertion and deletion operations performed in a dequeue. Consider all possible cases.
7. Write a code for deleting an element from a circular queue.
8. Write a program to implement a priority queue using a linked list.

Multiple Choice Questions

1. New elements in the queue are always inserted from where?
   - A. Front end
   - B. Middle
   - C. Rear end
   - D. Both (a) and (c)
2. A queue is a ________ data structure.
   - A. FIFO
   - B. LIFO
   - C. FILO
   - D. LILO
3. The overflow condition in the circular queue exists when ________.
   - A. FRONT = MAX – 1 and REAR = 0
   - B. FRONT = 0 and REAR = MAX – 1
   - C. FRONT = 0 and REAR = 0
   - D. FRONT = MAX – 1 and REAR = MAX – 1
4. If the elements *P*, *Q*, *R*, and *S* are placed in a queue and are deleted one by one, in what order will they be deleted?
   - A. PQRS
   - B. SRQP
   - C. PRQS
   - D. SRQP
5. A data structure in which elements are inserted or deleted from the front as well as the rear end is known as what?
   - A. Linear queue
   - B. Dequeue
   - C. Priority queue
   - D. Circular queue

6. A line outside a movie theater represents a ________.
   - A. Linked list
   - B. Array
   - C. Queue
   - D. Stack
7. In a queue, deletion is always done at the ________.
   - A. Top end
   - B. Back end
   - C. Front end
   - D. Rear end
8. In a priority queue, two elements with the same priority are processed on an FCFS basis.
   - A. False
   - B. True
9. The function that inserts the elements in a queue is called ________.
   - A. Push
   - B. Enqueue
   - C. Pop
   - D. Dequeue
10. Which of the implementations of queues is better when the size of the queue is not known in advance?
    - A. Linked list representation
    - B. Array representation
    - C. Both
    - D. None of the above

CHAPTER 6

SEARCHING AND SORTING

INTRODUCTION TO SEARCHING

As we all know, computer systems are often used to store large numbers. We require a search mechanism to retrieve a specific record from the large amounts of data stored in our computer systems. Searching means to find whether a particular data item exists in an array/list or not. The process of finding a particular value in a list or an array is called *searching*. If that particular value is present in the array, then the search is said to be successful, and the location of that particular value is returned by the searching process. If the value does not exist, however, or the value is not present in the array, then searching is said to be unsuccessful. There are many different searching algorithms, but three of the popular searching techniques are as follows:

- Linear or sequential search
- Binary search
- Interpolation search

We will discuss all these methods in detail in this chapter.

LINEAR OR SEQUENTIAL SEARCH

A *linear search* is also called a *sequential search*. This is a very simple technique used to search for a particular value in an array. It works by comparing the value of the key being searched for with every element of the array in a linear sequence until a match is found. A search will be unsuccessful if all the data elements are read and the desired element is not found. The following are some important points:

- It is the simplest way to search for an element in the list.
- It searches the data element sequentially, no matter whether the array is sorted or unsorted.

For example, let us take an array of ten elements, which is declared as follows:

```
int array[10] = {87, 25, 14, 39, 74, 1, 99, 12, 30, 67};
```

The value to be searched for in the array is declared as `VAL = 74`, so we search to find whether `74` exists in the array or not. If the value is present, then its position is returned. Here, the position of `VAL = 74` is `POS = 4` (index starting from zero). The process is shown below:

1. *Pass* 1 – `87` is compared with `74`. Since `87` is not equal to `74`, we will move to the next pass.

arr[0]	arr[1]	arr[2]	arr[3]	arr[4]	arr[5]	arr[6]	arr[7]	arr[8]	arr[9]
87	25	14	39	74	1	99	12	30	67

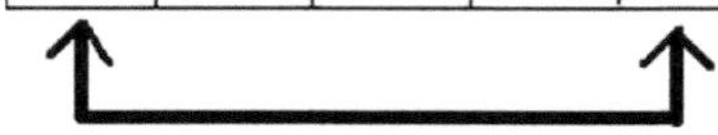

2. *Pass* 2 – `25` is compared with `74`. Since `25` is not equal to `74`, we will move to the next pass.

arr[0]	arr[1]	arr[2]	arr[3]	arr[4]	arr[5]	arr[6]	arr[7]	arr[8]	arr[9]
87	25	14	39	74	1	99	12	30	67

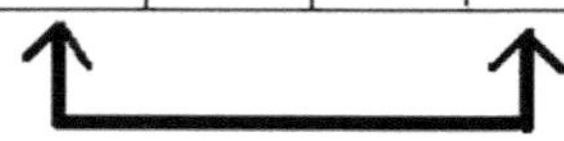

3. *Pass* 3 – `14` is compared with `74`. Since `14` is not equal to `74`, we will move to the next pass.

arr[0]	arr[1]	arr[2]	arr[3]	arr[4]	arr[5]	arr[6]	arr[7]	arr[8]	arr[9]
87	25	14	39	74	1	99	12	30	67

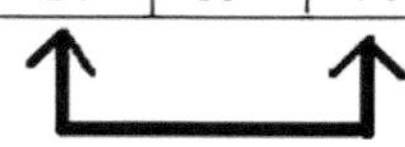

4. *Pass* 4 – `39` is compared with `74`. Since `39` is not equal to `74`, we will move to the next pass.

arr[0]	arr[1]	arr[2]	arr[3]	arr[4]	arr[5]	arr[6]	arr[7]	arr[8]	arr[9]
87	25	14	39	74	1	99	12	30	67

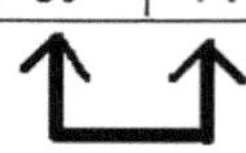

5. *Pass* 5 – `74` is compared with `74`. Since `74` is equal to `74`, we will return the position in which `74` is present, which, in this case, is `4`.v

arr[0]	arr[1]	arr[2]	arr[3]	arr[4]	arr[5]	arr[6]	arr[7]	arr[8]	arr[9]
87	25	14	39	74	1	99	12	30	67

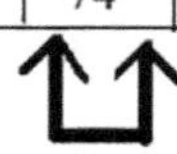

74 is found at POS = 4

In this way, a linear search is used to search for a particular value in the array. Now, let us understand it further with the help of an algorithm.

Practical Application:

A simple, real-life example of a linear search is a person searching for another person's contact number in a telephone directory. If the person could not recall the exact name of that person but knows that the name starts with *A and has confidence that name would be recalled if same appears written in front of them,* then they will start searching from the beginning of the telephone directory in the linear search order till they find the requisite name.

Let `ARR` be an array of `n` elements, `ARR[1]`, `ARR[2]`, `ARR[3]`, ..., `ARR[n]`, such that `VAL` is the element to be searched. Then, the algorithm will find the position (`POS`) of the value (`VAL`) in the array, `ARR`.

```
Step 1: START
Step 2: Set I = 0, POS = -1
Step 3: Repeat while I<N
IF (ARR[I] = VAL)
             POS = I
             PRINT POS
Go to Step 5
[End of IF]
[End of Loop]
Step 4: IF (POS = -1)
             PRINT "VALUE NOT FOUND, SEARCH UNSUCCESSFUL"
[End of IF]
Step 5: EXIT
```

In `Step 2` of the algorithm, we are initializing the value of `I` and `POS`. In `Step 3`, a `while` loop is executed in which a check is made, that is, whether a match is found between the current array element and `VAL`. If the match is found, then the position of that element is printed. In the last step, if all the elements have been compared and no match is found, the search will be unsuccessful, meaning the value is not present in the array.

Complexity of the Linear Search Algorithm

The execution time of a linear search is $O(n)$, where n is the number of elements in the array. The algorithm is called a linear search because its complexity can be expressed as a linear function, which means that the number of comparisons to find the target item increases linearly with the size of the data. The best case of a linear search is when the data element to be searched for is equal to the first element of the array. The worst case will happen when the data element to be searched for is equal to the last element in the array. In both cases, however, n comparisons have to be made.

Drawbacks of Linear Search

Let's look at the disadvantages of using a linear search:

- It is a very time-consuming process, as it works sequentially.
- It can be applied to only a small amount of data.

It is a very slow process as almost every data element is accessed in this process, especially when the data element is located near the end. Let us write a code to demonstrate practical implementation of linear search.

Write a program to search for an element in an array using a linear search technique:

```
# include <stdio.h>
# include <conio.h>
intlinear_search( intarr[], int n, int value ) ;
void main()
{
intarr[10], n, i, r, value ;
clrscr() ;
printf("***LINEAR SEARCH***") ;
printf("\nenter no of elements") ;
scanf("%d", &n) ;
printf("enter the elements of array") ;
   for( i=0 ; i<n ; i++ )
   {
printf("\nenter element %d", i+1) ;
scanf("%d", &arr[i]) ; //Accepting the elements of array
     }
printf("\nenter value to search") ;
scanf("%d", &value) ;
   r = linear_search(arr, n, value) ;
   if(r == -1)
printf("value not found") ;
   else
printf("%d value found at %d", value, r+1) ;
getch() ;
}
intlinearsearch( intarr[], int n, int value )
{
int i;
   for( i=0 ; i<n ; i++ )
    {
       if(arr[i] == value)
       return i ;
     }
  return (-1);
}
```

This is the output:

```
***LINEAR SEARCH***
Enter no of elements  9

Enter elements of array
Enter element 1:   8
Enter element 2:   2
Enter element 3:   9
Enter element 4:   10
Enter element 5:   1
Enter element 6:   3
Enter element 7:   23
Enter element 8:   69
Enter element 9:   17

Enter value to search   23
23 value found at 7th position
```

Frequently Asked Questions

Q1. Explain how a linear search technique is used to search for an element.

Answer.
Suppose that `ARR` is an array having `N` elements. `ITEM` is the value to be searched. Then, we have the following cases:

- *Case 1: Unsorted list* – `ITEM` is compared with every element of the array. If the element is found, then no further comparison is required. If all the elements are compared and checked, then `ITEM` is not found.
- *Case 2: Sorted list* – `ITEM` is greater than the first element and smaller than the last element of the list, so searching is performed by comparing each element in the list with `ITEM`; otherwise, `ITEM` is reported as `Not Found`.

BINARY SEARCH

A *binary search* is an extremely efficient searching algorithm compared to a linear search. A binary search works only when the array/list is already sorted. In a binary search, we first compare the value `VAL` with the data element in the middle position of the array. If the match is found, then the position POS of that element is returned; otherwise, if the value is less than that of the middle element, then we begin our search in the lower half of the array, and vice versa. We repeat this process on the lower and upper halves of the array.

Binary Search Algorithm

Let us now understand how the binary search algorithm works in an array:

1. Find the middle element of the array; n/2 is the middle element of the array containing n elements.
2. Now, compare the middle element of the array with the data element to be searched:
   a) If the middle element is the desired element, then the search is successful.
   b) If the data element to be searched for is less than the middle element of the array, then search only the lower half of the array, that is, those elements that are on the left side of the middle element.
   c) If the data element to be searched for is greater than the middle element of the array, then search only the upper half of the array, that is, those elements that are on the right side of the middle element.

   Repeat these steps until a match is found.

Practical Application:

A real-life application of a binary search is searching for a particular word in a dictionary. We first open the dictionary somewhere in the middle. Now, we will compare the desired word with the first word on that page. If the desired word comes after the first word on an open page, then we will look in the second half of the dictionary; otherwise, we will look in the first half. Then, we will again open a page in the second half and compare the desired word with the first word on that page, and the same process is repeated until we have found the desired word.

Let's look at the algorithm for a binary search:

```
Binary_Search(ARR, Lower_bound, Upper_bound, VAL)

Step 1: START
Step 2: Set BEG = lower_bound, END = upper_bound, POS = -1
Step 3: Repeat Steps 4 & 5 while BEG <= END
Step 4: Set MID = (BEG+END)/2
Step 5: IF (ARR[MID] = VAL)
            POS = MID
            PRINT POS
Go to Step 7
```

```
ELSE IF (ARR[MID] > VAL)
                Set END = MID - 1
ELSE
                Set BEG = MID + 1
[End of If]
     [End of Loop]
Step 6: IF (POS = -1)
                PRINT "VALUE NOT FOUND, SEARCH UNSUCCESSFUL"
    [End of IF]
Step 7: EXIT
```

In `Step 2` of the algorithm, we are initializing the values of `BEG`, `END`, and `POS`. In `Step 3`, a `while` loop is executed. In `Step 3`, the value of `MID` is calculated. In `Step 4`, we will check whether the value to be searched for is equal to the array value at `MID`. If the match is found, then the position of that element is printed. If the match is not found and the value to be searched for is less than that of the array value at `MID`, then `END` is modified; otherwise, if the value to be searched for is greater than that of the array value at `MID`, then `BEG` is modified. In the last step, if all the elements have been compared and no match is found, the search has been unsuccessful, meaning the value is not present in the array.

Let us now consider an example to search for a particular value in a sorted array, as shown below through following steps.

Consider an array of ten elements, declared as:

```
int array[10] = {0, 10, 20, 30, 40, 50, 60, 80, 90, 100};
```

The value to be searched for is declared as `VAL = 20`. Then, the algorithm will proceed as follows.

In the first pass, `BEG = 0` and `END = 10`. So, `MID = (BEG + END)/2`, meaning `(0 + 10)/2 = 5`.

Now, `VAL = 20` and `ARR[MID] = ARR[5] = 50`.

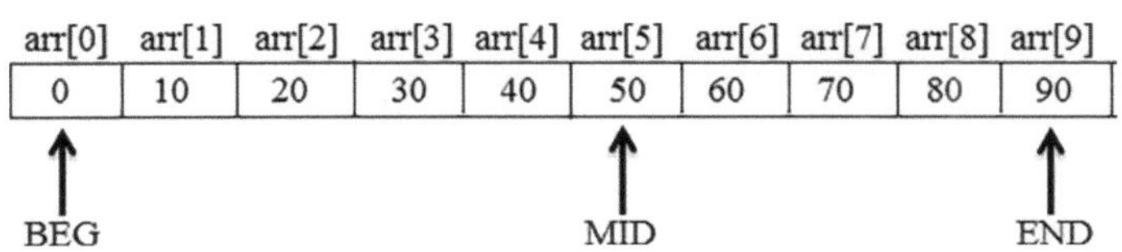

As `ARR[5] = 50 > VAL = 30`, we will now search for the value in the lower half of the array. So now the values of `END` and `MID` are modified, and we move to the next pass.

In the second pass, `END = MID - 1 = 4`. Here, `MID = (0 + 4)/2 = 2`.

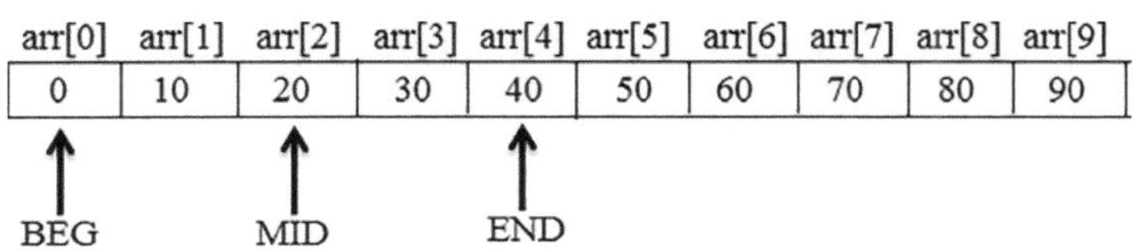

20 is found at POS = 2

Now, `VAL = 20` and `ARR[MID] = ARR[2] = 20`.
Hence, the search is successful, and `VAL = 20` is found at `POS = 2`.

Complexity of the Binary Search Algorithm

In a binary search algorithm, we can see that with each comparison, the size of the search area is reduced by half. So, we can claim that the efficiency of the binary search in the worst case is $O(log_{10}n)$, where n is the total number of elements in the array. The best case will happen when the value to be searched for is equal to the value of the array in the middle.

Drawbacks of Binary Search

Let's look at the disadvantages of using a binary search:

- A binary search requires that the data elements in the array be sorted; otherwise, it will not work.

A binary search cannot be used where there are many insertions and deletions of data elements in the array. Let us discuss the code to implement binary search algorithm.

Write a program to search for an element in an array using the binary search technique:

```
# include<stdio.h>
# include<conio.h>
intbinary_search (intarr[], int n, intval) ;
void main()
{
 intarr[10], n, i, r, val ;
 clrscr() ;
 printf("***BINARY SEARCH***") ;
 printf("\nenter no of elements") ;
 scanf("%d", &n) ;
 printf("enter the elements of array") ;
   for( i=0; i<n ; i++)
   {
     printf("\nenter element %d", i+1) ;
     scanf("%d", &arr[i]) ;}
 printf("\nenter value to search") ;
 scanf("%d", &val) ;
   r = binary_search(arr, n, val) ;
   if(r == -1)
      printf("value not found") ;
   else
      printf("%d value found at %d", val, r+1) ;
 getch() ;
}
intbinary_search(intarr[], int n, intval)
```

```
{
int start = 0, end = n-1, mid ;
   while(start <= end)
   {
      mid = (start+end)/2 ;
      if(arr[mid] == val)
      return mid;
      else if(arr[mid] <val)
      start=mid+1 ;
      else
      end=mid-1 ;
  }
  return (-1);
}
```

This is the output:

```
***BINARY SEARCH***
Enter no of elements  8


Enter elements of array
Enter element 1:   1
Enter element 2:   5
Enter element 3:   9
Enter element 4:   15
Enter element 5:   27
Enter element 6:   35
Enter element 7:   43
Enter element 8:   60


Enter value to search: 35
35 value found at 6th position
```

Frequently Asked Questions

Q2. What is a binary search? Explain.

Answer.

A binary search is one of the searching techniques that is used to find an element in an array. It works very efficiently with a sorted list. In a binary search, the element to be searched for is compared with the middle element of the array. If the value to be searched for is less than the middle element, we will search in the lower half of the array, and vice versa.

INTERPOLATION SEARCH

An *interpolation search*, also known as an *extrapolation search*, is a technique for searching for a particular value in an ordered array. This searching technique is more efficient than a binary search if the elements in the array are sorted. The technique of an interpolation search is similar to when we are searching for "Abhishek" in the telephone directory; we don't start in the middle, because we know that it will be near the extreme left, so we start from the front and work from there. That is the main idea of an interpolation search: instead of dividing the list into fixed halves, we cut it by an amount that seems most likely to succeed.

Practical Application:

If we want to search for "Ayush" in the directory, then we will always search in the extreme left of the directory.

Interpolation Search Algorithm

In each step of this searching technique, the remaining search area for the value to be searched for is calculated. The calculations are done on the values at the bounds of the search area and the value that is to be searched for. The value found at this position will now be compared with the value to be searched. If both values are equal, then the search is said to be successful. If both values are unequal, then, depending on the comparison done, the remaining search area is reduced to the part just before or after the initial position.

Consider an array, `ARR`, of `n` elements in which the elements are arranged in a sorted manner. Initially, `LOW` is set to `0`, and `HIGH` is set to `n-1`. Now, we are searching for a value, `VAL`, in `ARR` between `ARR[LOW]` and `ARR[HIGH]`. Then, in this case, `MID` will be calculated with the following formula:

```
MID = LOW + (HIGH - LOW) X ((VAL - ARR[LOW] / ARR[HIGH] - ARR[LOW]))
```

If the `VAL` is found at `MID`, then the search is complete; otherwise, if the value is lower than `ARR[MID]`, reset `HIGH = MID - 1`, and if the value is greater than `ARR[MID]`, reset `LOW = MID + 1`. Repeat these steps until the value is found.

Hence, we can say that the interpolation search is very much similar to the binary search technique. The main difference between the techniques is that in a binary search, the value selected is always the middle value of the list, and it discards half the values based on the comparison between the value to be searched for and the value found at the estimated position. Let us understand the interpolation search with the help of an algorithm as discussed below:

```
INTERPOLATION_SEARCH(ARR, Lower_bound, Upper_bound, VAL)
Step 1: START
Step 2: Set LOW = lower_bound, HIGH = upper_bound, POS = -1
Step 3: Repeat Steps 4 & 5 while LOW<= HIGH
Step 4: Set MID = LOW + (HIGH - LOW) X ((VAL - ARR[LOW] / ARR[HIGH] -
        ARR[LOW]))
Step 5: IF (ARR[MID] = VAL)
            POS = MID
            PRINT POS
Go to Step 7
ELSE IF (ARR[MID] > VAL)
            Set HIGH = MID - 1
ELSE
            Set LOW = MID + 1
[End of If]
[End of Loop]
Step 6: IF (POS = -1)
            PRINT "VALUE NOT FOUND, SEARCH UNSUCCESSFUL"
[End of IF]
Step 7: EXIT
```

For example, let's consider an array of seven numbers, declared as:

```
int array[] = {5, 16, 23, 34, 45, 56, 65} ;
```

This is shown in Figure 6.3. The value to be searched for is `45`. The algorithm will proceed as follows.

In the first pass, `LOW = 0`, `HIGH = 7 - 1 = 6`, and `VAL = 45`.

`ARR[LOW] = ARR[0] = 5` and `ARR[HIGH] = ARR[6] = 65`.

Now, `MID = LOW + (HIGH - LOW) X ((VAL - ARR[LOW]) / (ARR[HIGH] - ARR[LOW]))`, meaning `0 + (6 - 0) X ((45 - 5) / (65 - 5)`, and `0 + 6 X (40 / 60) = 4`.

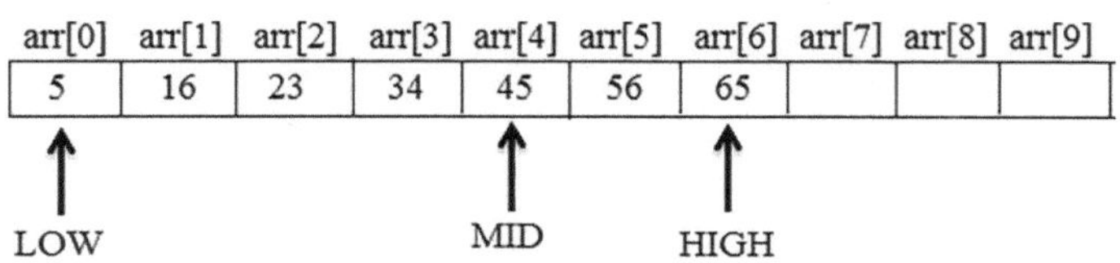

45 is found at POS = 4

If `(VAL == ARR[MID])`, then `45 == ARR[4] = 45`, and `45 = 45`.

Hence, the value is found.

Complexity of the Interpolation Search Algorithm

The interpolation search makes about $log_{10}(log_{10}\, n)$ comparisons when there are n elements in the list and the elements are uniformly distributed. The worst case will happen when the number of elements is increased exponentially; in that case, the algorithm can take up to $O(n)$ comparisons. The practical implementation of Interpolation search is discussed through following code.

Write a program to search for an element in an array using the interpolation search technique:

```
# include<stdio.h>
# include<conio.h>
intinterpolation_search (intarr[], int n, intval) ;
void main()
{
intarr[10], n, i, r, val ;
clrscr() ;
printf("***INTERPOLATION SEARCH***") ;
printf("\nenter no of elements") ;
scanf("%d", &n) ;
printf("enter the elements of array") ;
   for( i=0; i<n ; i++)
   {
    printf("\nenter element %d", i+1) ;
    scanf("%d", &arr[i]) ;
    }
printf("\nenter value to search") ;
scanf("%d", &val) ;
   r = interpolation_search(arr, n, val) ;
   if(r == -1)
printf("value not found") ;
   else
printf("%d value found at %d", val, r+1) ;
getch() ;
 }
intinterpolation_search(intarr[], int n, intval)
{
int low = 0, high = n-1, mid ;
   while(low <= high)
   {
      mid = low + (high - low) X ((val - arr[low]) / (arr[high] -
arr[high]))
      if(arr[mid] == val)
      return mid;
      else if(arr[mid] <val)
      low = mid+1 ;
      else
      high = mid-1 ;
  }
  return (-1);
}
```

This is the output:

```
***INTERPOLATION SEARCH***
Enter no of elements  7
Enter elements of array
Enter element 1:   15
Enter element 2:   25
```

```
Enter element 3:   35
Enter element 4:   45
Enter element 5:   55
Enter element 6:   65
Enter element 7:   75


Enter value to search: 35
35 value found at 3rd position
```

INTRODUCTION TO SORTING

Sorting refers to the process of arranging the data elements of an array in a specified order, that is, either in ascending or descending order. For example, it will be practically impossible for us to find a word in the dictionary if the words in it are not arranged in alphabetical order. This can also be true for dictionaries, book indexes, bank accounts, and so on. Hence, the convenience of having sorted data is unquestionable. Retrieval of information becomes much easier when the data is stored in some specified order. Therefore, sorting is a very important application in computer science.

Let us take an array that is declared and initialized as:

```
int array[] = {10, 25, 17, 8, 30, 3} ;
```

Then, the array after applying the sorting technique is:

```
array[] = {3, 8, 10, 17, 25, 30} ;
```

A sorting algorithm can be defined as an algorithm that puts the data elements of an array/list in a certain order, that is, either numerical order or any predefined order. There are many sorting algorithms that are available and are widely used according to the different environments required by the different sorting methods.

The two main categories of sorting methods are:

- *Internal sorting* – This refers to the sorting of the data elements stored in the computer's main memory.
- *External sorting* – This refers to the sorting of the data elements stored in the files. It is applied when the amount of data is large and cannot be stored in the main memory.

Types of Sorting Methods

The various sorting methods are:

- Selection sort
- Insertion sort
- Merge sort
- Bubble sort
- Quick sort

Let us discuss all of them in detail.

Selection Sort

Selection sort is a sorting technique that works by finding the smallest value in the array and placing it in the first position. After that, it then finds the second smallest value and places it in the second position. This process is repeated until the whole array is sorted. Thus, selection sort works by finding the smallest unsorted element remaining in the entire array and then swapping it with the element in the next position to be filled. It is a very simple technique, and it is also easier to implement than other sorting techniques. Selection sort is used for sorting files with large records.

Selection Sort Technique

Let us take an array, `ARR`, with `N` elements in it. The selection sort technique works as follows.

First of all, we will find the smallest value in the entire array, and we will place that value in the first position of the array. Then, we will find the second smallest value in the array, and we will place it in the second position of the array. We will repeat this process until the whole array is sorted.

In pass 1, we will find the position, `POS`, of the smallest value in the array of `N` elements and interchange `ARR[POS]` with `ARR[0]`. Hence, `ARR[0]` is sorted.

In pass 2, we will find the position, `POS`, of the smallest value in the array of `N-1` elements and interchange `ARR[POS]` with `ARR[1]`. Hence, `ARR[1]` is sorted.

We will continue this process until we get to pass N−1. In this pass, we will find the position, `POS`, of the smaller of the elements of `ARR[N-2]` and `ARR[N-1]` and interchange `ARR[POS]` with `ARR[N-2]`. Hence, `ARR[0]`, `ARR[1]`, …, `ARR[N-1]` is sorted.

Let us discuss it with the help of a detailed algorithm.

Consider an array, `ARR`, having `N` elements from `ARR[0]` to `ARR[N-1]`. `I` and `J` are the looping variables, and `POS` is the swapping variable:

```
SELECTION SORT(ARR, N)

Step 1: START
Step 2: Repeat Steps 3 & 4 for I = 1 to N - 1
Step 3: Call MIN(ARR, I, N, POS)
Step 4: Swap ARR[I] with ARR[POS]
[End of Loop]
Step 5: EXIT

MIN(ARR, I, N, POS)

Step 1:Set SMALLEST = ARR[I]
Step 2:Set POS = I
Step 3: Repeat Step 4 for J = I + 1 to N - 1
Step 4: IF (ARR[J] < SMALLEST)
            Set SMALLEST = ARR[J]
Set POS = J
[End of IF]
  [End of Loop]
Step 5: Return POS
```

As an example, let's sort the given array using selection sort:

arr[0]	arr[1]	arr[2]	arr[3]	arr[4]
4	14	29	11	35

Figure 6.4 shows the solution:

Pass	POS	Array[0]	Array[1]	Array[2]	Array[3]	Array[4]
1	4	4	14	29	11	35
2	3	4	11	29	14	35
3	3	4	11	14	29	35
4	3	4	11	14	29	35
5	4	4	11	14	29	35

FIGURE 6.4 Selection sort process.

Hence, after sorting, the new array is:

arr[0]	arr[1]	arr[2]	arr[3]	arr[4]
4	11	14	29	35

Complexity of the Selection Sort Algorithm

Selection sort is a simple technique of sorting. In this method, if there are n elements in the array, then $(n-1)$ comparisons or iterations are made. Thus, the selection sort technique has a complexity of $O(n^2)$. The practical implementation of selection sort algorithm is discussed through following code.

Write a program to sort an array using the selection sort method:

```
# include <stdio.h>
# include <conio.h>
void main()
{
int i, j, min, pos, arr[10], n, temp;
clrscr();
printf("enter no of elements in the array") ;
scanf("%d", &n) ;
printf("\nelements in the array are ") ;
 for( i=0 ; i<n ; i++ )
 {
   printf("\nenter element %d", i+1) ;
   scanf("%d", &arr[i]);
  }
printf("Selection Sort") ;
   for( i=1 ; i<n ; i++ )
```

```
    {
      min = arr[ i-1 ] ;
      pos = i-1;
      for( j=I ; j<n ; j++ )
       {
         if( arr[j] < min )
         {
              min = arr[j] ;
              pos = j ;
           }
        }
         if( pos != i-1 )//Swapping of variables is done
         {
           temp = arr[pos] ;
           arr[pos] = arr[i-1] ;
           arr[i-1] = temp ;
          }
       }
  printf("after sorting new array is") ;
      for( i=0 ; i<n ; i++ )
      {
        printf(" \t%d ", arr[i]) ;
      }
  getch() ;
  }
```

Here is the output:

```
Enter number of elements 5
Enter elements of array
Enter element 1    12
Enter element 2    78
Enter element 3    36
Enter element 4    94
Enter element 5    10


Selection Sort
After sorting new array is
10    12    36    78     94
```

Frequently Asked Questions

Q3. Define the selection sort technique.

Answer.

Selection sort is a sorting technique that works by finding the smallest element in the array and placing it in the first position. It then finds the second-smallest element and places it in the second position. This procedure is repeated until the whole array is sorted.

Insertion Sort

Insertion sort is another very simple sorting algorithm that works just like its name suggests, that is, it inserts each element into its proper position in the final list. To limit the wastage of memory, or to save memory, most implementations of an insertion sort work by moving the current element past the already sorted elements and repeatedly swapping or interchanging it with the preceding element until it is placed in its correct position.

Practical Application:

We usually use this technique when ordering a deck of cards while playing a game called bridge.

Insertion Sort Technique

Initially, there is only one element in the list, which is already sorted. Hence, we proceed to the next steps.

During the first iteration, the first and the second elements of the list are compared. The smaller value occupies the first position of the list.

During the second iteration, the first three elements of the list are compared. The smaller value will occupy the first position in the list. The second position will be occupied by the second smallest element, and so on.

This procedure is repeated for all the elements of the array up to $(n-1)$ iterations.

Here is the algorithm for an insertion sort:

```
INSERTION SORT(ARR, N)
Step 1: START
Step 2: Repeat Steps 3 to 6 for I = 1 to N - 1
Step 3: Set POS = ARR[I]
Step 4: Set J = I - 1
Step 5: Repeat while POS <= ARR[J]
            Set ARR[J + 1] = ARR[J]
            Set J = J - 1
[End of Inner while loop]
Step 6: Set ARR[J + 1] = POS
            [End of Loop]
Step 7: EXIT
```

In the previous algorithm, in `Step 2`, a `for` loop is executed, which will be repeated for every element in the array. In `Step 3`, we are storing the value of the `I`th element in `POS`. In `Step 5`, again, a loop is executed in which the new elements after sorting are placed. Finally, the element is stored at the `(J+1)`th position.

For example, consider the following array. Sort the given values in the array using the insertion sort technique:

arr[0]	arr[1]	arr[2]	arr[3]	arr[4]	arr[5]
39	54	10	28	95	7

Let's look at the solution, shown below

In pass 1, `ARR[0]` is sorted. Move to the next pass.

39	54	10	28	95	7

In pass 2, `39` and `54` are compared. `39 < 54`, so `ARR[0] = 39` and `ARR[1] = 54`.

39	54	10	28	95	7

In pass 3, `39`, `54`, and `10` are compared. `10 < 39` and `54`, so `ARR[0] = 10`. Now, `39 < 54`, hence `ARR[1] = 39` and `ARR[2] = 54`.

10	39	54	28	95	7

In pass 4, as `28 < 39` and `54`, so `ARR[1] = 28`.

10	28	39	54	95	7

In pass 5, `95` is greater than all the values, so there is no need for swapping.

10	28	39	54	95	7

In pass 6, `7` is the smallest value, so `ARR[0] = 7`.

Therefore, after sorting, the new array is:

7	10	28	39	54	95

Complexity of the Insertion Sort Algorithm

In an insertion sort, the best case will happen when the array is already sorted, and in that case, the running time of the algorithm is $O(n)$ (i.e., linear running time). The worst case will happen when the array is sorted in reverse order. Thus, in that case, the running time of the algorithm is $O(n^2)$ (i.e., quadratic running time). The practical implementation of insertion sort is discussed below through the following code.

Write a program to sort an array using the insertion sort method:

```
# include <stdio.h>
# include <conio.h>
void main()
{
 int i, j, min, arr[10], n, temp;
 clrscr();
 printf("enter no of elements in the array") ;
 scanf("%d", &n) ;
 printf("\nelements in the array are ") ;
 for( i=0 ; i<n ; i++ )
 {
   printf("\nenter element %d", i+1) ;
   scanf("%d", &arr[i]);
  }
 printf("Insertion Sort") for( i=1 ; i<n ; i++ )
  {
   temp = arr[i] ;
    j = i -1;
    while(( temp <arr[j] && (j >= 0))
     {
      arr[j + 1] = arr[j] ;
    j-- ;
     }
    arr[j + 1] = temp ;
        }
 printf("after sorting new array is") ;
 for( i=0 ; i<n ; i++ )
    {
      printf(" \t%d ", arr[i]) ;
    }
 getch() ;
}
```

Here is the output:

```
Enter number of elements 7
Enter elements of array
Enter element 1    58
Enter element 2    12
Enter element 3    20
Enter element 4    8
Enter element 5    11
Enter element 6    99
Enter element 7    63


Insertion Sort
After sorting new array is
811    12    20    58    63    99
```

Merge Sort

Merge sort is a sorting method that follows the divide and conquer approach. The divide and conquer approach is a very good approach in which *divide* means partitioning the array having *n* elements into two sub-arrays of *n*/2 elements each. If there are no elements present in the list/array or if an array contains only one element, however, then it is already sorted. If an array has more elements, then it is divided into two sub-arrays containing equal elements. *Conquer* is the process of sorting the two sub-arrays recursively using merge sort. Finally, the two sub-arrays are merged into one single sorted array.

Merge Sort Techniques

If the array has zero or one element in it, then there is no need to sort that array as it is already sorted. Otherwise, if there are more elements in the array, then divide the array into two sub-arrays containing equal elements.

Each sub-array is now sorted recursively using merge sort.

Finally, the two sub-arrays are merged into a single sorted array.

Here is the algorithm for merge sort:

```
MERGE SORT(ARR, BEG, END)

Step 1: START
Step 2: IF (BEG < END)
Step 3: Set MID = (BEG + END)/2
Call MERGE SORT (ARR, BEG, MID)
Call MERGE SORT (ARR, MID + 1, END )
             Call MERGE (ARR, BEG, MID, END)
[End of If]
Step 4: EXIT

MERGE(ARR, BEG, MID, END)

Step 1: START
Step 2: Set I = BEG, J = MID + 1, K = 0
Step 3: Repeat while (I <= MID)  && (J <= END)
  IF (ARR[I] > ARR[J])
             Set TEMP[K] = ARR[J]
             Set J = J + 1
             Set K = K + 1
ELSE IF (ARR[J] > ARR[I])
             Set TEMP[K] = ARR[I]
             Set I = I + 1
             Set K = K + 1
ELSE
             Set TEMP[K] = ARR[J]
             Set J = J + 1
             Set K = K + 1
             Set TEMP[K] = ARR[I]
             Set I = I + 1
             Set K = K + 1
    [End of If]
```

```
    [End of Loop]
Step 4: (Copying the remaining elements of left sub array if any)
Repeat while (I <= MID)
                 Set TEMP[K] = ARR[I]
                 Set I = I + 1
                 Set K = K + 1
          [End of Loop]
Step 5: (Copying the remaining elements of right sub array if any)
Repeat while (J <= END)
                 Set TEMP[K] = ARR[J]
                 Set I = I + 1
                 Set K = K + 1
            [End of Loop]
Step 6: Set IND = 0
Step 7: Repeat while (IND < K)
Set ARR[IND] = ARR[IND]
                 Set IND = IND + 1
            [End of Loop]
Step 8: EXIT
```

As an example, let's sort the following array using merge sort:

```
int array[] = { 40, 10, 86, 44, 93, 26, 69, 17 }
```

The solution is shown in Figure 6.6, which shows using the divide and conquer technique to merge the sub-arrays into one sorted array:

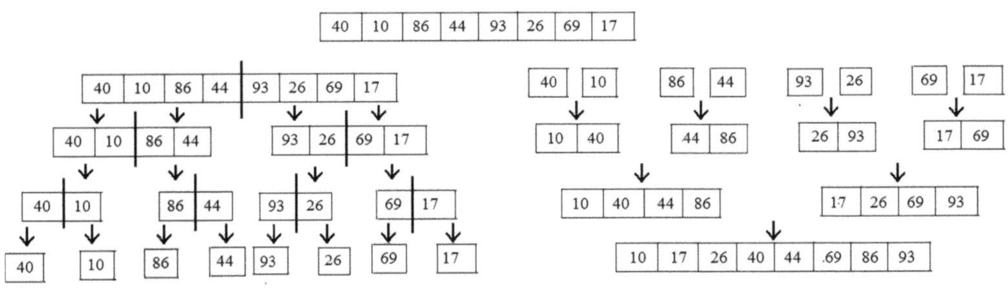

FIGURE 6.6 The merge sort process.

From the previous example, we can see how the merge sort algorithm works. First, the merge sort algorithm recursively divides the array into smaller sub-arrays. After dividing the array into smaller parts, we call the function `Merge()` to merge all the sub-arrays to form a single sorted array.

Complexity of the Merge Sort Algorithm

The running time of the merge sort algorithm is $O(n \log_{10} n)$. This runtime remains the same in the average as well as in the worst case of the merge sort algorithm. Although it has an optimal time complexity, sometimes this runtime can be $O(n)$. The practical implementation of Merge Sorting is discussed with the help of following code.

Write a program to sort an array using the merge sort method:

```
# include<stdio.h>
# include<conio.h>
void merge(intarr[], int i1, int j1, i2, int j2) ;
void merge_sort(intarr[], int beg, int end)
{
int mid ;
 if(beg < end)
 {
   mid=(beg+end)/2;
   merge_sort(arr, beg, mid) ;
   merge_sort(arr, mid+1, end) ;
   merge(arr, beg, mid, mid+1, end) ;
  }
}
void merge(intarr[], int i1, int j1, int i2, int j2)
{
  int temp[20] ;
  int i, j, k ;
   i = i1 ;
   j = i2 ;
   k = 0 ;
 while(i < j1 && j < j2)
 {
    if(arr[i] <arr[j])
   {
      temp[k] = arr[i] ;
      k++ ;
      i++ ;
  }
  else if(arr[i] >arr[j])
  {
    temp[k] = arr[j] ;
    k++ ;
    j++ ;
  }
else
{
   temp[k] = arr[j] ;
   k++ ;
   j++ ;
   temp[k] = arr[i] ;
  k++ ;
   i++ ;
 }
}
while(i < j1)
{
   temp[k] = arr[i] ;
   k++ ;
   i++ ;
}
while(j < j2)
```

```
{
  temp[k] = arr[j] ;
  k++ ;
  j++ ;
}
for(i = i1, j = 0 ; i <= j2 ; i++, j++)
{
  arr[i] = temp[j] ;
  }
 }
int main()
{
int i, n, arr[10] ;
clrscr() ;
printf("enter no of elements: ") ;
scanf("%d", &n) ;
printf("\nelements of array are:") ;
 for(i=0; i<n; i++)
 {
   printf("\nenter element %d", i+1) ;
   scanf("%d", &arr[i]) ;
  }
merge_sort(arr, 0, n-1) ;
printf("\nafter merge sort new array is:") ;
  for(i=0; i<n; i++)
  {
   printf("\t%d", arr[i]) ;
 }
return 0 ;
}
```

Here is the output:

```
Enter number of elements: 8
Enter elements of array:
Enter element 1    3
Enter element 2    15
Enter element 3    69
Enter element 4    32
Enter element 5    10
Enter element 6    87
Enter element 7    21
Enter element 8    45


After merge sort new array is:
3    10    15    21    32    45    69    87
```

Bubble Sort

Bubble sort, also known as *exchange sort*, is a very simple sorting method. It works by repeatedly moving the largest element to the highest position of the array. In bubble sort, we are comparing two elements at a time, and swapping is done if they are wrongly placed. If the element at a lower index or position is greater than the element at a higher index, then both elements are interchanged so that the smaller element is placed before the larger one. This process is repeated until the list becomes sorted. Bubble sort gets its name from the way that the smaller elements *bubble* to the top of the array. This sorting technique only uses comparisons to operate on the elements. Hence, we can also call it a *comparison sort*.

Bubble Sort Technique

The basic idea applied for a bubble sort is to let us assume that if an array, ARR, contains n elements, then the number of iterations required to sort the array will be $(n - 1)$.

During the first iteration, the largest value in the array is placed at the last position.

During the second iteration, the second largest value of the array is placed in the second last position.

During the third iteration, the third largest value of the array is placed in the third last position, and so on.

This procedure is repeated until all the elements in the array are scanned and are placed in their correct position, which means that the array is sorted.

Here is the algorithm for a bubble sort:

```
BUBBLE SORT(ARR, N)

Step 1: START
Step 2: Repeat Step 3 for I = 0 to N - 1
Step 3: Repeat for J = 0 to N - 1
Step 4: IF (ARR[J] > ARR[J+1])
             INTERCHANGE ARR[J] & ARR[J + 1]
[End of Inner Loop]
   [End of Outer Loop]
Step 5: EXIT
```

As an example, consider the following array. Sort the given values in the array using the bubble sort technique:

arr[0]	arr[1]	arr[2]	arr[3]	arr[4]
40	50	20	90	30

Let's go through the solution, discussed in following steps

In the given array, the number of elements in the array is 5, so the number of iterations will be $(n - 1) = 4$.

In pass 1, 40 and 50 are compared. Since 40 < 50, no swapping is done:

40	50	20	90	30

Then, 50 and 20 are compared. Since 50 > 20, swapping will be done:

40	50	20	90	30

Next, `50` and `90` are compared. Since 50 < 90, no swapping is done:

40	20	50	90	30

Finally, `90` and `30` are compared. Since 90 > 30, swapping is done:

40	20	50	90	30

At the end of the first pass, the largest element in the array is placed at the highest position in the array, but all the other elements are still unsorted:

40	20	50	30	90

Let us now proceed to pass 2.
In pass 2, `40` and `20` are compared. Since 40 > 20, swapping is done:

40	20	50	30	90

Then, `40` and `50` are compared. Since 40 < 50, no swapping will be done:

20	40	50	30	90

Next, `50` and `30` are compared. Since 50 > 30, swapping is done:

20	40	50	30	90

At the end of the second pass, the second largest element in the array is placed at the second last position in the array, but all the other elements are still unsorted:

20	40	30	50	90

Let us now proceed to pass 3.
In pass 3, `20` and `40` are compared. Since 20 < 40, no swapping is done:

20	40	30	50	90

Next, `40` and `30` are compared. Since 40 > 30, swapping will be done:

20	40	30	50	90

At the end of the third pass, the third largest element in the array is placed at the third largest position in the array, but all the other elements are still unsorted:

20	30	40	50	90

Let us now proceed to pass 4.

In pass 4, `20` and `40` are compared. Since 20 < 40, no swapping is done:

20	40	30	50	90

At the end of the fourth pass, we can see that all the elements in the list are sorted. Hence, after sorting, the new array will be:

20	40	30	50	90

Complexity of the Bubble Sort Algorithm

Bubble sort is the most inefficient sorting algorithm, and hence it is not commonly used. In the best case, the running time of bubble sort is $O(n)$, that is, when the array is already sorted. Otherwise, its level of complexity in average and worst cases is extremely poor, that is, $O(n^2)$. The practical implementation of Bubble sort method is discussed through following code.

Write a program to sort an array using the bubble sort method:

```
# include <stdio.h>
# include <conio.h>
void main()
{
int  i , n, temp, j, arr[10] ;
clrscr() ;
printf("\n enter the number of elements: ") ;
scanf ("%d", &n) ;
printf ("\ n enter the elements of array:") ;
  for(i=0 ; i<n ; i++)
  {
    printf("\nenter element %d", i+1) ;
    scanf("%d" , &arr[i]) ;
  }
  for(i=0 ; i<n ; i++)
  {
   for(j=0 ; j<=n ; j++)
    {
       if(arr[j] >arr[j+1])
        {
           temp= arr[j]
           arr[j] = arr[j+1] ;
           arr[j+1] = temp ;
         }
      }
    }
printf("\nThe Array Sorted In Ascending Order Is\n :") ;
for(i=0 ; i<n ; i++)
{
  printf("%d\t" , arr[i]) ;
  }
getch() ;
}
```

Here is the output:

```
Enter number of elements: 6
Enter elements of array:
Enter element 1    63
Enter element 2    78
Enter element 3    10
Enter element 4    9
Enter element 5    33
Enter element 6    47


The Array Sorted In Ascending Order Is:
9    10    33    47    63    78
```

Quick Sort

Quick sort, also known as *partition exchange sort* and developed by C. A. R. Hoare, is a widely used sorting algorithm that also uses the divide and conquer approach, as we discussed when talking about merge sort. Here, we will also divide a single unsorted array into its two smaller sub-arrays. The divide and conquer method means dividing the bigger problem into two smaller problems, and then those two smaller problems into smaller problems, and so on. Like merge sort, if there are no elements in the array or if an array contains only one element, then it is already sorted. The quick sort algorithm is faster than all the other sorting algorithms, which have a time complexity of $O(n \log_{10} n)$.

Working of Quick Sort

In quick sort, an element called a *pivot* is selected from the array elements. After choosing the pivot element, all the elements of the array are rearranged such that all the elements less than the pivot element will be on the left side, and all the elements greater than the pivot element will be placed on the right side of the pivot element. After rearranging all the elements, the pivot is now placed in its final position. Thus, this process is known as *partitioning*.

Now, the two sub-arrays obtained will be recursively sorted.

Quick Sort Technique

1. Initially, set the index of the first element to `LEFT` and `POS`. Similarly, set the index of the last element to `RIGHT`. Now, `LEFT = 0`, `POS = 0`, and `RIGHT = N - 1` (assuming *n* elements in the array).
2. We will start with the last element, which is pointed to by `RIGHT`, and we will traverse each element in the array from right to left, comparing each element with the first element pointed to by `POS`. Note that `ARR[POS]` should always be less than `ARR[RIGHT]`:
   - **a)** If `ARR[POS]` is less than `ARR[RIGHT]`, then continue comparing until `RIGHT = POS`. If `RIGHT = POS`, then it means that the pivot is placed in its correct position.
   - **b)** If `ARR[RIGHT]` is less than `ARR[POS]`, then swap the two values and go to the next step.

3. Set POS = RIGHT.
4. We will start from the first element, which is pointed to by LEFT, and we will traverse every element in the array from left to right, comparing each element with the element pointed to by POS. Note that ARR[POS] should always be greater than ARR[LEFT]:
   a) If ARR[POS] is greater than ARR[LEFT], then continue comparing until LEFT = POS. If LEFT = POS, then it means that the pivot is placed in its correct position.
   b) If ARR[LEFT] is greater than ARR[POS], then swap the two values and go to the previous step.
5. Set POS = LEFT.

Here is the algorithm for quick sort:

```
QUICK SORT(ARR, BEG, END)

Step 1: START
Step 2: IF (BEG < END)
Call PARTITION (ARR, BEG, END, POS)
Call QUICK SORT (ARR, BEG, POS - 1)
             Call QUICK (ARR, POS + 1, END)
[End of If]
Step 3: EXIT

PARTITION(ARR, BEG, END, POS)

Step 1: START
Step 2: Set LEFT = BEG, RIGHT = END, POS = BEG, TEMP = 0
Step 3: Repeat Steps 4 to 7 while TEMP = 0
Step 4: Repeat while ARR[RIGHT] >= ARR[POS]&& POS != RIGHT
             Set RIGHT = RIGHT - 1
[End of Loop]
Step 5: IF (POS = RIGHT)
            Set TEMP = 1
      ELSE IF (ARR[POS] > ARR[RIGHT])
             INTERCHANGE ARR[POS] with ARR[RIGHT]
             Set POS = RIGHT
   [End of If]
Step 6: IF TEMP = 0
             Repeat while ARR[POS] >= ARR[LEFT] && POS != LEFT
             Set LEFT = LEFT + 1
         [End of Loop]
Step 7: IF (POS = LEFT)
             Set TEMP = 1
ELSE IF (ARR[LEFT] > ARR[POS])
             INTERCHANGE ARR[POS] with ARR[LEFT]
             Set POS = LEFT
  [End of If]
      [End of If]
       [End of Loop]
Step 8: EXIT
```

As an example, let's sort the values given in the following array using the quick sort algorithm:

arr[0]	arr[1]	arr[2]	arr[3]	arr[4]	arr[5]
25	7	39	17	30	52

Let's go through the solution, as shown below

In step 1, the first element is chosen as the pivot. Now, set `POS = 0`, `LEFT = 0`, and `RIGHT = 5`:

25	7	39	17	30	52
POS, LEFT					RIGHT

In step 2, we traverse the list from right to left. Since `ARR[POS] < ARR[RIGHT]` (that is, `25 < 52`), `RIGHT = RIGHT - 1 = 4`.

25	7	39	17	30	52
POS, LEFT				RIGHT	

In step 3, since `ARR[POS] < ARR[RIGHT]` (that is, `25 < 30`), `RIGHT = RIGHT - 1 = 3`.

25	7	39	17	30	52
POS, LEFT			RIGHT		

In step 4, since `ARR[POS] > ARR[RIGHT]` (that is, `25 >17`), we will swap the two values and set `POS = RIGHT`.

17	7	39	25	30	52
LEFT			RIGHT, POS		

In step 5, we traverse the list from left to right. Since `ARR[POS] > ARR[LEFT]` (that is, `25 > 17`), `LEFT = LEFT + 1`.

17	7	39	25	30	52
	LEFT		RIGHT, POS		

In step 6, since `ARR[POS] > ARR[LEFT]` (that is, `25 > 7`), `LEFT = LEFT + 1`.

17	7	39	25	30	52
		LEFT	RIGHT, POS		

In step 7, since ARR[POS] < ARR[LEFT] (that is, 25 < 39), we will swap the values and set POS = LEFT.

17	7	25	39	30	52

LEFT, POS   RIGHT

In step 8, we traverse the list from right to left. Since ARR[POS] < ARR[RIGHT], RIGHT = RIGHT - 1.

17	7	25	39	30	52

LEFTPOS, RIGHT

Now, RIGHT = POS, so the process is over, and the pivot element of the array, 25, is placed in its correct position. Therefore, all the elements that are smaller than 25 are placed before it, and all the elements greater than 25 are placed after it. Hence, 17 and 7 are the elements in the left sub-array, and 39, 30, and 52 are the elements in the right sub-array, which are both sorted.

Complexity of the Quick Sort Algorithm

The running time efficiency of quick sort is $O(n\ log_{10}n)$ in the average and the best case. The worst case will happen if the array is already sorted and the leftmost element is selected as the pivot element. In the worst case, its efficiency is $O(n^2)$.

Write a program to sort an array using the quick sort method:

```
# include <stdio.h>
# include <conio.h>
# define size 100
int part(intarr[], int beg , int end) ;
void quick_sort( intarr[ ] , int beg , int end) ;
void main()
{
intarr[size], i, n ;
printf("\n enter the number of elements: ") ;
scanf ("%d", &n) ;
printf ("\ n enter the elements of array :") ;
   for(i=0 ; i<n ; i++)
    {
      printf("\nenter element %d", i+1) ;
      scanf("%d" , &arr[i]) ;
    }
quick_sort(arr , 0 , n-1) ;
printf("\nThe sorted array is :\n") ;
     for (i=0 ; i<n ; i++)
     {
       printf(" %d \t ", arr[i]) ;
     }
getch() ;
}
```

```
int part (int arr[], int beg, int end)
{
int left , right , temp , pos , flag ;
pos = left = beg ;
  right = end ;
  flag = 0 ;
while( flag ! = 1)
  {
   while ( ( a[pos] < = arr[right ] && (loc != right ))
   right-- ;
   if(pos == right)
   flag = 1 ;
   else if (arr[pos] >arr[right])
   {
      temp = arr[pos] ;
      arr[pos] = arr[right] ;
      arr[right] = temp ;
      pos  = right ;
  }
   if ( flag != 1)
{
   while (( arr[pos] >= arr[left] && (pos != left))
   left ++ ;
   if( pos == left )
   flag = 1 ;
   else if (arr[pos] <arr[left])
  {
    temp = arr[pos] ;
    arr[pos]= arr[left] ;
    arr[left]= temp ;
    pos = left ;
   }
  }
 }
return pos ;
}
void quick sort (int arr[], int beg, int end)
{
int pos ;
   if(beg < end)
   {
     pos = part(arr, beg , end) ;
     quick_sort(arr, beg, loc-1) ;
     quick_sort(arr, loc+1, end) ;
 }
}
```

Here is the output:

```
Enter number of elements: 5
Enter elements of array:
Enter element 1    52
Enter element 2    78
```

```
Enter element 3    16
Enter element 4    01
Enter element 5    86


The sorted array is :
01    16    52    78    86
```

EXTERNAL SORTING

External sorting is a sorting technique that is used when the amount of data is very large. When a large amount of data has to be sorted, it is not possible to bring it into main memory (RAM). Therefore, in that situation, a secondary memory needs to be used. At the same time, some portion of data is brought into the main memory from the secondary memory for sorting based on the availability of storage space in the main memory. After the data is sorted, it is sent back to the secondary memory. Now, the next portion of the data is brought into the main memory, and after sorting, it is sent back to the secondary memory. This procedure is repeated until all the data is sorted. Here, each portion is called a *segment*. The time required for sorting is longer because a lot of time will be spent transferring the data from secondary memory to main memory. The merge sort algorithm is widely and commonly used in external sorting, which has already been discussed.

External sorting is used in database applications for performing different kinds of operations, such as joins, unions, projections, and many more. It is also used to update a master file from a transaction file. For example, updating the company file based on new employees, existing employees, locations, and so on. Duplicate records or data can also be removed from external sorting.

SUMMARY

- The process of finding a particular value in a list or an array is called searching. If that particular value is present in the array, then the search is said to be successful, and the location of that particular value is retrieved by the searching process.
- Linear search, binary search, and interpolation search are the commonly used searching techniques.
- Linear search works by comparing the values to be searched for with every element of the array in a linear sequence until a match is found
- Binary search works efficiently when the list is sorted. In a binary search, we first compare the value, `VAL`, with the data element in the middle position of the array.
- Interpolation search, also known as extrapolation search, is a technique for searching for a particular value in an ordered array. In each step of this searching technique, the remaining search area for the value to be searched for is calculated. The calculations are done on the values at the bounds of the search area and the value that is to be searched.
- Sorting refers to the technique of arranging the data elements of an array in a specified order, that is, either in ascending or descending order.

- Selection sort is a sorting technique that works by finding the smallest value in the array and placing it in the first position. After that, it then finds the second smallest value and places it in the second position. This process is repeated until the whole array is sorted.
- Insertion sort works by moving the current data element past the already sorted data elements and repeatedly interchanging it with the preceding element until it is in the correct place.
- Merge sort is a sorting method that follows the divide and conquer approach. *Divide* means partitioning the array having n elements into two sub-arrays of $n/2$ elements each. *Conquer* is the process of sorting the two sub-arrays recursively using merge sort. Finally, the two sub-arrays are merged into one single sorted array.
- Bubble sort, also known as exchange sort, is a very simple sorting method. It works by repeatedly moving the largest element to the highest position of the array.
- Quick sort is an algorithm that selects a pivot element and rearranges the values in such a way that all the elements less than the pivot element appear before it and the elements greater than the pivot element appear after it.
- External sorting is a sorting technique that is used when the amount of data is very large.

EXERCISES

Theory Questions

1. Define sorting. Explain the importance of sorting.
2. What are the different types of sorting techniques? Discuss each of them in detail.
3. Define searching. Which searching technique will you use while searching for an element in an array?
4. Explain selection sort and merge sort with suitable examples. Show various stages.
5. How is linear search used to find an element? Explain the working of insertion sort with a suitable example.
6. Explain different types of searching techniques. Give a suitable example to illustrate binary search.
7. Why is quick sort known as "quick"?
8. Explain the concept of external sorting.
9. Differentiate between binary search and interpolation search. Give a suitable example.
10. Discuss the limitations and advantages of insertion sort.
11. Explain the working of bubble sort with a suitable example. Why is bubble sort called "bubble"?

Programming Questions

1. Write a program to implement the bubble sort technique.
2. Write an algorithm to implement the interpolation search technique.
3. Write an algorithm to perform a merge sort. Show various stages in merge sorting over the data: `11, 2, 9, 13, 57, 25, 17, 1, 90, 3`.
4. Write a program to implement an insertion sort.
5. Write a program to search for an element using the binary search technique.

6. Write a program to perform a comparison sort.
7. Write an algorithm to perform a partition exchange sort technique. Show various stages over the data: `24, 52, 98, 12, 45, 6, 59, 90.`
8. Write an algorithm/program to implement a linear search technique.

Multiple Choice Questions

1. A binary search algorithm cannot be applied to ___________.
   - A. A sorted array
   - B. A sorted linked list
   - C. An unsorted linked list
   - D. Binary trees
2. In a merge sort algorithm, which term refers to "sorting the sub-arrays recursively"?
   - A. Conquer
   - B. Combine
   - C. Divide
   - D. All of these
3. What is the time complexity of the bubble sort algorithm?
   - A. $O(log\ n)$
   - B. $O(n)$
   - C. $O(n.log\ n)$
   - D. $O(n^2)$
4. Which sorting algorithm is known as a partition and exchange sort?
   - A. Selection sort
   - B. Merge sort
   - C. Quick sort
   - D. Bubble sort
5. Which case would exist when the element to be searched for using linear search is equal to the first element of the array?
   - A. Best case
   - B. Worst case
   - C. Average case
   - D. None of these
6. Quick sort is faster than ___________.
   - A. Bubble sort
   - B. Selection sort
   - C. Insertion sort
   - D. All of the above
7. When the amount of data is very large, which type of sorting is preferred?
   - A. Internal sorting
   - B. External sorting
   - C. Both of these
   - D. None of these

8. Which of the searching techniques will be best when the value to be searched for is present in the middle?
   - A. Linear search
   - B. Interpolation search
   - C. Binary search
   - D. All of these
9. What is the complexity of a binary search algorithm?
   - A. $O(n^2)$
   - B. $O(log\ n)$
   - C. $O(n)$
   - D. $O(n\ log\ n)$
10. Does selection sort have a linear running time complexity?
    - A. Yes
    - B. No

CHAPTER 7

STACKS

INTRODUCTION

A *stack* is an important data structure that is widely used in many computer applications. It can be visualized with many examples that we are already familiar with from our day-to-day life. A very simple illustration of a stack is a pile of books where one book is placed on top of another, as in Figure 7.1. When we want to remove a book, we remove the topmost book first. Hence, we can add or remove an element (in this case, a book) only at or from one position, which is the topmost position. Similarly, there are many daily life examples in which we can see how the stack is implemented. Hence, we observe that whenever we talk about a stack, we see that the element at the last position will be served first. Thus, a stack can be described as a *last-in, first-out* (*LIFO*) data structure; that is, the element that is inserted last will be the first one to be taken out.

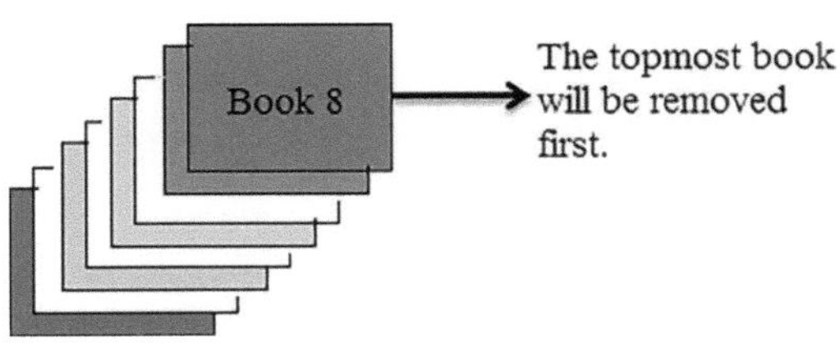

FIGURE 7.1 Stack of books.

Now, let us discuss stacks in detail.

DEFINITION OF A STACK

A stack is a linear collection of data elements in which the element inserted last will be the element taken out first (so, a stack is a LIFO data structure). It is an abstract data structure, somewhat similar to queues. Unlike queues, a stack is open only from one end. The stack is a linear data structure in which the insertion, as well as the deletion of an element, is done only from the end called `TOP`, as shown in Figure 7.2. One end is always closed, and the other end is used to insert and remove data.

Stacks can be implemented by using arrays or linked lists. We will discuss the implementation of stacks using arrays and linked lists in this section.

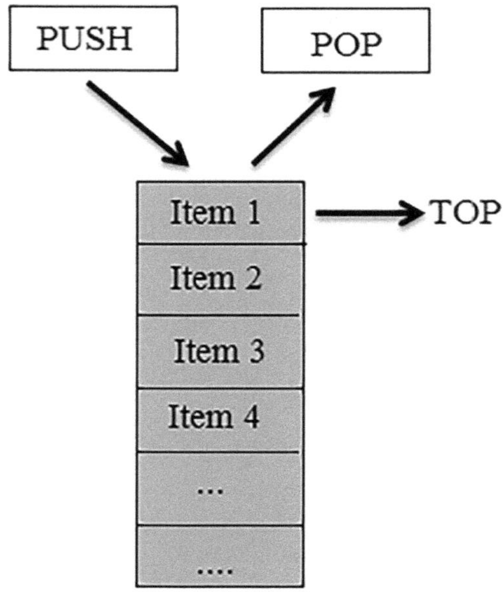

FIGURE 7.2 Representation of a stack.

Practical Application:

A real-life example of a stack is a pile of dishes where one dish is placed on top of another. When we want to remove a dish, we remove the topmost dish first.

Another real-life example of a stack is a pile of disks where one disk is placed on top of another. When we want to remove a disk, we remove the topmost disk first.

OVERFLOW AND UNDERFLOW IN STACKS

Let us discuss both overflow and underflow in stacks in detail:

- *Overflow in stacks* – An *overflow* condition occurs when we try to insert elements in the stack, but the stack is already full. Hence, the new elements cannot be inserted into the stack. If an attempt is made to insert a value in a stack that is already full, an overflow message is printed. It can be checked by the following formula:

  $TOP = MAX - 1$, where MAX is the size of the stack.

- *Underflow in stacks* – An *underflow* condition occurs when we try to remove elements from the stack, but the stack is already empty. Hence, no deletions can take place from the stack. If an attempt is made to delete a value from a stack that is already empty, an underflow message is printed. It can be checked by the following formula:

$$TOP = NULL$$

Frequently Asked Questions

Q1. Define a stack. What are the types of operations performed on stacks?

Answer.

A stack is a linear data structure in which the insertion as well as deletion of an element is done only from the end called `TOP`. It is LIFO in nature. The different operations that can be performed on stacks are:

- Push operation
- Pop operation
- Peek operation

OPERATIONS ON STACKS

There are three basic operations that can be performed on stacks: push, pull, and peek. Let's look at each in detail next.

Push Operations

A *push operation* is the process of adding new elements to the stack. Before inserting any new element in the stack, however, we must always check for the overflow condition, which occurs when we try to insert an element in a stack that is already full. An overflow condition can be checked as follows: $TOP = MAX - 1$, where MAX is the size of the stack. Hence, if the overflow condition is true, then an overflow message is displayed on the screen; otherwise, the element is inserted into the stack.

For example, let us take a stack that has five elements in it. Suppose we want to insert another element, `10`, into it; then `TOP` will be incremented by 1. Thus, the new element is inserted at the position pointed to by `TOP`. Figure 7.3 shows the stack before the push operation occurs:

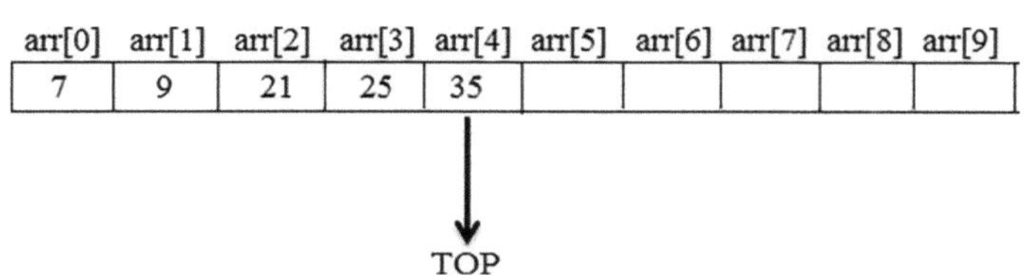

FIGURE 7.3 Stack before inserting a new element.

Figure 7.4 shows the stack after inserting `10` into it:

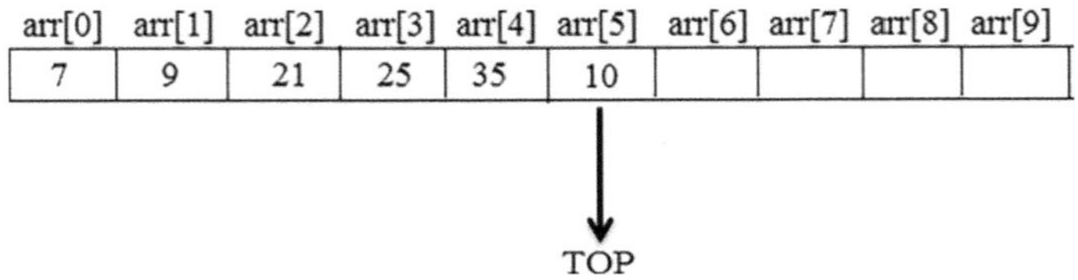

FIGURE 7.4 Stack after inserting a new element.

Let's look at the algorithm for a push operation in a stack:

```
Step 1: START
Step 2: IF TOP = MAX - 1
          Print OVERFLOW ERROR
Go to Step 5
[End of If]
Step 3:Set TOP = TOP + 1
Step 4: Set STACK[TOP] = ITEM
Step 5: EXIT
```

In the previous algorithm, first, we check for the overflow condition. In `Step 3`, `TOP` is incremented so that it points to the next location. Finally, the new element is inserted in the stack at the position pointed to by `TOP`.

Pop Operations

A *pop operation* is the process of removing elements from the stack. Before deleting an element from the stack, however, we must always check for the underflow condition, which occurs when we try to delete an element from a stack that is already empty. An underflow condition can be checked as follows: *TOP = NULL*. Hence, if the underflow condition is true, then an underflow message is displayed on the screen; otherwise, the element is deleted from the stack.

For example, let us take a stack that has five elements in it, as shown in Figure 7.5. Suppose we want to delete an element, `35`, from a stack; then `TOP` will be decremented by 1. Thus, the element is deleted from the position pointed to by `TOP`.

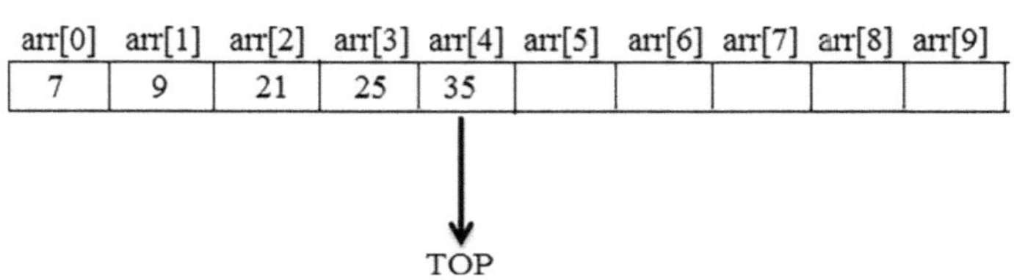

FIGURE 7.5 Stack before deleting an element.

Figure 7.6 shows the stack after deleting `35` from it:

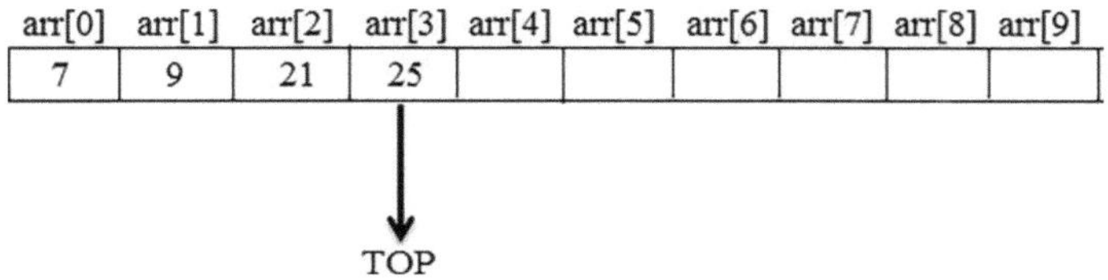

FIGURE 7.6 Stack after deleting an element.

Let's look at the algorithm for a pop operation in a stack:

```
Step 1: START
Step 2: IF TOP = NULL
            Print UNDERFLOW ERROR
            Go to Step 5
[End of If]
Step 3: Set ITEM = STACK[TOP]
Step 4:Set TOP = TOP - 1
Step 5: EXIT
```

In the previous algorithm, first, we check for the underflow condition, that is, whether the stack is empty or not. If the stack is empty, then no deletion takes place; otherwise, `TOP` is decremented to the previous position in the stack. Finally, the element is deleted from the stack.

Peek Operations

A *peek operation* returns the value of the topmost element of the stack. It does so without deleting the topmost element of the array. The peek operation first checks for the underflow condition. An underflow condition can be checked as follows: *TOP = NULL*. Hence, if the underflow condition is true, then an underflow message is displayed on the screen; otherwise, the value of the element is returned, as shown in Figure 7.7. The practical implementation of various operations on stack is discussed below through a menu driven code.

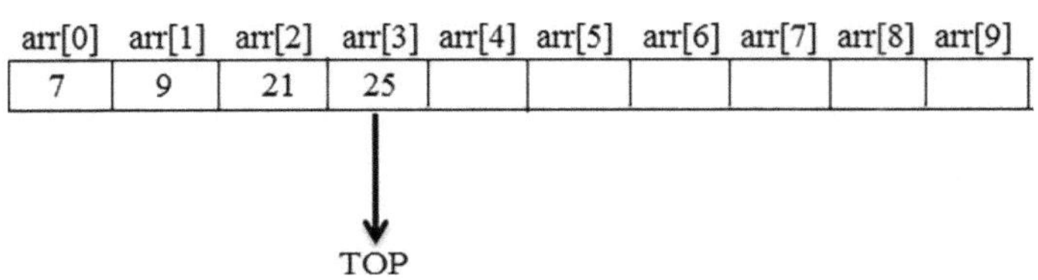

FIGURE 7.7 Stack returning the topmost value.

Let's first look at the algorithm for a peek operation in a stack:

```
Step 1: START
Step 2: IF TOP = NULL
            Print UNDERFLOW ERROR
            Go to Step 4
[End of If]
Step 3: Return STACK[TOP]
Step 4: EXIT
```

Write a menu-driven program for stacks performing all the operations:

```
# include<stdio.h>
# include<conio.h>
# define MAX 10
void push(int stack[], int item) ;
void pop(int stack[], int item) ;
int peek(int stack[]) ;
void display(int stack[]) ;
int top = -1 ;
void main()
{
 int stack[MAX], item, ch ;
 while(1)
 {
  clrscr() ;
  printf("\n***MENU***") ;
  printf("\n1. PUSH") ;
  printf("\n2. POP") ;
  printf("\n2. PEEK") ;
  printf("\n4. DISPLAY") ;
  printf("\n5. EXIT") ;
  printf("\nenter your choice: ") ;
  scanf("%d", &ch) ;
  switch(ch)
  {
    case 1 :
    printf("enter value to push: ") ;
    scanf("%d", &item) ;
    push(stack, item) ;
    break ;

    case 2 :
    pop(stack) ;
    break ;

   case 3 :
    item = peek(stack) ;
    if(item != 1)
   printf("The value at the top of stack is %d", item) ;
    break ;
```

```
        case 4 :
        display(stack) ;
        break ;

        case 5 :
        printf("!!Exit!!") ;
        exit(0) ;

        default :
        printf("wrong choice") ;
       }
   }
 }

 void push(int stack[], int item)
 {
 //Case 1 is for inserting an element in the stack
 if(top == MAX - 1)
   {
     printf("\noverflow error") ;
   }
   else
  {
    top++ ;
    stack[top] = item ;
    printf("Success!!") ;
    getch() ;
  }
 }

 void pop(int stack[], int item)
 {
 //Case 2 is for deleting an element from the stack
 if(top == -1)         {
      printf("\nunderflow error") ;
 }
 else
 {
    item = stack[top] ;
    top -- ;
    printf("deleted value is %d", item) ;
 }
 getch();
 }

 int peek(int stack[])
 {
 //Case 3 is for displaying the topmost element of the stack
  if(top == -1)
  {
   printf("Empty Stack") ;
  }
  else
```

```
     return(stack[top]) ;
    }

    void display(int stack[])
       int i ;
    //Case 4 is for displaying the elements of the stack
       front = front + 1 ;
       for(i = rear ; i > front ; i--)
       {
        printf("\n%d", stack[i]) ;
     }
    getch() ;
    }
```

Here is the output:

```
***MENU***
1: PUSH
2: POP
3: PEEK
4: DISPLAY
5: EXIT

Enter your choice: 1
Enter value to insert: 125
Success!!
```

IMPLEMENTATION OF STACKS

Stacks can be represented by two data structures: arrays and linked lists. Now, let us discuss both of them in detail.

Implementation of Stacks Using Arrays

Stacks can be easily implemented using arrays. Initially, the TCP pointer of the stack points to the first position or location of the array. As we insert new elements into the stack, the TOP keeps on incrementing, always pointing to the position where the next element will be inserted. The representation of a stack using an array is shown in Figure 7.8:

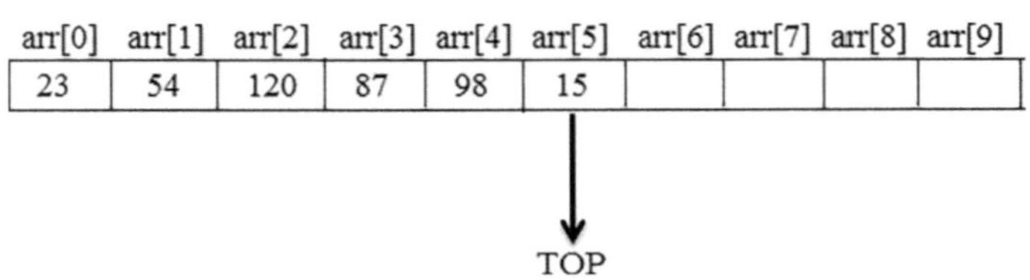

FIGURE 7.8 Array representation of a stack.

Implementation of Stacks Using Linked Lists

We already know that in linked lists, dynamic memory allocation takes place, that is, the memory is allocated at runtime. In the case of arrays, memory is allocated at the start of the program. If we are aware of the maximum size of the stack in advance, then the implementation of stacks using arrays will be efficient. If the size is not known in advance, then we will use the concept of a linked list in which dynamic memory allocation takes place. As we know, a linked list has two parts: the first part contains the information of the node, and the second part stores the address of the next element in the linked list. Similarly, we can also implement a linked stack. Now, the `START` pointer in the linked list will become the `TOP` pointer in a linked stack. All insertion and deletion operations will be done at the node pointed to by `TOP` only, as shown in Figure 7.9.

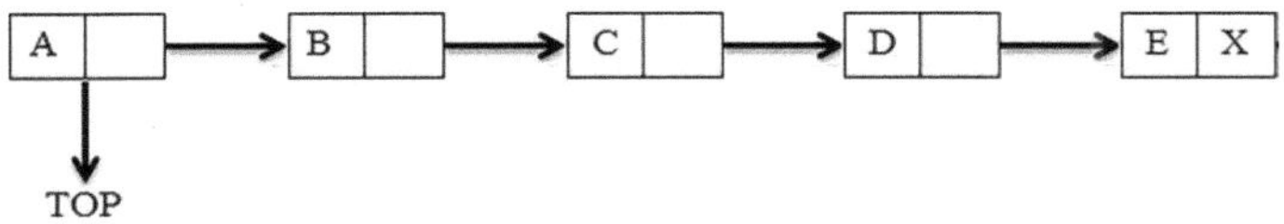

FIGURE 7.9 Linked representation of a stack.

Push Operation in Linked Stacks

A push operation is the process of adding new elements to an already existing stack. The new elements in the stack will always be inserted at the topmost position of the stack. Initially, we will check whether *TOP = NULL*. If the condition is true, then the stack is empty; otherwise, the new memory is allocated for the new node. We will understand it further with the help of an algorithm:

```
Step 1: START
Step 2: Set NEW NODE -> INFO = VAL
          IF TOP = NULL
Set NEW NODE -> NEXT = NULL
              Set TOP  = NEW NODE
ELSE
     Set NEW NODE -> NEXT = TOP
              Set TOP  = NEW NODE
 [End of If]
Step 3: EXIT
```

For example, consider a linked stack with four elements, as shown in Figure 7.10. A new element is to be inserted into the stack.

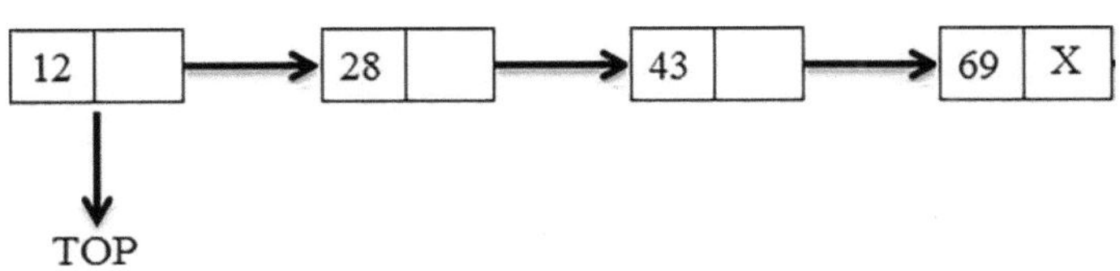

FIGURE 7.10 Linked stack before insertion.

After inserting the new element in the stack, the updated stack becomes as shown in Figure 7.11:

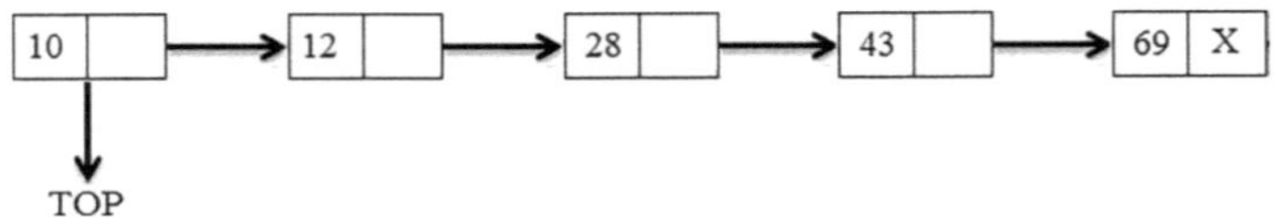

FIGURE 7.11 Linked stack after inserting a new element.

Pop Operation in Linked Stacks

A pop operation is the process of removing elements from an already existing stack. The elements from the stack will always be deleted from the node pointed to by `TOP`. Initially, we will check whether *TOP = NULL*. If the condition is true, then the stack is empty, which means we cannot delete any elements from it. Therefore, an underflow error message is displayed on the screen. We will understand it further with the help of an algorithm and code to discuss practical implementation of push and pop functions:

```
Step 1: START
Step 2: IF TOP = NULL
           Print UNDERFLOW ERROR
[End of If]
Step 3: Set TEMP = TOP
Step 4: Set TOP = TOP -> NEXT
Step 5: FREE TEMP
Step 6: EXIT
```

For example, consider a linked stack with five elements, as shown in Figure 7.12. An element is to be deleted from the stack.

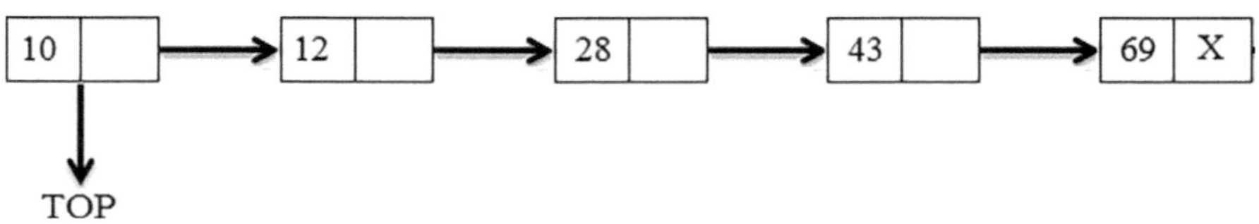

FIGURE 7.12 Linked stack before deletion.

After deleting an element from the stack, the updated stack becomes as shown in Figure 7.13:

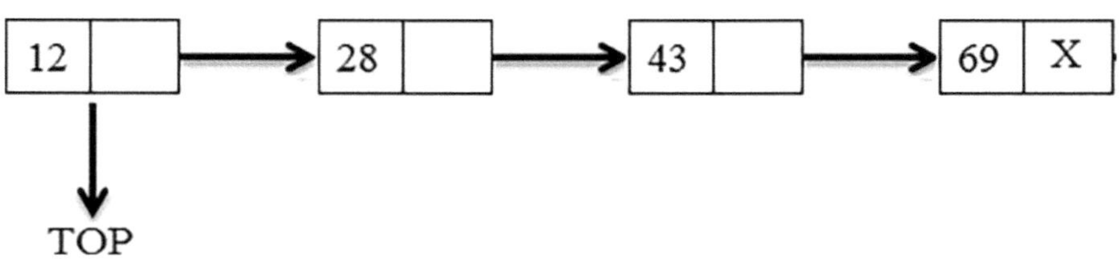

FIGURE 7.13 Linked stack after deleting the topmost element.

Write a menu-driven program implementing a linked stack performing push and pop operations:

```
# include<stdio.h>
# include<conio.h>
struct node
{
  int info ;
  struct node *next ;
 } *top = NULL ;
void push(int item) ;
void pop() ;
void display() ;
void main()
{
 int item, choice ;
 while(1)
 {
  clrscr() ;
  printf("\n***MENU***") ;

  printf("\n1. PUSH") ;
  printf("\n2. POP") ;
  printf("\n3. DISPLAY") ;
  printf("\n4. EXIT") ;
  printf("\nenter your choice: ") ;
  scanf("%d", &choice) ;
  switch(choice)
  {
    case 1 :
    printf("enter value to insert: ") ;
    scanf("%d", &item) ;
    push(item) ;
    break ;


    case 2 :
    pop() ;
    break ;


    case 3 :
    display() ;
    break ;


    case 4 :
    printf("!!Exit!!") ;
    exit(0) ;
```

```
    default :
    printf("wrong choice") ;
   }
 }
}


void push(int item)
{
  struct node *Newnode ;
//Case 1 is for inserting an element in the linked stack
  Newnode = (struct node*)malloc(sizeof(struct node)) ;
  Newnode -> info = item ;
if(top == NULL)
  {
    Newnode -> next = NULL ;
    top = Newnode
  }
  else
 {
Newnode -> next = top ;
   top = Newnode ;
  }
   printf("Success!!") ;
   getch() ;
 }
}


void pop()
{
  struct node *temp ;
//Case 2 is for deleting an element from the linked stack
  temp = top ;
  if(top == NULL)
   {
     printf("\nEmpty stack") ;
  }
else
{
  top = top -> next ;
  printf("deleted value is %d", temp -> info) ;
  free(temp) ;
 }
getch() ;
}


void display()
{
  struct node *ptr ;
```

```
//Case 3 is for displaying the elements in the linked stack
  if(top == NULL)
  {
   printf("\nEmpty stack") ;
 }
 ptr = top ;
 while(ptr != NULL)
 {
   printf("%d", ptr -> info) ;
   ptr = ptr -> next ;
   }
getch() ;
}
```

Here is the output:

```
***MENU***
1: PUSH
2: POP
3: DISPLAY
4: EXIT


Enter your choice: 2
Empty stack
```

APPLICATIONS OF STACKS

In this section, we will discuss various applications of stacks. The topics that will be covered in this section are the following:

- Polish and reverse Polish notations and their need
- Conversion from an infix expression to a postfix expression
- Conversion from an infix expression to a prefix expression
- Evaluation of a postfix expression
- Evaluation of a prefix expression
- Parenthesis balancing

Now, let us understand each one of them in detail.

Polish and Reverse Polish Notations and Their Need

Polish notation refers to a notation in which the operator is placed before the operands. Polish notation was named after the Polish mathematician Jan Łukasiewicz. We can also say that the transformation of an expression into a form is called a Polish notation. An algebraic expression can be represented in three forms. All these forms refer to the relative position of operators with respect to the operands:

- *Prefix form* – In an expression, if the operator is placed before the operands, that is, `+XY`, then it is said to be in prefix form.
- *Infix form* – In an expression, if the operator is placed in the middle of operands, that is, `X+Y`, then it is said to be in infix form.
- *Postfix form* – In an expression, if the operator is placed after the operands, that is, `XY+`, then it is said to be in postfix form.

Reverse Polish notation often refers to the postfix notation or suffix notation. It refers to the notation in which the operator is placed after its two operands, that is, `XY + AF BC*`.

It is comparatively easy for a computer system to evaluate an expression in Polish notation as the system need not check for priority-wise execution of various operators (such as the BODMAS rule), as all the operators in a prefix or postfix expression will automatically occur in their order of priority.

Conversion from an Infix Expression to a Postfix Expression

In any expression, we observe that there are two types of parts/components clubbed together: *operands* and *operators*. Operators indicate the operation to be carried out, and operands are the ones on which the operators operate. Operators have their own priority of execution. For simplicity of the algorithm, we will use only addition (+), subtraction (-), modulus (%), multiplication (*), and division (/) operators. The precedence of these operators is given as follows: *, ^, /, and % are given higher priority, whereas + and - are given lower priority.

The order of evaluation of these operators can be changed by using parentheses. For example, the expression `X * Y + Z` can be solved, as first `X * Y` will be done and then the result is added to `Z`. If the same expression is written with parentheses as `X * (Y + Z)`, now `Y + Z` will be evaluated first, and then the result is multiplied by `X`.

We can convert an infix expression to a postfix expression using a stack. First, we start to scan the expression from left to right. In an expression, there may be some operators, operands, and parentheses. Hence, we have to keep in mind some of the basic rules, which are:

- Each time we encounter an operand, it is added directly to the postfix expression.
- Each time we get an operator, we should always check the top of the stack to check the priority of the operators.
- If the operator at the top of the stack has higher precedence or the same precedence as that of the current operator, then it is repeatedly popped out from the stack and added to the postfix expression. Otherwise, it is pushed into the stack.
- Each time an opening parenthesis is encountered, it is directly pushed into the stack, and similarly, if a closing parenthesis is encountered, we will repeatedly pop out from the stack and add the operators in the postfix expression. Also, the opening parenthesis is deleted from the stack.

Now, let us understand it with the help of an algorithm, in which the first step is to push an opening parenthesis into the stack and also add a closing parenthesis at the end of the infix expression. The algorithm is repeated until the stack becomes empty:Further , we will also discuss practical implementation of conversion from infix expression to postfix expression with the help of a code.

```
Step 1: START
Step 2: Add ")" parenthesis to the end of infix expression.
Step 3: Push "(" parenthesis on the stack.
Step 4: Repeat the steps until each character in the infix expression
        is scanned.
          a) IF "(" parenthesis is found, push it onto the stack.
          b) If an operand is encountered, add it to the postfix
             expression.
          c) IF ")" parenthesis is found, then follow these steps -
             - Continually pop from the stack and add it to the postfix
               expression until a "(" parenthesis is encountered.
             - Eliminate the "(" parenthesis.
          d) If an operator is found, then follow these steps -
             - Continually pop from the stack and add it to the postfix
               expression which has same or high precedence than the
               current operator.
             - Push the current operator to the stack.
Step 5: Continually pop from the stack to the postfix expression until
        the stack becomes empty.
Step 6: EXIT
```

For example, let's convert the following infix expressions into postfix expressions.

a) (A + B) * C / D

b) [((A +B) * (C – D)) + (F – G)]

Here is the solution for the first conversion:

Character	Stack	Expression
(	(	
A	(	A
+	(+	A
B	(+	AB
)		AB+
*	*	AB+
C	*	AB+C
/	/	AB+C*
D		AB+C*D/
		Answer = AB+C*D/

Here is the solution for the second conversion:

Character	Stack	Expression
[	[	
(	[(	
(	[((	
A	[((	A
+	[((+	A
B	[((+	AB
)	[(	AB+
*	[(*	AB+
(	[(* (	AB+
C	[(* (	AB+C
-	[(* (-	AB+C
D	[(* (-	AB+CD
)	[(*	AB+CD-
)	[	AB+CD-*
+	[+	AB+CD-*
(	[+ (	AB+CD-*
F	[+ (	AB+CD-*F
-	[+ (-	AB+CD-*F
G	[+ (-	AB+CD-*FG
)	[+	AB+CD-*FG-
]		AB+CD-*FG-+
		Answer = AB+CD-*FG-+

Write a program to convert an infix expression to a postfix expression:

```
# include<stdio.h>
# include<conio.h>
# include<ctype.h>
# include<string.h>
# define SIZE 50
char stack[SIZE];
int top = -1;
void push( char stack[] ,char);
char pop(char stack[]);
void inftopostf(char src[],char trg[]);
int prior(char);
int main()
{
```

```
   char inf[50], postf[50];
   clrscr();
   printf(" \n Enter infix expression of your choice::");
   gets(inf);
   rev( inf);
   strcpy(postf, " ");
   inftopostf( inf, postf);
   printf("\n The postfix expression is:");
   puts(postf);
   getch();
  return 0;
 }
 void inftopostf(char src[],char trg[])
 {
  int p =0,q=0;
  char tmp;
  strcpy( trg , " ");
 while(src[p]!='\0')
 {
   if(src[p]=='(' )
 {
    push(stack, src[p]);
    p++;
 }
 else if(src[p]== ')')
 {
  while((top !=-1)&&(stack[top]!='(')
 {
    trg[q] = pop(stack);
    q++;
 }
 if(top=-1)
 {
    printf(" \n invalid expression");
    exit(1);
 }
  tmp = pop(stack);
  p++;
 }
 else if( isdigit(src[p]) || isalpha(src[p]))
 {
   trg[q]=src[p];
   p++;
 }
 else
 {
   printf("\n wrong element ");
   exit(1);
 }
 }
 while((top!=1)&& (stack[top]!='('))
```

```
{
  trg[q] = pop(stack);
  q++;
}
trg[q] ='\0';
}
int prior(char opr)
{
 if(opr== '/' || opr== '*' || opr == '%')
  return 1;
 else if (op == '+' || opr == '-')
  return 0;
}
void push(char stack[], char item)
{
 if(top == SIZE-1)
   printf("\n overflow of stack");
else
 {
   top++;
   stack[top] = item;
}
}
}
char pop( char stack[])
{
  char item = ' ';
  if(top==-1)
  printf(" \n underflow of stack");
else
  item = stack[top];
  top--;
}
 return item;
}
```

Here is the output:

```
Enter infix expression of your choice: X / Y - Z
The postfix expression is: XY/Z-
```

Frequently Asked Questions

Q2. Convert the following infix expression into a postfix expression:

(A + B) ^ C – (D * E) / F

Answer.

Character	Stack	Expression
(	(	
A	(	A
+	(+	A
B	(+	AB
)		AB+
^	^	AB+
C	^	AB+C
-	-	AB+C^
(	-(	AB+C^
D	-(	AB+C^D
*	-(*	AB+C^D
E	-(*	AB+C^DE
)	-	AB+C^DE*
/	-/	AB+C^DE*
F	-/	AB+C^DE*
		Answer = AB+C^DE*F/-

Conversion from an Infix Expression to a Prefix Expression

We can convert an infix expression to its equivalent prefix expression with the help of the following algorithm:

```
Step 1: START
Step 2: Reverse the infix expression. Also, interchange left and right
        parenthesis on reversing the infix expression.
Step 3: Obtain the postfix expression of the reversed infix expression.
Step 4: Reverse the postfix expression so obtained in Step 3. Finally,
        the expression is converted into prefix expression.
Step 5: EXIT
```

For example, let's convert the following infix expressions into prefix expressions:

a) `(X - Y) / (A + B)`

b) `(X - Y / Z) * (A / B - C)`

Let's go through the solutions:

a) After reversing the given infix expression, we get `((B + A) / Y - X)`.
Now, we find the postfix expression of `(B + A) / (Y - X)`.

Character	Stack	Expression
(	(	
(	((	
B	((	B
+	((+	B
A	((+	BA
)	(	BA+
/	(/	BA+
Y	(/	BA+Y
-	(-	BA+Y/
X	(-	BA+Y/X
)		BA+Y/X-
		Answer = BA+Y/X-

Now, when we reverse the postfix expression obtained, we get `-X/Y+AB`.
Hence, the prefix expression is `-X/Y+AB`.

b) After reversing the given infix expression, we get `(C - B / A) * (Z / Y - X)`.
Now, we find the postfix expression of `(C - B / A) * (Z / Y - X)`.

Character	Stack	Expression
(	(	
C	(	C
-	(-	C
B	(-	CB
/	(-/	CB
A	(-/	CBA
)		CBA/-
*	*	CBA/-
(	*(	CBA/-
Z	*(	CBA/-Z
/	*(/	CBA/-Z
Y	*(/	CBA/-ZY
-	*(-	CBA/-ZY/
X	*(-	CBA/-ZY/X
)	*	CBA/-ZY/X-
		Answer = CBA/-ZY/X-*

Now, when we reverse the postfix expression obtained, we get `*-X/YZ-/ABC`.

Hence, the prefix expression is `*-X/YZ-/ABC`. The practical implementation of conversion from infix expression to polish notations is discussed below through following code.

Write a program to convert an infix expression to a prefix expression:

```
# include<stdio.h>
# include<conio.h>
# include<ctype.h>
# include<string.h>
# define SIZE 50
char stack[SIZE];
int top = -1;
void  rev(char string[ ]);
void push( char stack[] ,char);
char pop(char stack[]);
void inftopostf(char src[],char trg[]);
int prior(char);
char inf[50], postf[50] , temp[50];
int main()
{
clrscr();
printf(" \n Enter infix expression of your choice::");
gets(inf);
rev( inf);
strcpy(postf, " ");
inftopostf( temp, postf);
printf("\n The postfix expression :");
puts(postf);
strcpy(temp, " ");
rev(postf);
printf("\n prefix expression is :");
puts(temp);
getch();
return 0;
}
void rev(char string[])
{
 int length , p=0;q=0;
 length = strlen(string);
 q = length -1;
while(q>=0)
{
  if(string[q]== '( ')
   temp[p] = ')';
  else if(string[q]== ')')
   temp[p] = '(';
 else
   temp[p] = string[q];
   p++;
   q--;
}
temp[p] = '\0';
}
void inftopostf(char src[],char trg[])
```

```
{
 int p =0,q=0;
 char tmp;
 strcpy( trg , " ");
while(src[p]!='\0')
{
   if(src[p]=='(' )
   {
    push(stack, src[p]);
    p++;
}
else if(src[p]== ')')
{
  while((top !=-1)&&(stack[top]!='(')
{
   trg[q] = pop(stack);
   q++;
}
if(top=-1)
{
   printf(" \n invalid expression");
   exit(1);
}
tmp = pop(stack);
p++;
}
else if( isdigit(src[p]) || isalpha(src[p]))
{
  trg[q]=src[p];
  q++;
  p++;
}
else if(src[p] =='+' || src[p]=='-' || src[p]==' ' || src[p]== '/' ||
src[p] == '%')
{
while((top!=1)&&(stack[top]!='(')&&(prior(stack[top]>prior(src[p])))
{
  trg[p]= pop(stack);
  q++;
}
push (stack,src[p]);
 p++;
}
else
{
  printf("\n wrong element ");
  exit(1);
}
}
while((top!=1)&& (stack[top]!='('))
{
 trg[q] = pop(stack);
 q++;
}
```

```
trg[q] ='\0';
}
int prior(char opr)
{
 if(opr== '/' || opr== '*' || opr == '%')
 return 1;
else if (op == '+' || opr == '-')
 return 0;
}
void push(char stack[], char item)
{
 if(top == SIZE-1)
 printf("\n overflow of stack");
else
{
  top++;
  stack[top] = item;
}
}
}
char pop( char stack[])
{
  char item = ' ';
  if(top==-1)
   printf(" \n underflow of stack");
else
  item = stack[top];
  top--;
}
 return item;
}
```

Here is the output:

```
Enter infix expression of your choice: W+X-Y*Z
The prefix expression: WX+YZ*-
Prefix expression is: -+WX*YZ
```

Evaluation of a Postfix Expression

With the help of stacks, any postfix expression can easily be evaluated. Every character in the postfix expression is scanned from left to right. The steps involved in evaluating a postfix expression are given in the following algorithm:

```
Step 1: START
Step 2: IF an operand is encountered, push it onto the stack.
Step 3: IF an operator "op1" is encountered, then follow these steps -
        a) Pop the two topmost elements from the stack, where X is the
           topmost element and Y is the next top element below X.
```

```
        b) Evaluate X op1 Y.
        c) Push the result onto the stack.
Step 4: Set the result equal to the topmost element of the stack.
Step 5: EXIT
```

For example, let's evaluate the following postfix expressions:

a) `2 3 4 + * 5 6 7 8 + * + +`
b) `T F T F AND F F F XOR OR AND T XOR AND OR`

Here is the solution for the first expression:

Character	Stack	Operation
2	2	PUSH 2
3	2, 3	PUSH 3
4	2, 3, 4	PUSH 4
+	2, 7	POP 4, 3 ADD(4 + 3 = 7) PUSH 7
*	14	POP 7, 2 MUL(7 * 2 = 14) PUSH 14
5	14, 5	PUSH 5
6	14, 5, 6	PUSH 6
7	14, 5, 6, 7	PUSH 7
8	14, 5, 6, 7, 8	PUSH 8
+	14, 5, 6, 15	POP 8, 7 ADD(8 + 7 = 15) PUSH 15
*	14, 5, 90	POP 15, 6 MUL(15 * 6 = 90) PUSH 90
+	14, 95	POP 90, 5 ADD(90 + 5 = 95) PUSH 95
+	109	POP 95, 14 ADD(95 + 14 = 109) PUSH 109
	Answer = 109	

Here is the solution for the second expression:

Character	Stack	Operation
T	T	PUSH T
F	T, F	PUSH F
T	T, F, T	PUSH T
F	T, F, T, F	PUSH F
AND	T, F, F	POP F, T AND(F AND T = F) PUSH F
F	T, F, F, F	PUSH F
F	T, F, F, F, F	PUSH F
F	T, F, F, F, F, F	PUSH F
XOR	T, F, F, F, T	POP F, F XOR(F XOR F = T) PUSH T
OR	T, F, F, T	POP T, F OR(T OR F = T) PUSH T
AND	T, F, F	POP T, F AND(T AND F = F) PUSH F
T	T, F, F, T	PUSH T
XOR	T, F, F	POP T, F XOR(T XOR F = F) PUSH F
AND	T, F	POP F, F AND(F AND F = F) PUSH F
OR	T	POP F, T OR(F OR T = T) PUSH T
	Answer = T	

Write a program for the evaluation of a postfix expression:

```
# include<stdio.h>
# include<conio.h>
# include<ctype.h>
# define SIZE 50
float stack[SIZE];
int top = -1;
void push( float stack[] , float item);
float pop(float stack[]);
float postf(char express[]);
```

```
int main()
{
float item;
char express[50];
clrscr();
printf(" \n Enter postfix expression of your choice::");
gets(express);
  item = postf(express)
printf("\n Value of postfix expression = %.2f",item);
getch();
return 0;
}
float postf( char express[])
{
 int i=0;
float  opr1,opr2, value;
while(express[i]!=0)
{
if(isdigit(express[i]))
{
  push(stack, (float)(express[i]-'0'));
else
{
opr2 = pop(stack);
opr1 = pop(stack);
switch(express[i])
{
case '+' :
value = opr1+opr2;
break;


case '-' :
value = opr1- opr2;
break;


case '*' :
value = opr1*opr2;
break;


case '/' :
value = opr1/opr2;
break;


case '%' :
value =(int) opr1%(int)opr2;
break;
}
push(stack, value);
}
i++;
}
return(pop(stack));
}
```

```
void push(float stack[] , float item)
{
  if(top == SIZE-1)
 {
  printf(" \n overflow of stack");
else
{
top++;
   stack[top] = item;
}
}
float pop(float stack[])
{
  float item =-1;
if(top ==-1)
printf("\n underflow of stack");
else
{
  item = stack[top];
top--;
}
 return item;
}
```

Here is the output:

```
Enter postfix expression of your choice: 8 7 6 * 5 + 1 + /
Value of postfix expression = 6.00
```

Frequently Asked Questions

Q3. Evaluate the following postfix expression:

2 3 4 * 6 / +

Answer.

Character	Stack
2	2
3	2, 3
4	2, 3, 4
*	2, 12
6	2, 12, 6
/	2, 2
+	4
	Answer = 4

Evaluation of a Prefix Expression

There are a variety of techniques for evaluating a prefix expression. The simplest of all the techniques is explained in the following algorithm:

```
Step 1: START
Step 2: Accept the prefix expression.
Step 3: Repeat the steps 4 to 6 until all the characters have been
        scanned.
Step 4: The prefix expression is scanned from the right.
Step 5: IF an operand is encountered, push it onto the stack.
Step 6: IF an operator is encountered, then follow these steps -
        a) Pop two elements from the operand stack.
        b) Apply the operator on the popped operands.
        c) Push the result onto the stack.
Step 7: EXIT
```

For example, let's evaluate the following prefix expressions:

a) `+ -  4 6 * 9 /10 50`
b) `+ * * + 2 3 4 5 + 6 7`

Here is the solution for the first expression:

Character	Stack	Operation
50	50	PUSH 50
10	50, 10	PUSH 10
/	5	POP 10, 50 DIV(50 / 10 = 5) PUSH 5
9	5, 9	PUSH 9
*	45	POP 9, 5 MUL(5 * 9 = 45) PUSH 45
6	45, 6	PUSH 6
4	45, 6, 4	PUSH 4
-	45, 2	POP 4, 6 SUB(6 - 4 = 2) PUSH 2
+	47	POP 2, 45 ADD(45 + 2 = 47) PUSH 47
	Answer = 47	

Here is the solution for the second expression:

Character	Stack	Operation
7	7	PUSH 7
6	7, 6	PUSH 6
+	13	POP 6, 7 ADD(7 + 6 = 13) PUSH 13
5	13, 5	PUSH 5
4	13, 5, 4	PUSH 4
3	13, 5, 4, 3	PUSH 3
2	13, 5, 4, 3, 2	PUSH 2
+	13, 5, 4, 5	POP 2, 3 ADD(3 + 2 = 5) PUSH 5
*	13, 5, 20	POP 5, 4 MUL(4 * 5 = 20) PUSH 20
*	13, 100	POP 20, 5 MUL(5 * 20 = 100) PUSH 100
+	113	POP 100, 13 ADD(13 + 100 = 113) PUSH 113
	Answer = 113	

Write a program for the evaluation of a prefix expression:

```
#include<stdio.h>
#include<conio.h>
#include<string.h>
int stack[10];
int top=-1;
void push(int);
int pop();
void main()
{
char pref[20];
int length,item,i,op1,op2,result;
clrscr();
printf("Enter the prefix Expression :");
gets(pref);
length=strlen(pref);
for(i=length-1;i>=0;i--)
```

```
{
switch (get(pref[i]))
{
case 0:
item=pref[i]-'0';
push(item);
break;


case 1:
op1=pop();
op2=pop();
switch(pref[i])
{
case '+':
res=opr1+opr2;
break;
case '-':
result=op1-op2;
break;
case '*':
result=op1*op2;
break;
case '/':
result=op1/op2;
break;
}
push(result);
}
}
printf("Result is %d",stack[0]);
getch();
}


void push(int item)
{
stack[++top]=item;
}
int pop()
{
return(stack[top--]);
}
int get(char ch)
{
if(ch=='+'||ch=='-'||ch=='*'||ch=='/')
return 1;
else
return 0;
}
```

Here is the output:

```
Enter the prefix Expression: +-846
Result is 10
```

Parenthesis Balancing

Stacks can be used to check the validity of parentheses in any arithmetic or algebraic expression. We are already aware that in a valid expression, the parentheses (the brackets) occur in pairs; that is, if a parenthesis is opening, then it must be closed in an expression. Otherwise, the expression would be invalid. For example, `(X + Y - Z` is invalid, but `(X + Y - Z)` is a valid expression. Hence, there are some key points that must be kept in mind:

- Each time an opening parenthesis is encountered, it should be pushed onto the stack.
- Each time a closing parenthesis is encountered, the stack is examined.
- If the stack is already empty, then the closing parenthesis does not have an opening parenthesis, and hence, the expression is invalid.
- If the stack is not empty, then we will pop the stack and check whether the popped element corresponds to the closing parenthesis.
- When we reach the end of the stack, the stack must be empty. Otherwise, one or more opening parentheses do not have a corresponding closing parenthesis, and, therefore, the expression will become invalid.

For example, let's check whether the following given expressions are valid or not:

a) `((A - B) * Y`
b) `[(A + B) - {X + Y} * [C - D]]`

Here is the solution for the first expression:

	Symbol	Stack
1.	(	(
2.	(	(, (
3.	A	(, (
4.	-	(, (
5.	B	(, (
6.	)	(
7.	*	(
8.	Y	(
9.		(

As the stack is not empty, the expression is not a valid expression.
Here is the solution for the second expression:

	Symbol	Stack
1.	[	[
2.	(	[, (
3.	A	[, (

	Symbol	Stack
4.	+	[, (
5.	B	[, (
6.	)	[
7.	-	[
8.	{	[, {
9.	X	[, {
10.	+	[, {
11.	Y	[, {
12.	}	[
13.	*	[
14.	[	[, [
15.	C	[, [
16.	-	[, [
17.	D	[, [
18.	]	[
19.	]	

As the stack is empty, the given expression is a valid expression.

Write a program to implement parenthesis balancing:

```
#include<stdio.h>
#include<conio.h>
#include<string.h>
# define SIZE 30
int stack[SIZE];
int top=-1;
void push(char);
char pop();
void main()
{
char express[SIZE], val;
 int i , temp=1;
clrscr();
printf("Enter the Expression :");
gets(express);
for(i=0; i<strlen(express); i++)
{
  if (express[i]== '(' || express[i]== '{' || express[i]== '[')
push(express[i]);
if(express[i]== ')' || express[i]== '}' || express[i]== ']' )
if(top== -1)
flag = 0;
```

```
else
{
   Val = pop( );
if (express[i]== ')' && (val == '{' || val== '['))
temp = 0;
if (express[i]== '}' && (val == '(' || val== '['))
temp = 0;
if (express[i]== '}' && (val == '(' || val== '{'))
 temp = 0;
}
}
if(top>=0)
temp=0;
if(temp==1)
 printf("\n expression is valid");
else
printf("\n expression is invalid");
}
void push(char c)
{
 if(top == (SIZE-1))
printf(" \noverflow of stack");
else
{
top = top+1;
  stack[top] = c;
}
}
char pop( )
{
 if(top==-1)
 printf("\n underflow of stack");
 else
 return(stack[top--]);
}
```

Here is the output:

```
Enter the Expression: (X - (Y * Z))
Expression is valid
```

SUMMARY

- A stack is a linear collection of data elements in which the element inserted last will be the element taken out first (meaning a stack is a LIFO data structure). A stack is a linear data structure, in which the insertion as well as the deletion of an element is done only from the end called TOP.
- In computer memory, stacks can be implemented by using either arrays or linked lists.
- The overflow condition occurs when we try to insert elements into the stack, but the stack is already full.

- The underflow condition occurs when we try to remove elements from the stack, but the stack is already empty.
- The three basic operations that can be performed on the stacks are push, pop, and peek operations.
- A push operation is the process of adding new elements to the stack.
- A pop operation is the process of removing elements from the stack.
- A peek operation is the process of returning the value of the topmost element of the stack.
- Polish notation refers to a notation in which the operator is placed before the operands.
- Infix, prefix, and postfix notations are three different but equivalent notations of writing algebraic expressions.

EXERCISES

Theory Questions

1. What is a stack? Give a real-life example.
2. What do you understand by a stack overflow and a stack underflow?
3. What is a linked stack, and how is it different from a linear stack?
4. Discuss the various operations that can be performed on stacks.
5. Explain the terms *Polish notation* and *reverse Polish notation*.
6. In what ways can a stack be implemented?
7. What are the various applications of a stack? Explain in detail.
8. Why is a stack known as a LIFO structure?
9. What are the different notations to represent an algebraic expression? Which one is mostly used in computers?
10. Explain the concept of *linked stacks* and also discuss how insertion and deletion take place in them.
11. Draw the stack structure when the following operations are performed one after another on an empty stack:
    - **A.** Push `1, 2, 6, 17, 100`
    - **B.** Pop three numbers
    - **C.** Peek
    - **D.** Push `50, 23, 198, 500`
    - **E.** Display
12. Convert the following infix expressions to their equivalent postfix expressions:
    - **A.** `A + B + C - D * E / F`
    - **B.** `[A - C] + {D * E}`
    - **C.** `[X / Y] % (A * B) + (C % D)`
    - **D.** `[(A - C + D) % (B - H + G)]`
    - **E.** `18 / 9 * 3 - 4 + 10 / 2`
13. Check the validity of the given algebraic expressions.
    - **A.** `(([A - V - D] + B)`
    - **B.** `[(X - {Y * Z})]`
    - **C.** `[A + C + E)`

14. Convert the following infix expressions to their equivalent prefix expressions.

    A. `8 / 9 * 3 - 4 + 10 / 2`
    B. `X * (Z / Y)`
    C. `[(A + B) - (C + D)] * E`

15. Evaluate the following postfix expressions:

    A. `1 2 3 * * 4 5 6 7 + + * *`
    B. `12 4 / 45 + 2 3 * +`

Programming Questions

1. Write a program to implement a stack using arrays.
2. Write a program to convert an infix expression to a postfix expression.
3. Write a function that performs a peek operation on the stack.
4. Write a program to implement a stack using a linked list.
5. Write a program to input two stacks and compare their contents.
6. Write a program to convert the expression `x + y` into `xy+`.
7. Write a program to evaluate a postfix expression.
8. Write a program to evaluate a prefix expression.
9. Write a program to convert `b - c` into `-bc`.
10. Write a function that performs a push operation in a linked stack.

Multiple Choice Questions

1. New elements in the stack are always inserted from the__________.

    A. Front end
    B. Top end
    C. Rear end
    D. Both (a) and (c)

2. A stack is a _______ data structure.

    A. FIFO
    B. LIFO
    C. FILO
    D. LILO

3. The overflow condition in the stack exists when:

    A. *TOP = NULL*
    B. $TOP = MAX$
    C. $TOP = MAX - 1$
    D. None of the above

4. The function that inserts the elements in a stack is called ________.

    A. `push()`
    B. `peek()`
    C. `pop()`
    D. None of the above

5. Disks piled up one above the other represent a _______.
   A. Queue
   B. Stack
   C. Tree
   D. Linked list
6. Reverse Polish Notation is another name for ________.
   A. A postfix expression
   B. A prefix expression
   C. An infix expression
   D. All of the above
7. Stacks can be represented by ________.
   A. A linked list only
   B. Arrays only
   C. Both of the above
   D. None of the above
8. If the numbers `10`, `45`, `13`, `50`, and `32` are pushed onto a stack, what does the pop operation return?
   A. `10`
   B. `45`
   C. `50`
   D. `32`
9. Which data structure is required for implementing a parenthesis balancer?
   A. Queue
   B. Tree
   C. Stack
   D. Heap
10. What will the postfix representation of the expression `(2 – b) * (a + 10) / (c * 8)` be?
   A. `8 a * c 10 + b 2 – * /`
   B. `/ 2 a c * + b 10 * 9 –`
   C. `2 b – a 10 + * c 8 * /`
   D. `10 a + * 2 b – / c 8 *`

CHAPTER 8

TREES

INTRODUCTION

In earlier chapters, we learned about various data structures, such as arrays, linked lists, stacks, and queues. These are all linear data structures. Although linear data structures are flexible, it is quite difficult to use them to organize data into a hierarchical representation. Hence, to overcome this limitation, we can create a new data structure, which is called a tree.

A *tree* is a data structure that is defined as a set of one or more nodes that allows us to associate a parent-child relationship. In trees, one node is designated as the root node or parent node, and all the remaining nodes can be partitioned into non-empty sets, each of which is a subtree of the root. Unlike natural trees, a tree data structure is upside down, having a root at the top and leaves at the bottom. There is no parent of the root node; it can only have child nodes. On the contrary, leaf nodes (or leaves) have no child nodes. When there are no nodes in the tree, then the tree is known as a null tree or an empty tree. Trees are widely used in various day-to-day applications. The recursive programming of trees makes the programs optimized and easily understandable. Trees are also used to represent the structure of mathematical formulas. Figure 8.1 represents a tree in which *A* is the root node of the tree. *X*, *Y*, and *Z* are the child nodes of the root node, *A*. They also form the subtrees of the tree. *B*, *C*, *Y*, *D*, and *E* are the leaf nodes of the tree, as they have no children.

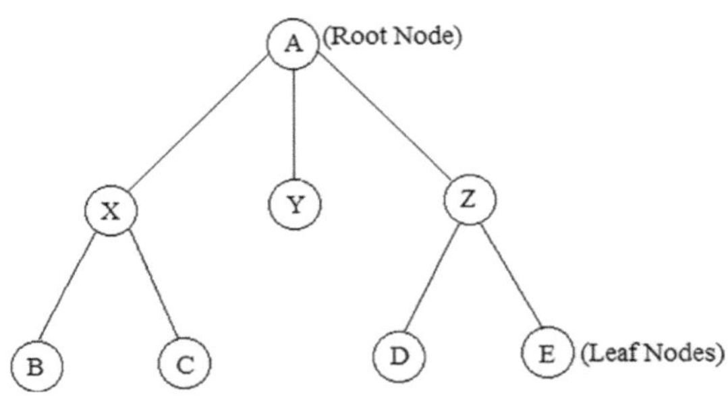

FIGURE 8.1 A tree.

Practical Application:

The members of a family can be visualized as a tree in which the root node can be visualized as a grandparent, whose two children can be visualized as the child nodes. Then, the grandchildren form the left and the right subtrees of the tree.

Trees are used to organize information in database systems and also to represent the syntactic structure of the source programs in compilers.

DEFINITIONS

Let's look at the definitions of the main words used when discussing trees:

- *Node* – A node is the main component of the tree data structure. It stores the actual data along with the links to the other nodes, as shown in Figure 8.2.

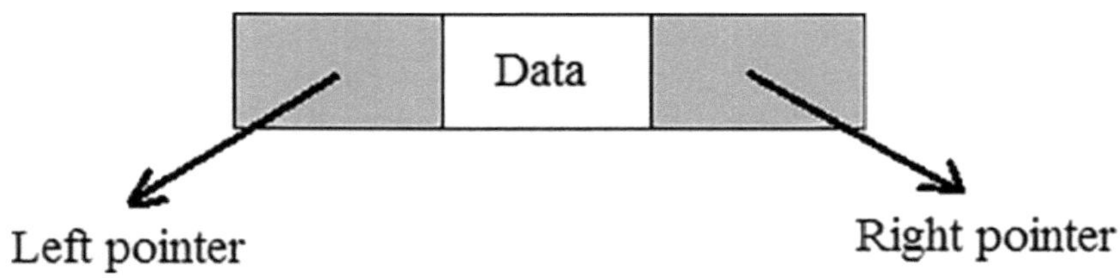

FIGURE 8.2 Structure of a node.

- *Root* – The root node is the topmost node of the tree. It does not have a parent node. If the root node is empty, then the tree is empty.
- *Parent* – The parent of a node is the immediate predecessor of that node. In Figure 8.3, *X* is the parent of the *Y* and *Z* nodes.

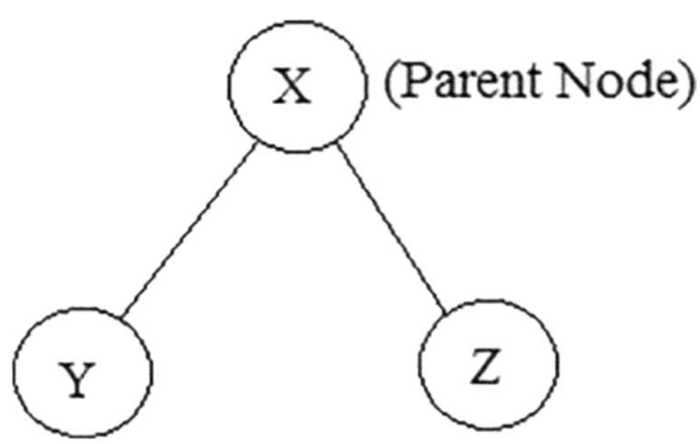

FIGURE 8.3 Parent node.

- *Child* – The child nodes are the immediate successors of a node. They must have a parent node. A child node placed at the left side is called the left child, and similarly, a child node placed at the right side is called the right child. In Figure 8.4, *Y* is the left child of *X*, and *Z* is the right child.

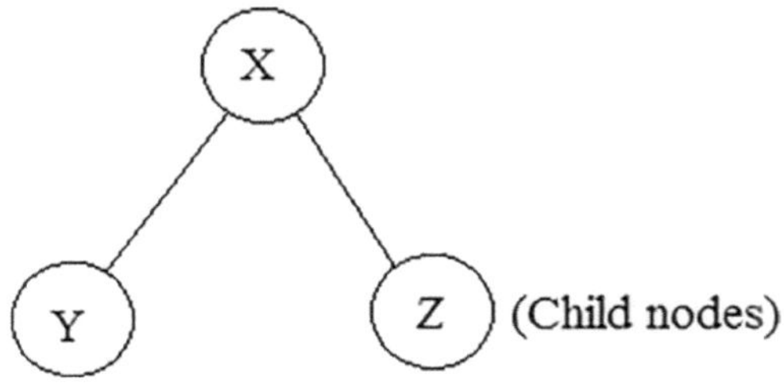

FIGURE 8.4 Child nodes.

- *Leaf/terminal nodes* – A leaf node does not have any child nodes.
- *Subtrees* – In Figure 8.5, nodes *B*, *X*, and *Y* form the left subtree of root *A*. Similarly, nodes *C* and *Z* form the right subtree of *A*.

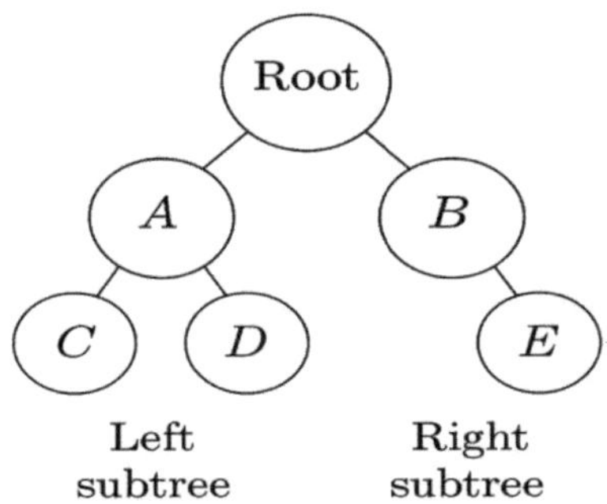

FIGURE 8.5 Subtrees.

- *Path* – This is a unique sequence of consecutive edges that is required to be followed to reach the destination from a given source. In Figure 8.6, the path from root node *A* to *Y* is given as *A–B*, *B–Y*.

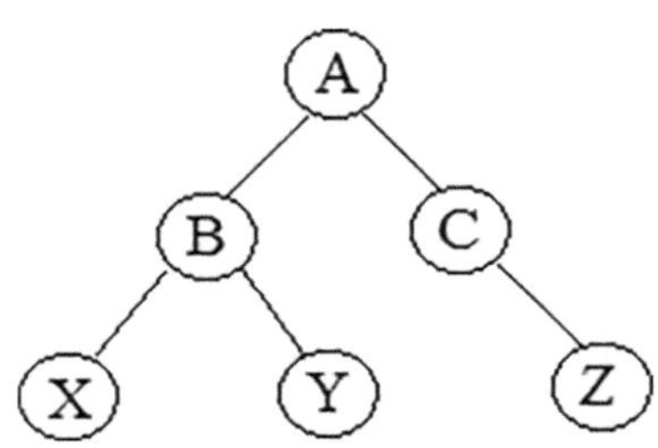

FIGURE 8.6 Path.

- *Level number of a node* – Every node in the tree is assigned a level number, a shown in Figure 8.7. The root is at level 0, the children of the root node are at level 1, and so on.

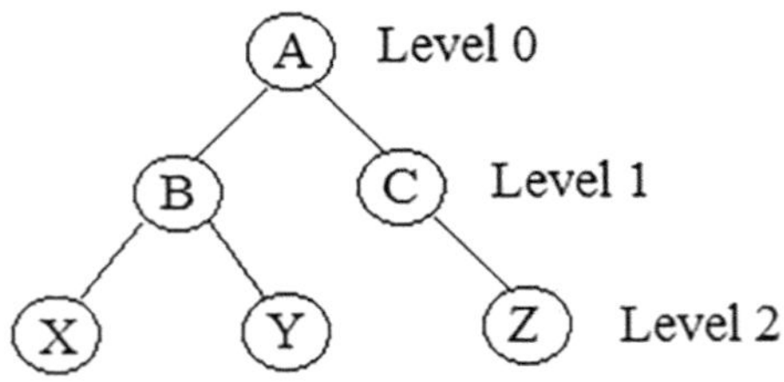

FIGURE 8.7 Node level numbers.

- *Height* – The height of the tree is the maximum level of the node + 1. The height of a tree containing a single node will be 1. Similarly, the height of an empty tree will be 0.
- *Ancestors* – The ancestors of a node are any predecessor nodes on the path between the root and the destination. There are no ancestors for the root node. Nodes *A* and *B* are the ancestors of node *X*.
- *Descendants* – The descendants of a node are any successor nodes on the path between the given source and the leaf node. There are no descendants of the leaf node. Here, *B*, *X*, and *Y* are the descendants of node *A*.
- *Siblings* – The child nodes of a given parent node are called siblings. *X* and *Y* are the siblings from *B* in Figure 8.8.

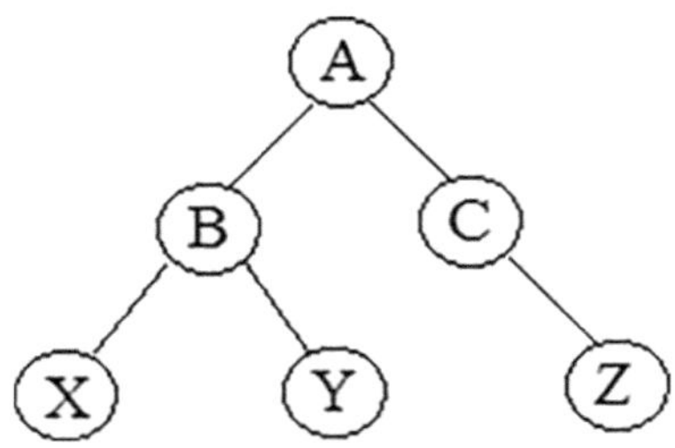

FIGURE 8.8 Siblings.

- *Degree of a node* – This is equal to the number of children that a node has.
- *Out-degree of a node* – This is equal to the number of edges leaving that node.
- *In-degree of a node* – This is equal to the number of edges arriving at that node.
- *Depth* – This is given as the length of the path from the root node to the destination node.

BINARY TREES

A *binary tree* is a collection of nodes where each node contains three parts: left pointer, right pointer, and data item. The left pointer points to the left subtree, and the right pointer points to the right subtree. The topmost element of the binary tree is known as the root node. The root pointer points to the root node. As the name suggests, a binary tree can have at most two children, meaning a parent can have zero, one, or two children. Also, if *root* = *NULL*, then it means that the tree is empty.

Figure 8.9 represents a binary tree. Here, *A* represents the root node of the tree. *B* and *C* are the children of root node *A*. Nodes *B*, *D*, *E*, *F*, and *G* constitute the left subtree. Similarly, nodes *C*, *H*, *I*, and *J* constitute the right subtree. Nodes *G*, *E*, *F*, *I*, and *J* are the terminal/leaf nodes of the binary tree as they have no children. Hence, node *A* has two successors, *B* and *C*. Node *B* has two successors, *D* and *G*. Similarly, node *D* also has two successors, *E* and *F*. Node *G* has no successors. Node *C* has only one successor, *H*. Node *H* has two successors, *I* and *J*. Since nodes *E*, *F*, *G*, *I*, and *J* have no successors, they are said to have empty subtrees.

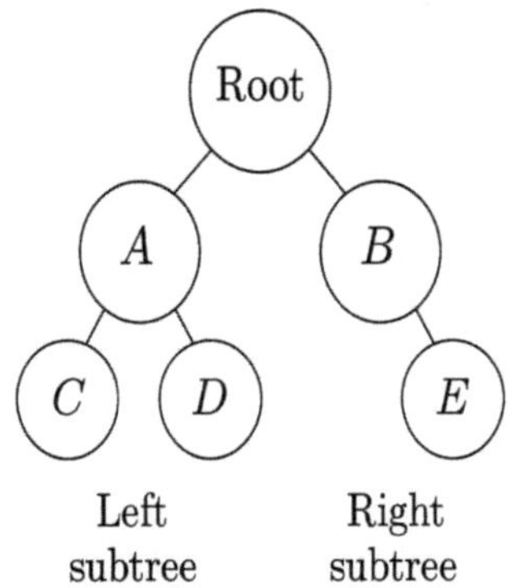

FIGURE 8.9 A binary tree.

Types of Binary Trees

There are two types of binary trees: complete and extended.

A *complete binary tree* is a type of binary tree that obeys/satisfies two properties:

a) First, every level in a complete binary tree except the last one must be completely filled.

b) Second, all the nodes in the complete binary tree must appear left as much as possible.

In a complete binary tree, the number of nodes at level n is 2^n. Also, the total number of nodes in a complete binary tree of depth d is equal to the sum of all nodes present at each level between 0 and d. Figure 8.10 shows two examples of complete binary trees.

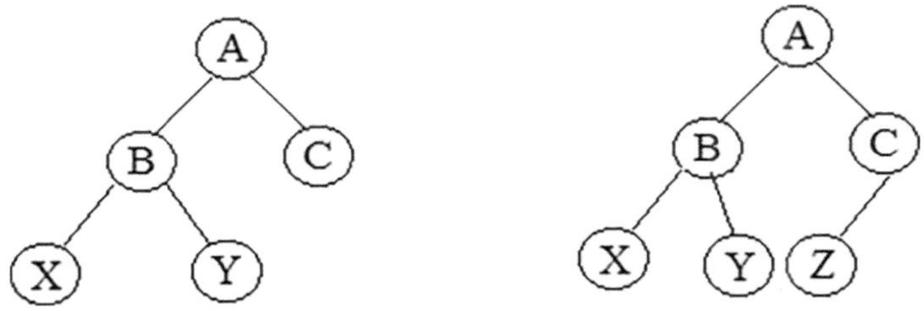

FIGURE 8.10 Complete binary trees.

Extended binary trees are also known as *2T-trees*. A binary tree is said to be an extended binary tree if and only if every node in the tree has either zero children or two children. In an extended binary tree, nodes having two children are known as *internal nodes*. Nodes having no children are known as *external nodes*. In Figure 8.11, the internal nodes are represented by *I*, and the external nodes are represented by *E*.

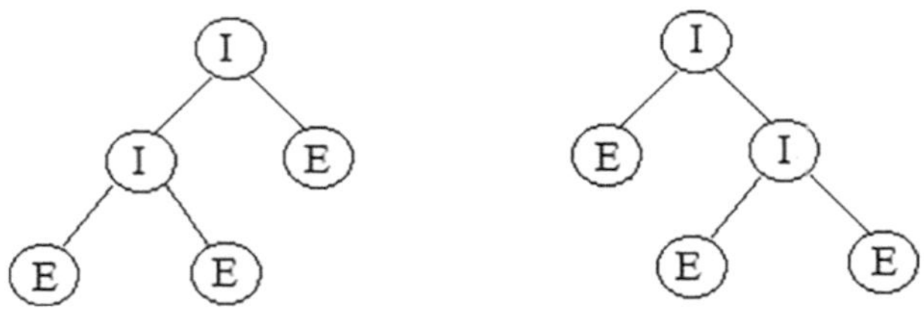

FIGURE 8.11 Extended binary trees.

Memory Representation of Binary Trees

Binary trees can be represented in a computer's memory in either of the following ways:

- Array representation
- Linked representation

Now, let us discuss both of them in detail.

Array Representation of Binary Trees

Here, a binary tree is represented using an array in the computer's memory. It is also known as *sequential representation*. Sequential representation of binary trees is done using *one-dimensional* (1*D*) arrays. This type of representation is static and hence it is inefficient, as the size must be known in advance, and it requires a lot of memory space. The following rules are used to decide the location of each node in the memory:

a) The root node of the tree is stored in the first location.

b) If the parent node is present at location k, then the left child is stored at location $2k$, and the right child is stored at location $(2k + 1)$.

c) The maximum size of the array is given as $(2^h - 1)$, where h is the height of the tree.

For example, look at the binary tree given in Figure 8.12, which also shows its array representation in memory.

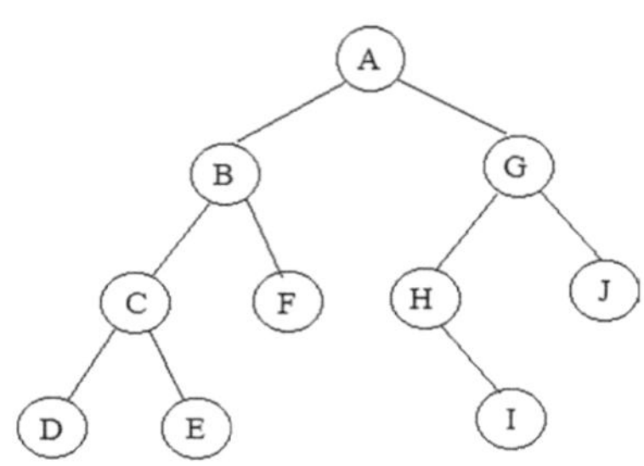

A	B	G	C	F	H	J	D	E				I		
1	2	3	4	5	6	7	8	9	10	11	12	13	14	15

FIGURE 8.12 Binary tree and its array representation.

Linked Representation of Binary Trees

A binary tree can also be represented using a linked list in a computer's memory. This type of representation is dynamic, as memory is dynamically allocated (that is, when it is needed), and thus it is efficient and avoids wastage of memory space. In linked representation, every node has three parts:

- The 1st part is called the left pointer, which contains the address of the left subtree.
- The 2nd part is called the data part, which contains the information of the node.
- The 3rd part is called the right pointer, which contains the address of the right subtree.

The structure of the node is declared as follows:

```
struct node
{
struct node *leftchild ;
int information ;
struct node *rightchild ;
}
```

A representation of a node is shown in Figure 8.2. When there are no children of a node, the corresponding pointer fields are *NULL*.

For example, look at the binary tree given in Figure 8.13, which also shows its linked representation in memory.

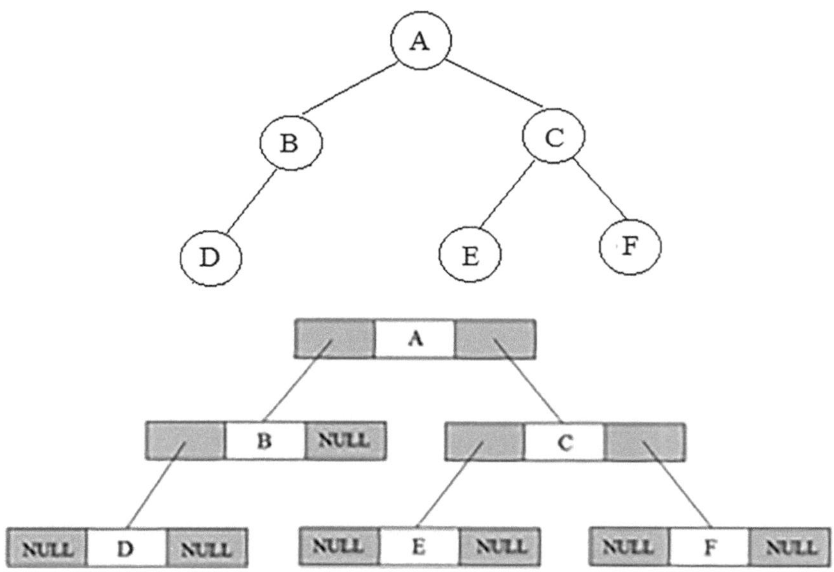

FIGURE 8.13 Binary tree and its linked representation.

BINARY SEARCH TREE

A *binary search tree* (*BST*) is a variant of a binary tree. The special property of a BST is that all the nodes in the left subtree have a value less than that of the root node. Similarly, all the nodes in the right subtree have a value more than that of the root node. Hence, a BST is also known as an *ordered binary tree*, because all the nodes in a BST are ordered. Also, the left and the right subtrees are also BSTs, and thus, the same property is applicable to every subtree in the BST. Figure 8.14 shows a BST in which all the keys are ordered.

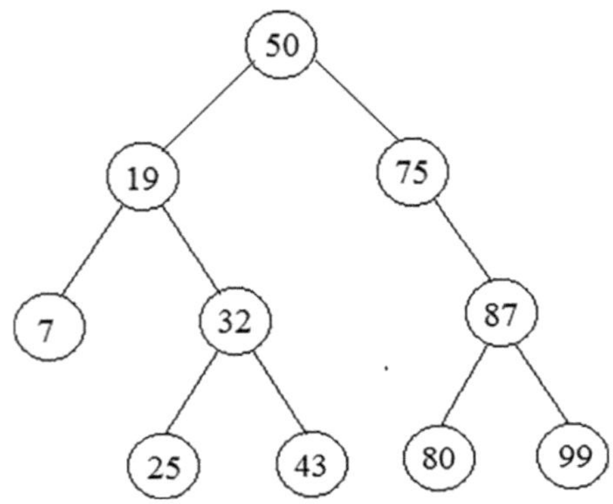

FIGURE 8.14 Binary search tree

Now, in the previous figure, the root node is 50. The left subtree of the root node consists of nodes 19, 7, 32, 25, and 43. We can see that all these nodes have smaller values than the root node, and hence they constitute the left subtree. Similarly, the right subtree of the root node consists of nodes 75, 87, 80, and 99. Here also, we can see that all these nodes have higher values than the root node, and hence they constitute the right subtree. Each of the subtrees is also ordered. Thus, it becomes easier to search for an element in the tree, and as a result, time is also reduced by a great margin. BSTs are very efficient regarding searching for an element. These trees are already sorted in nature. Thus, these trees have a low time complexity. Various operations can be performed on BSTs, which will be discussed in the upcoming section.

Operations on Binary Search Trees

In this section, we will discuss different operations that are performed on BSTs:

- Searching for a node/key
- Inserting a node/key
- Deleting a node/key
- Deleting the entire binary search tree
- Finding the mirror image of the binary search tree
- Finding the smallest node
- Finding the largest node
- Determining the height of the binary search tree

Now, let us discuss all of these operations in detail.

Searching for a Node/Key

The searching operation is one of the most common operations performed in the BST. This operation is performed to find whether a given key exists in the tree or not. The searching operation starts at the root node. First, it will check whether the tree is empty or not. If the tree is empty, then the node/key for which we are searching is not present in the tree, and the algorithm terminates there by displaying the appropriate message. If the tree is not empty and the nodes are present in it, then the search function checks the node/value to be searched and compares it with the key value of the current node. If the node/key to be searched is less than the key value of the current node, then we will recursively call the left child node. On the other hand, if the node/key to be searched is greater than the key value of the current node, then we will recursively call the right child node. Now, let us look at the algorithm for searching for a key in the BST:

```
SEARCH(ROOT, VALUE)
Step 1: START
Step 2: IF(ROOT == NULL)
Return NULL
Print "Empty Tree"
ELSE IF(ROOT -> INFO == VALUE)
Return ROOT
ELSE IF(ROOT -> INFO > VALUE)
SEARCH(ROOT -> LCHILD, VALUE)
ELSE IF(ROOT -> INFO < VALUE)
SEARCH(ROOT -> RCHILD, VALUE)
ELSE
Print "Value not found"
           [End of IF]
          [End of IF]
         [End of IF]
        [End of IF]
Step 3: END
```

In the previous algorithm, first, we check whether the tree is empty or not. If the tree is empty, then we return `NULL`. If the tree is not empty, then we check whether the value stored at the current node (`ROOT`) is equal to the node/key we want to search for or not. If the value of the `ROOT` node is equal to the key value to be searched, then we return the current node of the tree, that is, the `ROOT` node. Otherwise, if the key value to be searched is less than the value stored at the current node, we recursively call the left subtree. If the key value to be searched is greater than the value stored at the current node, then we recursively call the right subtree. Finally, if the value is not found, then an appropriate message is printed on the screen.

For example, let's say we have been given a BST. We want to search for the node with a value of `20` in the BST.

Initially, the BST is given as shown in Figure 8.15:

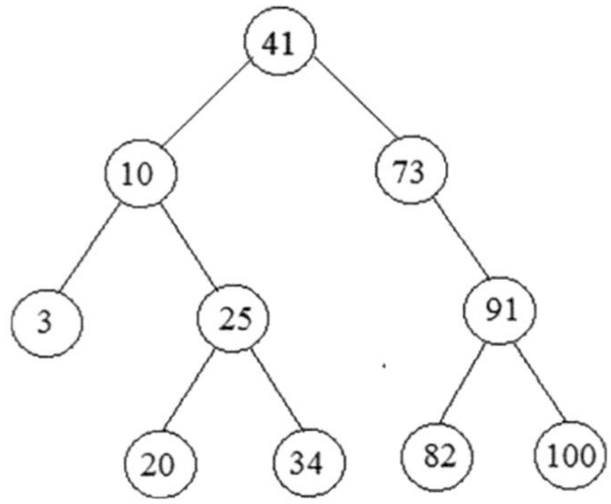

FIGURE 8.15 Sample binary search tree.

First, the root node, `41`, is checked, as shown in Figure 8.16.

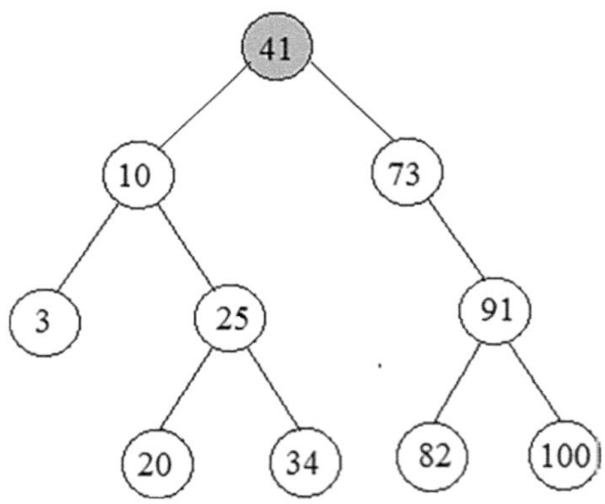

FIGURE 8.16 Traversing through root node

The value stored at the root node is not equal to the value to be searched, but we know that 20 < 41; thus, we will traverse the left subtree, as shown in Figure 8.17.

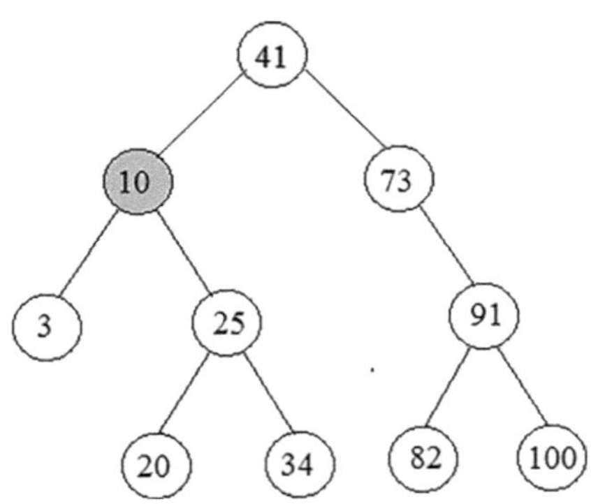

FIGURE 8.17 Traversing through left subtree

We know that `10` is not the value to be searched, but 20 > 10; thus, we will now traverse the right subtree with respect to `10`, as shown in Figure 8.18.

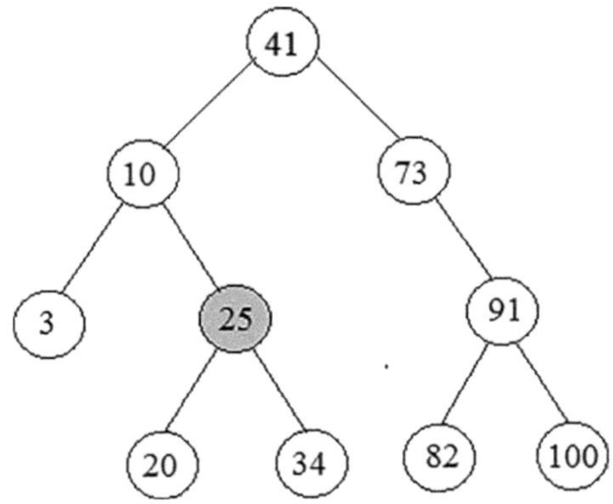

FIGURE 8.18 Traversing through right subtree of the left subtree of root

Again, `25` is not the value to be searched, but 20 < 25; thus, we will now traverse the left subtree with respect to `25`, as shown in Figure 8.19.

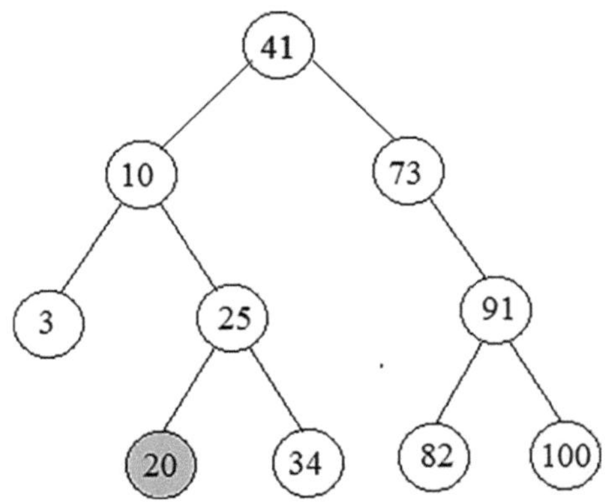

FIGURE 8.19 Searching for a node with a value of 20 in the binary search tree.

Finally, a node with a value of `20` is successfully found in the BST.

Inserting a Node/Key

The insertion operation is performed to insert a new node with the given value into the BST. The new node is inserted at the correct position following the BST constraint. It should not violate the properties of the BST. The insertion operation also starts at the root node. First, it will check whether the tree is empty or not. If the tree is empty, then we will allocate the memory for the new node. If the tree is not empty, then we will compare the key value to be inserted with the value stored in the current node. If the node/key to be inserted is less than the key value of the current node, then the new node is inserted in the left subtree. On the other hand, if the node/key to be inserted is greater than the key value of the current node, then the new node is inserted in the right subtree. Now, let us discuss the algorithm for inserting a node in the BST:

```
INSERT(ROOT, VALUE)
Step 1: START
Step 2: IF(ROOT == NULL)
 Allocate memory for ROOT node
 Set ROOT -> INFO = VALUE
Set ROOT -> LCHILD = ROOT -> RCHILD = NULL
        [End of IF]
Step 3: IF(ROOT -> INFO > VALUE)
 INSERT(ROOT -> LCHILD, VALUE)
        ELSE
INSERT(ROOT -> RCHILD, VALUE)
        [End of IF]
Step 4: END
```

In the previous algorithm, first, we check whether the tree is empty or not. If the tree is empty, then we will allocate memory for the `ROOT` node. In `Step 3`, we are checking whether the key value to be inserted is less than the value stored at the current node; if so, we will simply insert the new node in the left subtree. Otherwise, the new child node is inserted into the right subtree.

For example, let's say we have been given a BST. We want to insert a new node with the value of `7` into the BST.

Initially, the BST is given as shown in Figure 8.20:

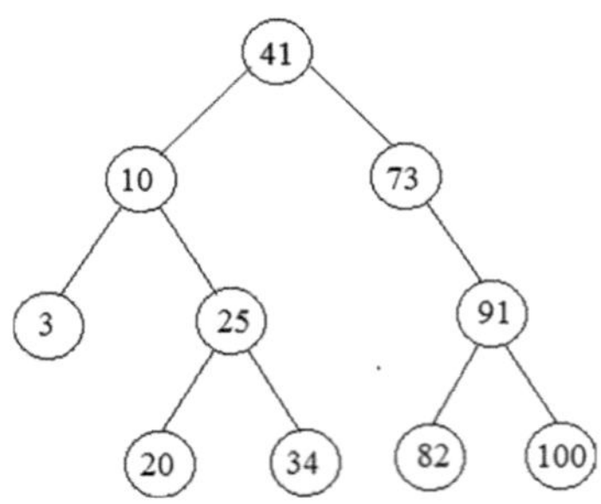

FIGURE 8.20 Sample BST.

First, we check whether the tree is empty or not. So, we will check the root node, `41`, as shown in Figure 8.21. As the root node is not empty, we will begin the insertion process.

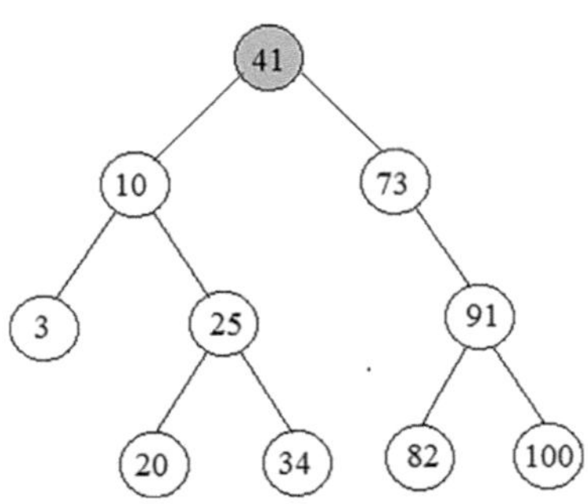

FIGURE 8.21 Traversing through root node.

We know that 7 < 41; thus, we will traverse the left subtree to insert the new node, as shown in Figure 8.22.

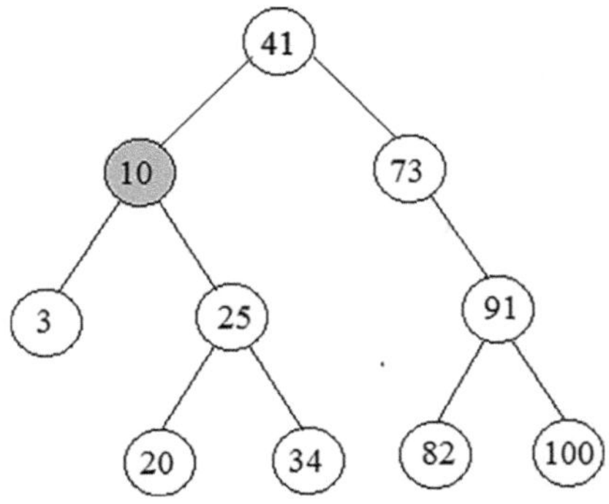

FIGURE 8.22 Traversing through left subtree.

We know that 7 < 10; thus, we will again traverse the left subtree to insert the new node,as shown in Figure 8.23.

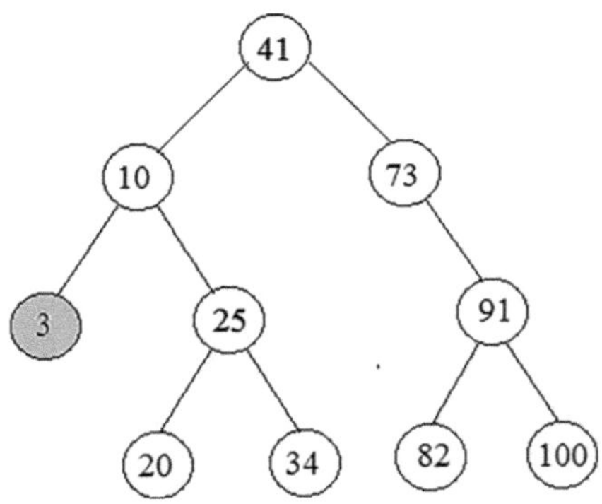

FIGURE 8.23 Traversing through left most child.

Now, we know that 7 > 3; thus, the new node with the value of 7 is inserted as the right child of the parent node 3, as shown in Figure 8.24.

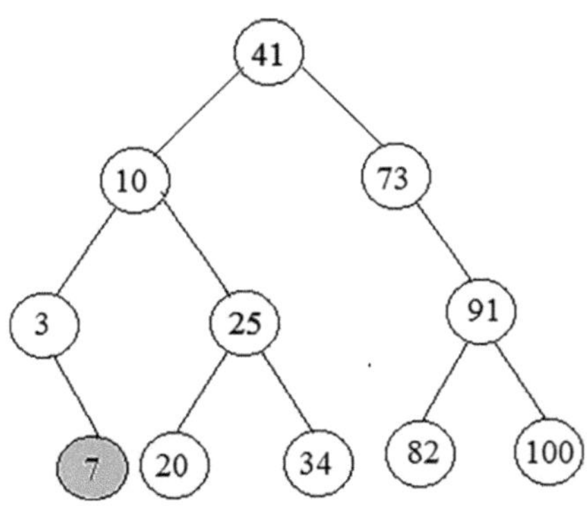

FIGURE 8.24 Inserting a new node with the value of 7 in the binary search tree.

Finally, the new node with the value of `7` is inserted as a right child in the BST.

Deleting a Node/Key

Deleting a node/key from a BST is the most crucial process. We should be careful when performing the deletion operation; while deleting the nodes, we must be sure that the property of the BSTs is not violated so that we don't lose necessary nodes during this process. The deletion operation is divided into three cases, as follows.

Case 1: Deleting a Node Having No Children

This is the simplest case of deletion, as we can directly remove or delete a node that has no children. Look at the BST given in Figure 8.25 as an example and see how deletion is done in this case.

We have been given a BST. We want to delete the node with the value of `61` from the BST.

Initially, the BST is given as follows:

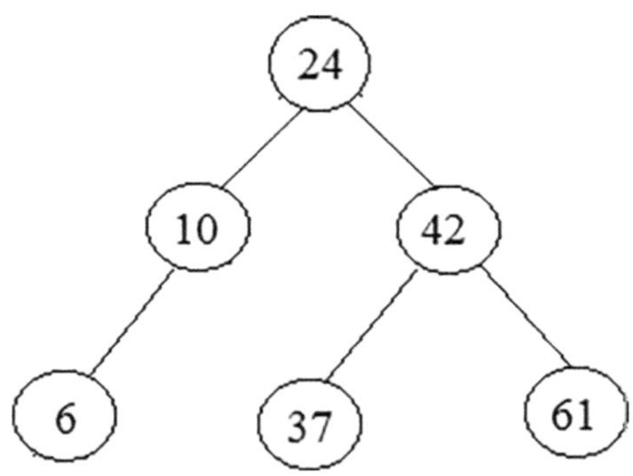

FIGURE 8.25 Sample BST.

First, we will check whether the tree is empty or not by checking the root node, `24`, as shown in Figure 8.26.

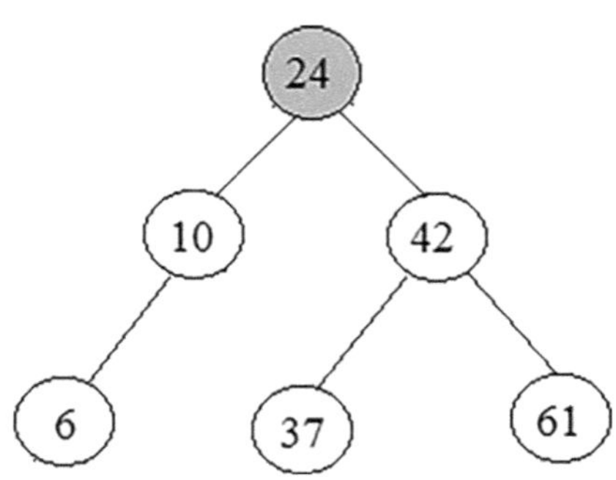

FIGURE 8.26 Traversing through root node.

As the root node is present, we will compare the value to be deleted with the value stored at the current node. As 61 > 24, we will recursively traverse the right subtree, as shown in Figure 8.27.

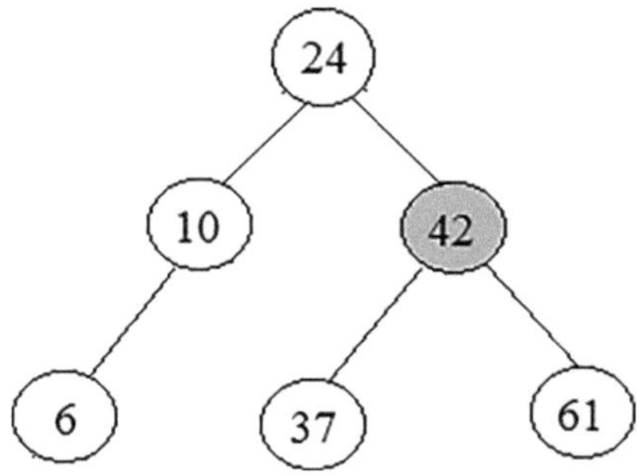

FIGURE 8.27 Traversing through right subtree to search for key to be deleted.

Again, we will compare the value to be deleted with the value stored at the current node. As 61 > 42, thus, we will recursively traverse the right subtree, as shown in Figure 8.28.

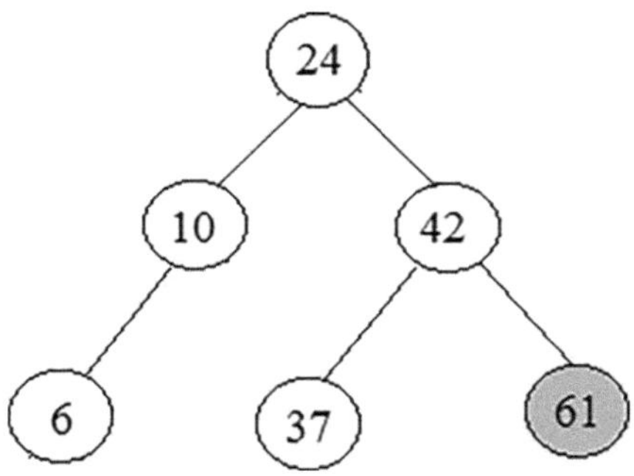

FIGURE 8.28 Traversing through right most child to locate node to be deleted.

Finally, a node having the value `61` is found and deleted from the BST, as shown in Figure 8.29.

FIGURE 8.29 Deleting the node with the value 61 from the binary search tree.

Case 2: Deleting a Node Having One Child

In this case of deletion, the node that is to be deleted, the parent node, is simply replaced by its child node. Look at the BST given in Figure 8.30 and see how deletion is done in this case.

We have been given a BST. We want to delete a node with the value `10`.
Initially, the binary search tree is given as follows:

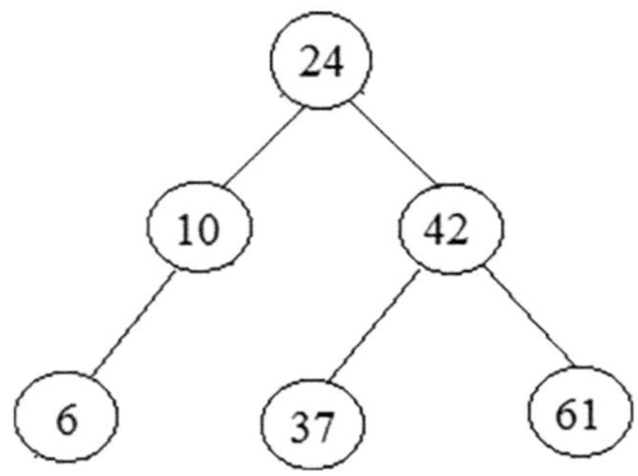

FIGURE 8.30 Sample BST.

First, we will check whether the tree is empty or not by checking the root node, `24`, as shown in Figure 8.31.

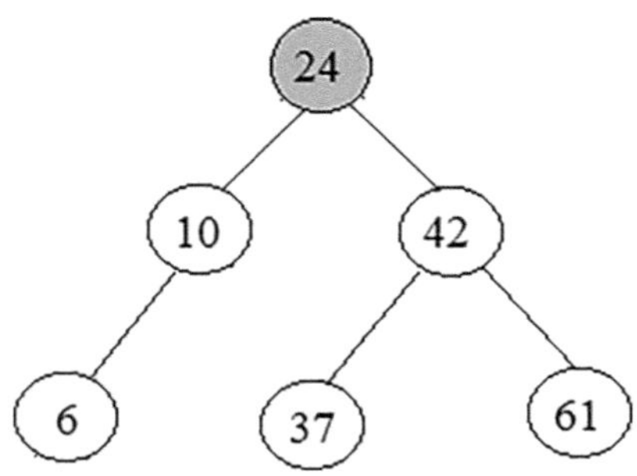

FIGURE 8.31 Tarversing through root nodeAs the root node is present, we will compare the value to be deleted with the value stored at the current node As *10 < 24*, we will recursively traverse the left subtree, as shown in Figure 8.32.

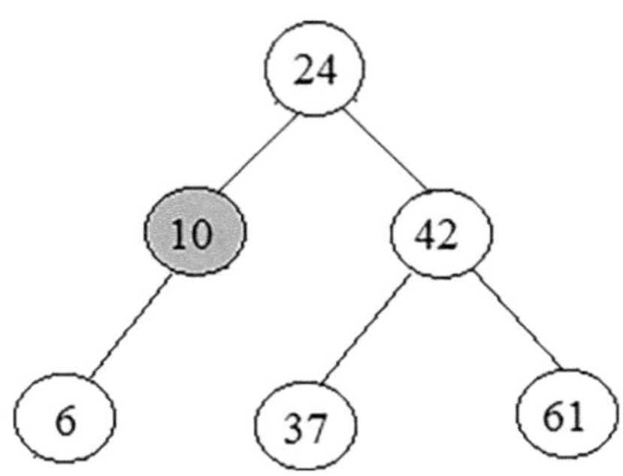

FIGURE 8.32 Traversing through left subtree to locate the node to be deleted.

Now, as the node to be deleted is found and has one child, `6`, the node to be deleted is now replaced with its child node, and the actual node is deleted, as shown in Figure 8.33.

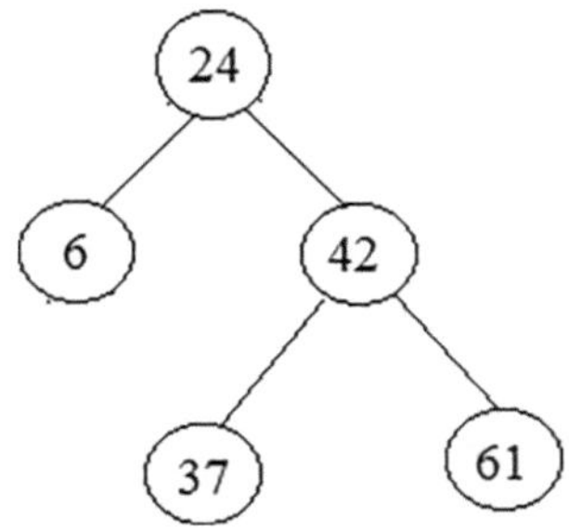

FIGURE 8.33 Deleting the node with value 10 from the binary search tree.

Case 3: Deleting a Node Having Two Children

In this case, the node that is to be deleted is simply replaced by its in-order predecessor (the largest value in the left subtree), or by its in-order successor (the smallest value in the right subtree). The in-order predecessor or in-order successor can be deleted using either of the two cases. Now, look at the BST shown in Figure 8.34 and see how the deletion will take place in this case.

We have been given a BST. We want to delete the node with the value `42`.

Initially, the BST is given as follows:

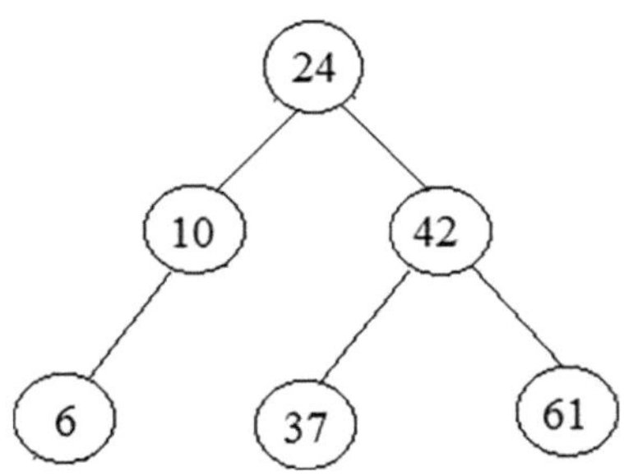

FIGURE 8.34 Sample BST.

First, we will check whether the tree is empty or not by checking the root node, `24`, as shown in Figure 8.35.

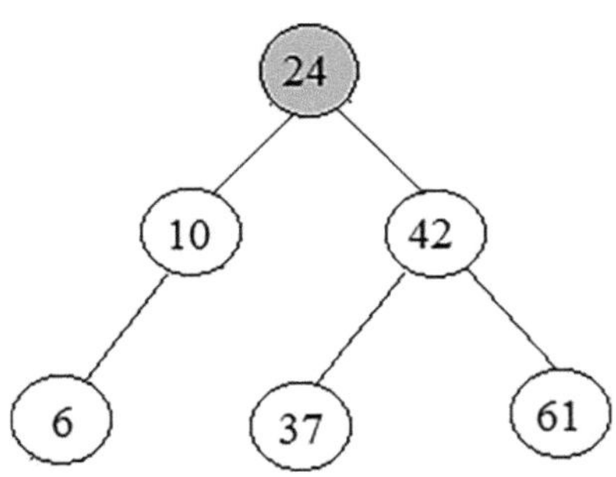

FIGURE 8.35 Traversing through root node.

As the root node is present, we will compare the value to be deleted with the value stored at the current node. As 42 > 24, we will recursively traverse the right subtree, as shown in Figure 8.36.

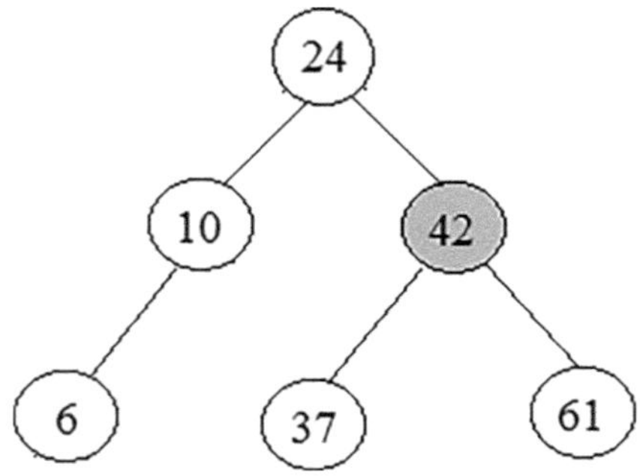

FIGURE 8.36 Traversing through right subtree to locate the node to be deleted.

As the node to be deleted is found and has two children, we will find the in-order predecessor of the current node (`42`) and replace the current node with its in-order predecessor so that the actual node `42` is deleted, as shown in Figure 8.37.

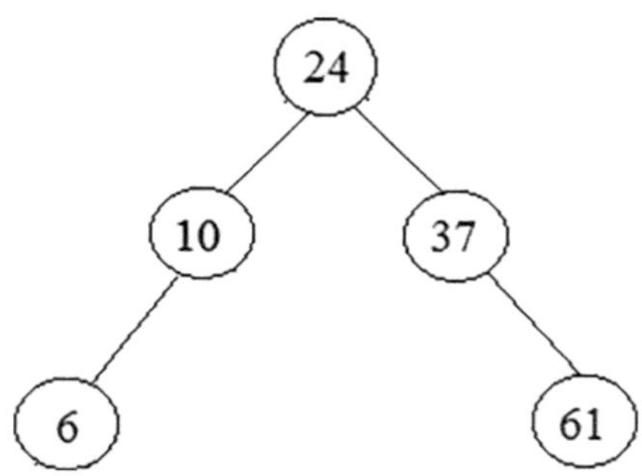

FIGURE 8.37 Deleting the node with the value 42 from the binary search tree.

Now, let us discuss the algorithm for deleting a node from the BST:

```
DELETE_NODE(ROOT, VALUE)
Step 1: START
Step 2: IF(ROOT == NULL)
 Print "Error"
          [End of IF]
Step 3:IF(ROOT -> INFO > VALUE)
 DELETE_NODE(ROOT -> LCHILD, VALUE)
         ELSE IF(ROOT -> INFO < VALUE)
DELETE_NODE(ROOT -> RCHILD, VALUE)
         ELSE
IF(ROOT -> LCHILD = NULL & ROOT -> RCHILD = NULL)
            FREE(ROOT)
          ELSE IF(ROOT -> LCHILD & ROOT -> RCHILD)
```

```
TEMP = FIND_LARGEST(ROOT -> LCHILD)
                          OR
TEMP = FIND_SMALLEST(ROOT -> RCHILD)
Set ROOT -> INFO = TEMP -> INFO
FREE(TEMP)
         ELSE
IF(ROOT -> LCHILD != NULL)
Set TEMP = ROOT -> LCHILD
Set ROOT -> INFO = TEMP -> INFO
FREE(TEMP)
ELSE
Set TEMP = ROOT -> RCHILD
Set ROOT -> INFO = TEMP -> INFO
FREE(TEMP)
          [End of IF]
         [End of IF]
        [End of IF]
Step 4: END
```

In the previous algorithm, first, we check whether the tree is empty or not. If the tree is empty, then the node to be deleted is not present. Otherwise, if the tree is not empty, we will check whether the node/value to be deleted is less than the value stored at the current node. If the value to be deleted is less than the value stored at the current node, we will recursively call the left subtree. If the value to be deleted is greater than the value stored at the current node, we will recursively call the right subtree. Now, if the node to be deleted has no children, then the node is simply freed. If the node to be deleted has two children (that is, both a left and right child), then we will find the in-order predecessor by calling (`TEMP = FIND_LARGEST(ROOT -> LCHILD)`) or the in-order successor by calling (`TEMP = FIND_SMALLEST(ROOT -> RCHILD)`) and replace the value stored at the current node with that of the in-order predecessor or in-order successor. Then, we will simply delete the initial node of either the in-order predecessor or the in-order successor. Finally, if the node to be deleted has only one child, the value stored at the current node is replaced by its child node, and the child node is deleted.

Deleting the Entire Binary Search Tree

It is very easy to delete the entire BST. First, we will delete all the nodes present in the left subtree, followed by the nodes present in the right subtree. Finally, the root node is deleted, and the entire tree is deleted.

Here is the algorithm for deleting the entire BST:

```
DELETE_BST(ROOT)
Step 1: START
Step 2: IF(ROOT != NULL)
          DELETE_BST(ROOT -> LCHILD)
          DELETE_BST(ROOT -> RCHILD)
          FREE(ROOT)
        [End of IF]
Step 3: END
```

Finding the Mirror Image of the Binary Search Tree

This is an exciting operation to perform in the BST. The mirror image means interchanging the right subtree with the left subtree at each and every node of the tree, as shown in Figure 8.38.

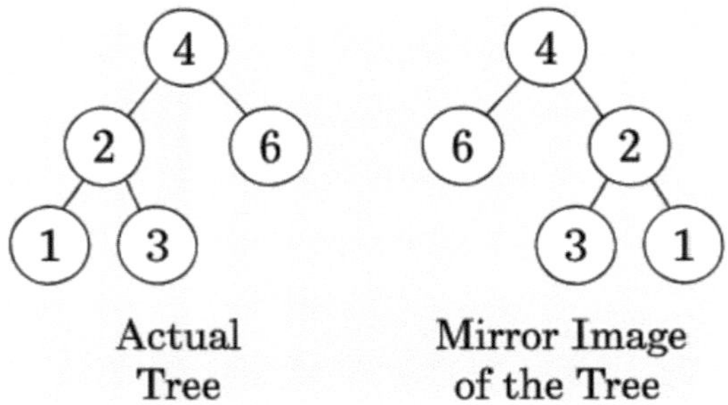

FIGURE 8.38 Binary search tree and its mirror image.

Here is the algorithm for finding the mirror image of the BST:

```
MIRROR_IMAGE(ROOT)
Step 1: START
Step 2:IF(ROOT != NULL)
          MIRROR_IMAGE(ROOT -> LCHILD)
          MIRROR_IMAGR(ROOT -> RCHILD)
          Set TEMP = ROOT -> LEFT
          ROOT -> LEFT = ROOT -> RIGHT
          Set ROOT -> RIGHT = TEMP
        [End of IF]
Step 3: END
```

Finding the Smallest Node

We know that it is the basic property of the BST that the smallest value always occurs in the extreme left of the left subtree. If there is no left subtree, then the value of the root node will be the smallest. Hence, to find the smallest value in the BST, we will simply find the value of the node present at the extreme left of the left subtree.

Here is the algorithm for finding the smallest node in the BST:

```
SMALLEST_VALUE(ROOT)
Step 1: START
Step 2: IF(ROOT = NULL OR ROOT -> LCHILD = NULL)
          Return NULL
        ELSE
          Return SMALLEST_VALUE(ROOT)
        [End of IF]
Step 3: END
```

Finding the Largest Node

We know that it is the basic property of the BST that the largest value always occurs in the extreme right of the right subtree. If there is no right subtree, the value of the root node will be the largest. Hence, to find the largest value in the BST, we will simply find the value of the node present at the extreme right of the right subtree.

Here is the algorithm for finding the largest node in the BST:

```
LARGEST_VALUE(ROOT)
Step 1: START
Step 2: IF(ROOT = NULL OR ROOT -> RCHILD = NULL)
          Return NULL
        ELSE
          Return LARGEST_VALUE(ROOT)
        [End of IF]
Step 3: END
```

Determining the Height of the Binary Search Tree

The height of the BST can easily be determined. We will first calculate the height of the left subtree and the right subtree. Whichever height is greater, we add 1 to that height; that is, if the height of the left subtree is greater, then 1 is added to the height of the left subtree. Similarly, if the height of the right subtree is greater, then 1 is added to the height of the right subtree.

Here is the algorithm for determining the height of the BST:

```
CALCULATE_HEIGHT(ROOT)
Step 1: START
Step 2: IF ROOT = NULL
          Print "Can't find height of the tree."
        ELSE
          Set LHEIGHT = CALCULATE_HEIGHT(ROOT -> LCHILD)
          Set RHEIGHT = CALCULATE_HEIGHT(ROOT -> RCHILD)
        IF(LHEIGHT < RHEIGHT)
Return (RHEIGHT) + 1
        ELSE
          Return (LHEIGHT) + 1
         [End of IF]
        [End of IF]
Step 3: END
```

Binary Tree Traversal Methods

Traversing is the process of visiting each node in the tree exactly once in a particular order. We all know that a tree is a non-linear data structure, and therefore, a tree can be traversed in various ways. There are three types of traversals, which are:

- Pre-order traversal
- In-order traversal
- Post-order traversal

Now, we will discuss all of these traversals in detail.

Pre-Order Traversal

In pre-order traversal, the following operations are performed recursively at each node:

1. Visit the root node.
2. Traverse the left subtree.
3. Traverse the right subtree.

The term *pre* in *pre-order* determines that the root node is accessed before accessing any other node in the tree. Hence, it is also known as a *DLR traversal*, meaning *Data Left Right*. Therefore, in a DLR traversal, the root node is accessed first, followed by the left subtree and then the right subtree. Now, let us see an example of pre-order traversal.

Let's say we want to find the pre-order traversal of the binary tree of the word *EDUCATION*, shown in Figure 8.39.

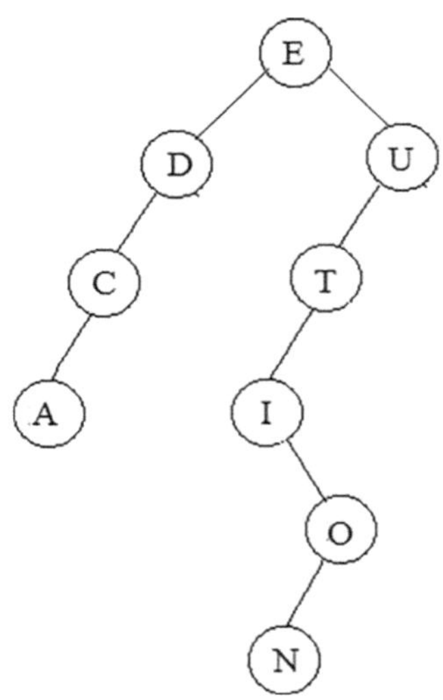

FIGURE 8.39 Binary tree of the word EDUCATION.

The pre-order traversal of the previous binary tree is:

E D C A U T I O N

Now, let us look at the function for pre-order traversal:

```
void pre-order(struct BST * root)
{
  if(root != NULL)
  {
    printf("%d", root -> info) ;
    pre-order(root ->lchild) ;
    pre-order(root ->rchild) ;
  }
}
```

In-Order Traversal

With in-order traversal, the following operations are performed recursively at each node:

1. Traverse the left subtree.
2. Visit the root node.
3. Traverse the right subtree.

The term *in* in *in-order* determines that the root node is accessed between the left and the right subtrees. Hence, it is also known as an *LDR traversal*, meaning *Left Data Right*. Therefore, in an LDR traversal, the left subtree is traversed first, followed by the root node and then the right subtree. Now, let us see an example of an in-order traversal.

Let's say we want to find the in-order traversal of the binary tree of the word EDUCATION, as shown in Figure 8.40.

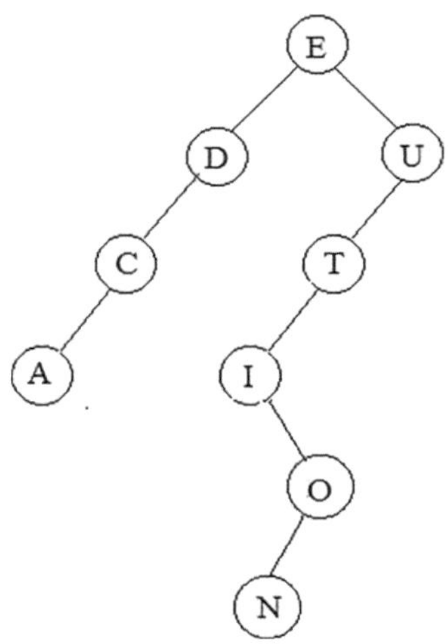

FIGURE 8.40 Binary tree of the word EDUCATION.

The in-order traversal of the previous binary tree is:

A C D E I N O T U

Now, let us look at the function for an in-order traversal:

```
void in-order(struct BST * root)
{
  if(root != NULL)
  {
    in-order(root ->lchild) ;
    printf("%d", root -> info) ;
    in-order(root ->rchild) ;
  }
}
```

Post-Order Traversal

In a post-order traversal, the following operations are performed recursively at each node:

1. Traverse the left subtree.
2. Traverse the right subtree.
3. Visit the root node.

The term *post* in *post-order* determines that the root node will be accessed last after the left and the right subtrees. Hence, it is also known as an *LRD traversal*, meaning *Left Right Data*. Therefore, in an LRD traversal, the left subtree is traversed first, followed by the right subtree and then the root node. Now, let us see an example of a post-order traversal.

Let's say we want to find the post-order traversal of the binary tree of the word EDUCATION, as shown in Figure 8.41.

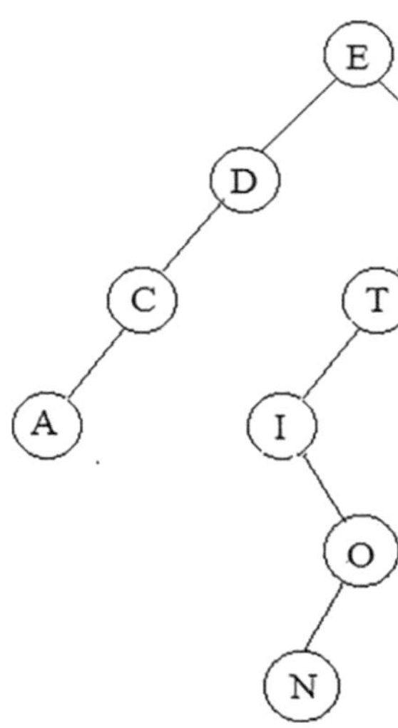

FIGURE 8.41 Binary tree of the word EDUCATION.

The post-order traversal of the previous binary tree is:

A C D N O I T U E

Now, let us look at the function of the post-order traversal and it's practical implementation through a program as discussed below:

```
void post-order(struct BST * root)
{
  if(root != NULL)
  {
    post-order(root ->lchild) ;
    post-order(root ->rchild) ;
    printf("%d", root -> info) ;
  }
}
```

Write a program to create a binary search tree and perform different operations on it:

```
#include<stdio.h>v
#include<malloc.h>
#include<conio.h>
struct node
{
 int info;
 struct node *lchild;
 struct node *rchild;
};
struct node *BST;

void create_BST(struct node *);
struct node *insert_value(struct node*,int);
void in-order_traversal(struct node *);
void post-order_traversal(struct node*);
void pre-order_traversal(struct node *);
struct node *delete_value(struct node *,int);
intcalculate_height(struct node*);
struct node *delete_BST(struct node *);

int main()
{

 struct node *root;
 int choice, item;
create_BST(BST);
clrscr();
 while(1)
```

```
    {
  printf("\n\t***MENU***");
     printf("\n 1. Insert Value ");
     printf("\n 2. In-order traversal of the tree");
     printf("\n 3. post-order traversal of the tree");
     printf("\n 4. Pre-order traversal of the tree");
     printf("\n 5. Delete value");
  printf("\n 6. Calculate height");
  printf("\n 7. Delete BST);
     printf("\n 8. Exit);
     printf("\nEnter your choice : ");
     scanf(" %d",&choice);
     switch(choice)
    {
   case 1:
    printf("\n Enter the new value to be inserted: ");
    scanf("%d", &item);
    BST = insert_value(BST, item);
  break;

     case 2:
   printf("\n In-order traversal of the tree is : ");
   in-order_traversal(BST);
   break;

  case 3:
   printf("\n Post-order traversal of the tree is : ");
   post-order_traversal( BST);
   break;

  case 4:
   printf("\n Pre-order traversal of binary tree is : ");
   pre-order_traversal(BST);
   break;

  case 5:
    printf("\n Enter the value to delete : ");
    scanf(" %d",&item);
    BST=delete_value(BST,item);
    break;

  case 6:
  printf("\nHeight of tree is %d",calculate_height(BST);
  break;

  case 7;
  BST = delete_BST(BST);
  break;

     case 8:
  printf("Wrong Choice!!!");
     exit(0);
  }
  }
  getch();
```

```
return0;
}

void create_BST(struct node*BST)
{
 BST = NULL:
 }

struct node *insert_value(struct node *BST, int item)
{
struct node *root , *node_root , *parent_root;
  root=(struct node*)malloc(sizeof(struct node));
  root->info = item;
  root->lchild = NULL;
  root->rchild= NULL;
  if(BST == NULL)
{
BST = root;
BST->lchild == NULL;
BST->rchild == NULL;
}
else
{
parent_root = NULL:
node_root = BST;
while(node_root!=NULL)
{
parent_root= node_root;
   if(item < (node_root->info))
node_root=node_root->lchild;
else
node_root=parent_root->rchild;
}
if(item < (parent_root->info))
parent_root->lchild = root;
else
parent_root->rchild = root;
}
return BST;
}

void in-order_traversal(struct node *BST)
{
 if(BST != NULL)
 {
  in-order_traversal(BST->lchild);
  printf("%d\t",BST->info);
  in-order_traversal(BST->rchild);
 }
}

void post-order_traversal(struct node *BST)
{
 if( BST!= NULL)
```

```
  {
   post-order_traversal(BST->lchild);
   post-order_traversal(BST->rchild);
   printf("%d\t", BST ->info);
  }
 }

 void pre-order_traversal(struct node *BST)
 {
  if(BST!= NULL)
  {
   printf("%d\t",BST->info);
   pre-order_traversal(BST->lchild);
   pre-order_traversal(BST->rchild);
  }
 }

 struct tree *delete_value(struct node*BST,int item)
 {
  struct node *current,*parent,*succ,*psucc,*ptr;
 if(BST->lchild==NULL)
  {
 printf("\n The tree is empty");
 return(BST);
 }
 parent = BST;
 current = BST->lchild;
 while(current !=NULL && item!= current->info)
 {
 parent = current;
 current = (item<current -> info) ? current ->lchild: current
 ->rchild);
 }
 if(current == NULL)
 {
 printf("\n Item to be deleted is not in the binary tree);
   return(BST);
  }
 if(current ->lchild == NULL)
 ptr= current ->rchild;
 else if (current ->rchild == NULL)
 ptr = c->lchild;
 else
 {
 //Finding in-order successor and its parent
 psucc= current;
 current = current ->lchild;
 while(succ->lchild!=NULL)
 {
 psucc = succ;
 succ = succ->lchild;
 }
 if(current == psucc)
 {
```

```
 //Case A
succ->lchild = current ->rchild;
}
else
{
  //case B
succ->lchild = current ->lchild;
psucc->lchild= succ->rchild;
succ->rchild = current ->rchild;
}
ptr= succ;
}
if(parent ->lchild == current)
 parent ->lchild =ptr;
 else
parent ->rchild =ptr;
free(current);
return BST;
}

intcalculate_height(struct node*BST)
{
intlheight, rheight;
 if(BST== NULL)
return 0;
else
{
lheight = calculate_height(BST->lchild);
rheight = calculate_height(BST->rchild);
  if(lheight>rheight)
return (lheight+1);
else
  return(rheight+1);
}
}

struct node *delete_BST(struct  node*BST)
{
 if(BST!= NULL)
{
delete_BST(BST->lchild);
delete_BST(BST->rchild);
free(BST);
}
}
```

Here is the output:

```
  ***MENU***
Insert value
In-order traversal of the tree
Post-order traversal of the tree
```

```
Pre-order traversal of the tree
Delete value
Calculate height
Delete BST
Exit
Enter your choice: 1
Enter the value to be inserted: 95
Enter the value to be inserted: 20
Enter your choice: 4
Pre-order traversal of binary tree is: 95  20
Enter your choice : 5
Enter the value to delete : 20
Enter your choice : 8
Wrong Choice!!!
```

Creating a Binary Tree Using Traversal Methods

A binary tree can be constructed if we are given at least two of the traversal results. One traversal should always be an in-order traversal, and the second should either be a pre-order traversal or a post-order traversal. An in-order traversal determines the left and right child nodes of the binary tree. Pre-order or post-order traversal determines the root node of the binary tree. Hence, there are two different ways of creating a binary tree:

- In-order and pre-order traversal
- In-order and post-order traversal

Now, we have pre-order and in-order traversal sequences. The following steps are followed to construct a binary tree:

1. The pre-order traversing sequence is used to determine the root node of the binary tree. The first node in the pre-order sequence will be the root node.
2. The in-order traversing sequence is used to determine the left and the right subtrees of the binary tree. Keys toward the left side of the root node in the in-order sequence form the left subtree. Similarly, keys toward the right side of the root node in the in-order sequence form the right subtree.
3. Now, each element from the pre-order traversing sequence is recursively selected, and the left and the right subtrees are created from the in-order traversing sequence.

Let's look at this with an example. Let's say we want to create a binary tree from the following traversing sequences:

- *In-order* – A C D E I N O T U
- *Pre-order* – E D C A U T I O N

Now, we will construct the binary tree:

1. The first node in the pre-order sequence is the root node of the tree. Hence, *E* is the root node of the binary tree:

2. Now, we can easily determine the left and right subtrees from the in-order sequence. Keys toward the left side of the root node (that is, *A*, *C*, and *D*) form the left subtree. Similarly, elements on the right side of the root node (that is, *I*, *N*, *O*, *T*, and *U*) form the left subtree:

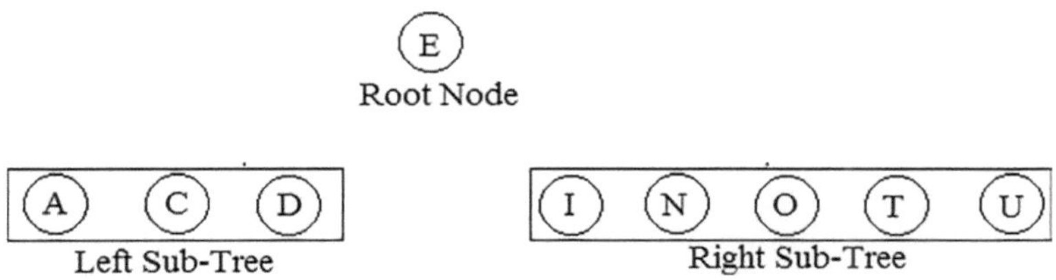

3. Now, the left child of the root node will be the first node in the pre-order traversing sequence after the root node, *E*. Thus, *D* is the left child of the root node, *E*:

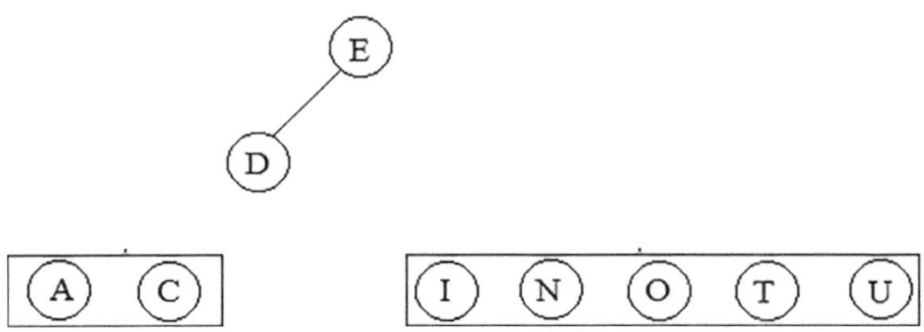

4. Similarly, the right child of the root node will be the first node in the pre-order traversing sequence after the nodes of the left subtree. Thus, *U* is the right child of the root node, *E*:

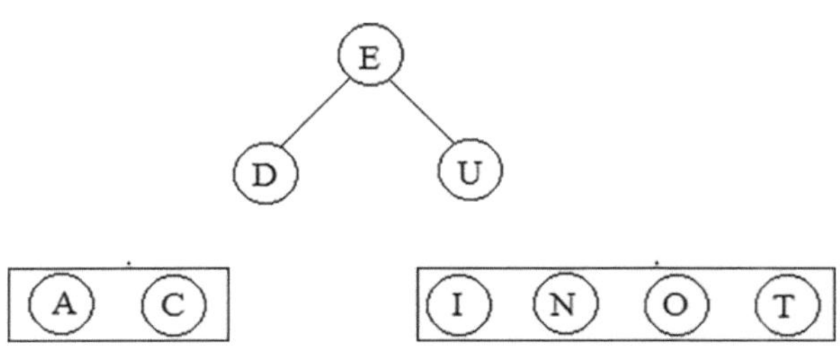

5. In the in-order sequence, *A* and *C* are on the left side of *D*. So, *A* and *C* will form the left subtree of *D*:

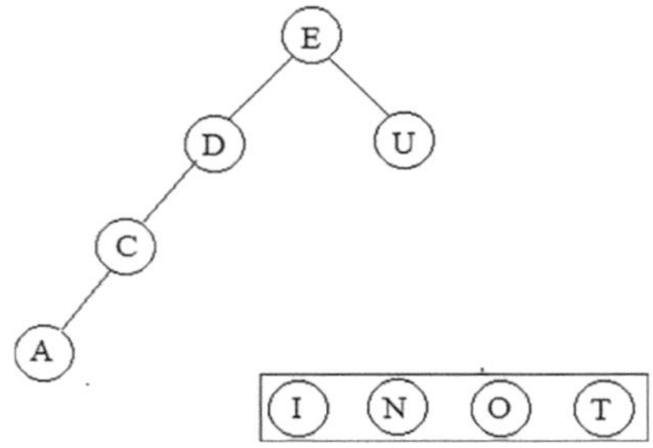

6. The next elements in the pre-order sequence are *T* and *I*. In the in-order sequence, *T* and *I* are on the left side of *U*. So, *T* and *I* will form the left subtree of *U*:

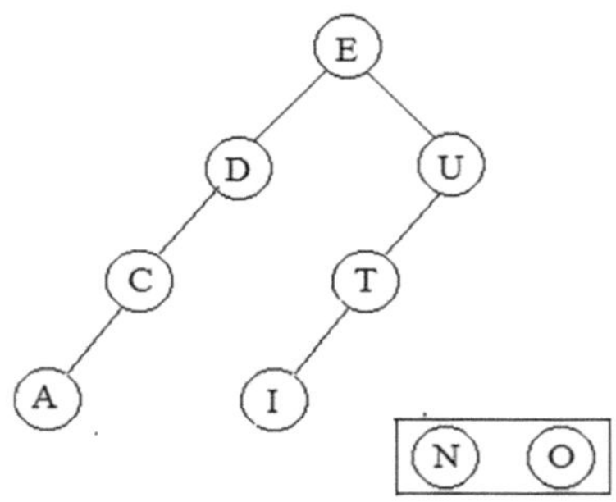

7. The next element in the pre-order sequence is *O*. In the in-order sequence, *O* is on the right side of *I*. So, *O* will form the right subtree of *I*. The last element in the pre-order sequence is *N*, which is on the left side of *O* in the in-order sequence. Thus, *N* will form the left subtree of *O*:

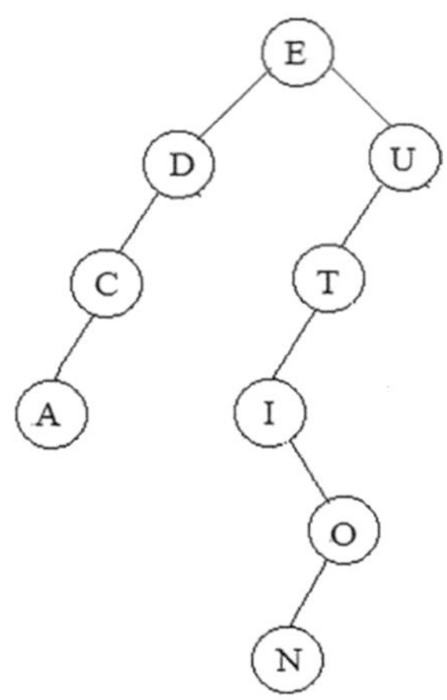

Finally, the binary tree is created from the given traversing sequences.

Frequently Asked Questions

Q1. Create a binary tree from the following traversing sequences:

- *In-order* – d b e a f c g
- *Pre-order* – a b d e c f g

Answer.

1. *a* is the root node of the binary tree:

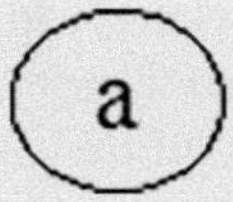

2. *d*, *b*, and *e* are on the left side of the a node in the in-order sequence. Hence, *d*, *b*, and *e* are the left subtrees of root a. Also, *d* is the left subtree of *b*, and *e* is the right subtree of *b*:

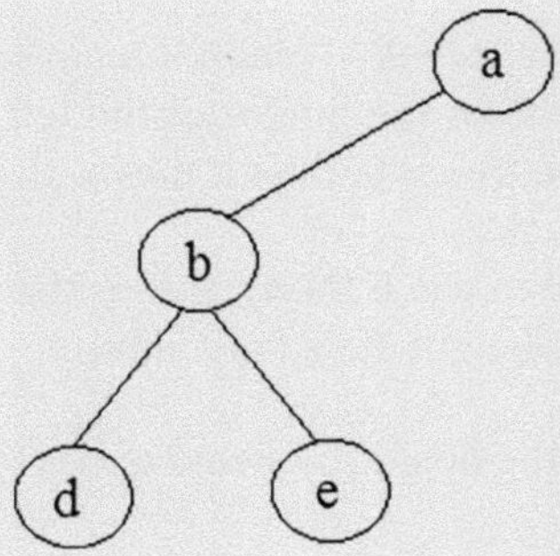

3. f, *c*, and *g* are on the right side of root a in the in-order sequence. Hence, *f*, *c*, and *g* are the right subtrees of root *a*. Also, *f* is the left subtree of *c*, and *g* is the right subtree of *c*:

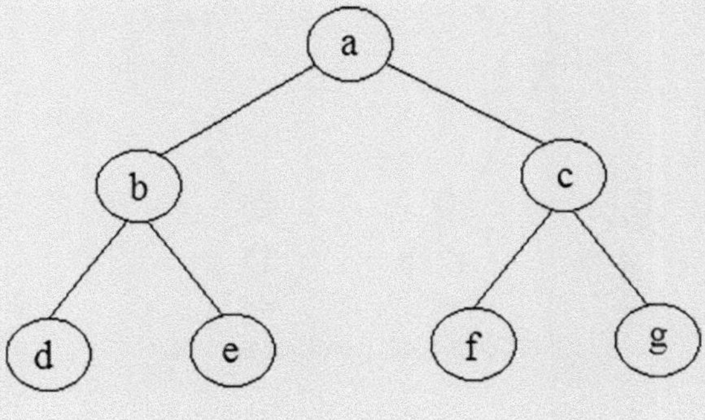

AVL TREES

The *AVL tree* was invented by Adelson-Velski and Landis in 1962. It is so named in honor of its inventors and was the first balanced BST. It is a self-balancing BST. The AVL tree is also known as a *height-balanced tree* because of its property that the heights of the two subtrees of a node can differ at most by one. AVL trees are very efficient in performing searching, insertion, and deletion operations, as they take $O(\log n)$ time to perform all these operations.

Need for AVL Trees

AVL trees are very similar to BSTs, but with a small difference. AVL trees have a special variable, known as a balance factor, associated with them. Every node in the AVL tree has a balance factor associated with it. The balance factor is determined by subtracting the height of the right subtree from the height of the left subtree. Thus, a node with a balance factor of –1, 0, or 1 is said to be a height-balanced tree. The primary need for the height-balanced tree is that the process of searching becomes very fast. This balancing condition also ensures that the depth of the tree is $O(\log n)$. The balance factor is calculated as follows:

$$Balance\ Factor = Height(Left\ subtree) - Height(Right\ subtree)$$

- If the balance factor of the tree is –1, then it means that the height of the right subtree of that node is one more than the height of the left subtree of that node.
- If the balance factor of the tree is 0, then it means that the height of the left and the right subtrees of a node are equal.
- If the balance factor of the tree is 1, then it means that the height of the left subtree of that node is one more than the height of its right subtree.

Thus, the overall benefit of the height-balanced tree is to assist in fast searching. An example is shown in Figure 8.42.

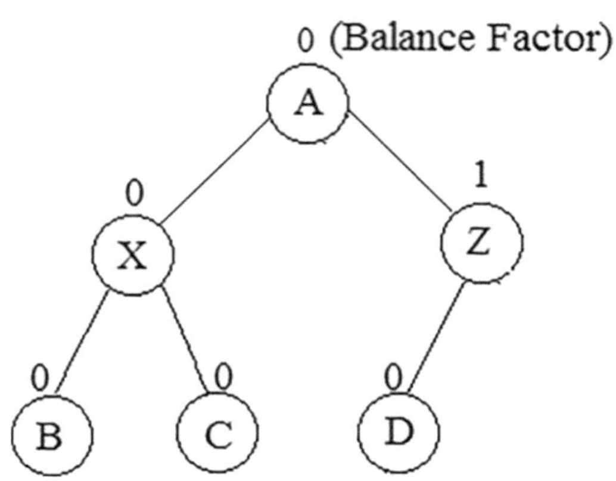

FIGURE 8.42 Balanced AVL tree.

Operations on an AVL Tree

In this section, we will discuss two operations that are performed on the AVL trees:

- Searching for a node in an AVL tree
- Inserting a new node in an AVL tree

The process of searching for a node in an AVL tree is the same as for a BST.

The process of inserting a new node in an AVL tree is quite similar to that of BSTs. The new node is always inserted as a terminal/leaf node in the AVL tree, but the insertion of a new node can disturb the balance of the AVL tree, as the balance factor may be disturbed. Thus, for the tree to remain balanced, the insertion process is followed by a *rotation* process. The rotation process is usually done to restore the balance factor of the tree. If the balance factor of each node is –1, 0, or 1 after the insertion process, then the rotation is not required, as the tree is already balanced; otherwise, rotation is required. Now, let us look at an example and see how insertion is done without rotations.

In the AVL tree shown in Figure 8.43, let's say we want to insert a new node with the value 60. Initially, the AVL tree is as shown here:

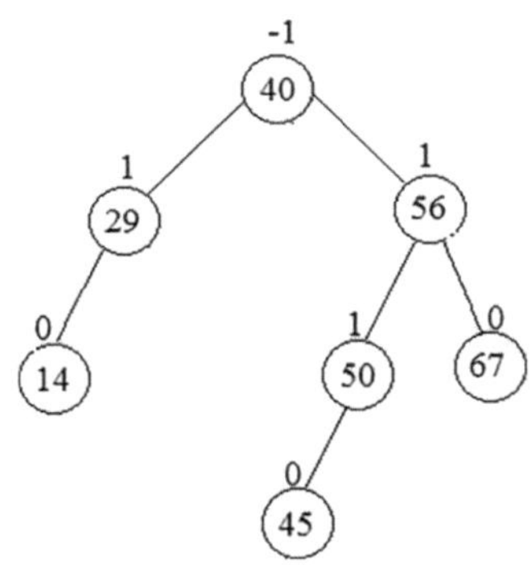

FIGURE 8.43 AVL tree before insertion.

Now, we will insert 60 into the AVL tree, as shown in Figure 8.44:

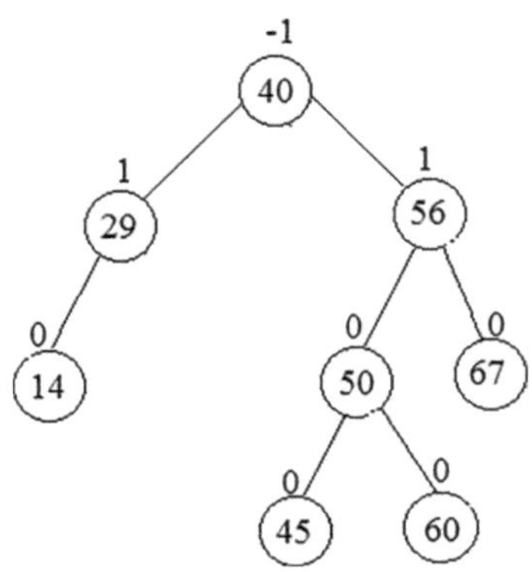

FIGURE 8.44 AVL tree after inserting 60.

Hence, after insertion, there are no nodes in the tree that are unbalanced. Thus, there is no need to apply rotation here. Now, we will discuss how the rotation process is performed in AVL trees.

AVL Rotations

Rotation is done when the balance factor of the node becomes disturbed after inserting a new node. We know that a new node that is inserted will always have a balance factor of 0, as it will be the leaf node. Hence, the nodes whose balance factors will be disturbed are the ones that lie in the path from the root node to the newly inserted node. So, we will perform the rotation process only on those nodes whose balance factors will be disturbed. In the rotation process, our first task is to find the critical node in the AVL tree. The critical node is the nearest ancestor node from the newly inserted node to the root node that does not have a balance factor of –1, 0, or 1. First, let us understand the concept of the critical node with the help of an example.

Let's say we want to find the critical node in an AVL tree.

Initially, the AVL tree is as shown in Figure 8.45:

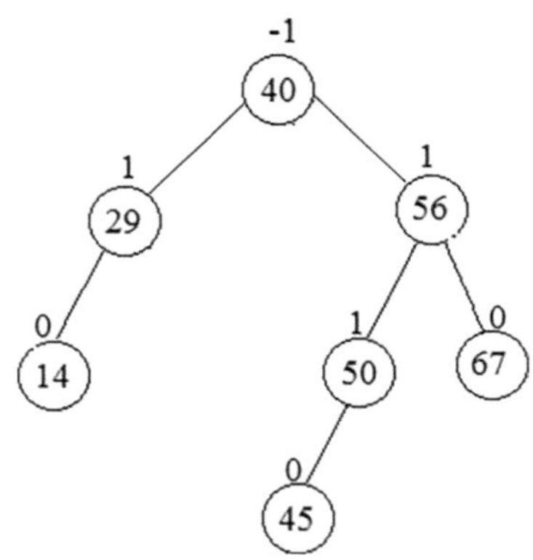

FIGURE 8.45 AVL tree.

Now, we will insert a new node with the value 42 in the tree, as shown in Figure 8.46.

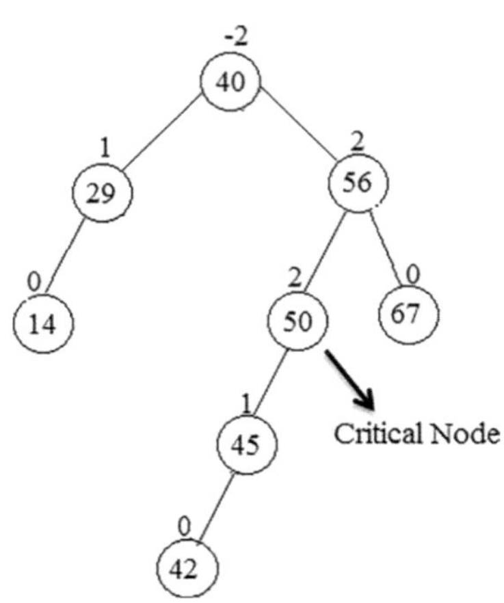

FIGURE 8.46 AVL tree.

After inserting 42 into the AVL tree, we can see that there are three nodes in the tree that have balance factors equal to –2, 2, and 2. The critical node is the one that is nearest to the newly inserted node with a disturbed balance factor. We can see that 50 is the nearest node to 42, and 50 has a balance factor of 2. Thus, 50 is the critical node in this AVL tree. To restore the balance factor of the previous AVL tree, rotations are performed.

There are four types of rotations:

1. *Left-left rotation* (*LL rotation*) – A new node is inserted in the left subtree of the left subtree of the critical node.
2. *Right-right rotation* (*RR rotation*) – A new node is inserted in the right subtree of the right subtree of the critical node.
3. *Right-left rotation* (*RL rotation*) – A new node is inserted in the left subtree of the right subtree of the critical node.
4. *Left-right rotation* (*LR rotation*) – A new node is inserted in the right subtree of the left subtree of the critical node.

Now, let us discuss all of these rotations in detail.

LL Rotation

LL rotation is also known as left-left rotation, as the new node is inserted in the left subtree of the left subtree of the critical node. It is a single rotation. Let us take an example and perform an LL rotation in it.

Initially, the AVL tree is as shown in Figure 8.47:

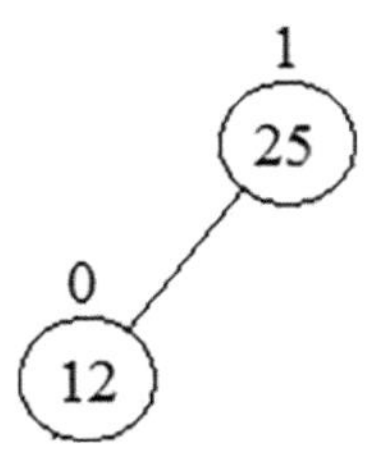

FIGURE 8.47 Sample AVL Tree.

Let's insert a new node, 5, in the AVL tree, as shown in Figure 8.48.

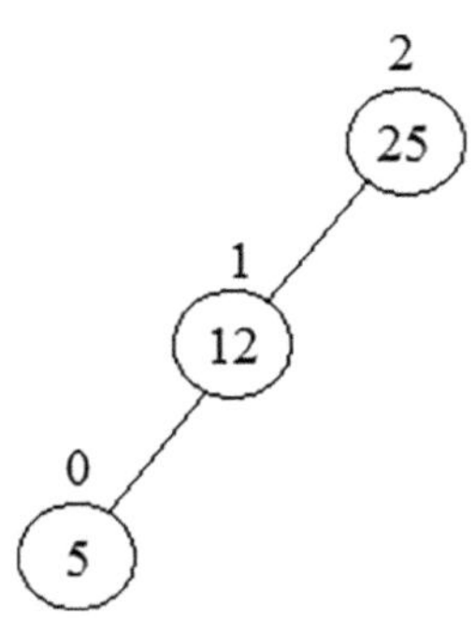

FIGURE 8.48 Insertion of key as left most child resulting to skewed tree.

After inserting 5, the balance factor of 25 is disturbed. Thus, 25 is the critical node. Hence, we will apply the LL rotation to restore the balance factor of the tree. After rotation, node 12 becomes the root node, and nodes 5 and 25 become the left and the right child of the tree, respectively, as shown in Figure 8.49.

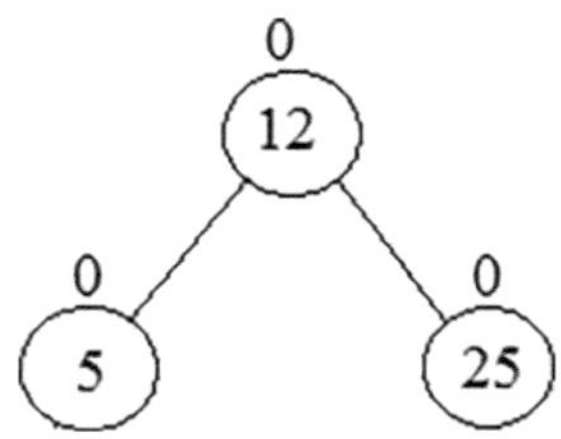

FIGURE 8.49 LL rotation in an AVL tree.

Therefore, the LL rotation is performed, and the balance factor of each node is also restored.

RR Rotation

RR rotation is also known as right-right rotation, as the new node is inserted in the right subtree of the right subtree of the critical node. It is also a single rotation. Let us take an example and perform an RR rotation in it.

Initially, the AVL tree is as shown in Figure 8.50:

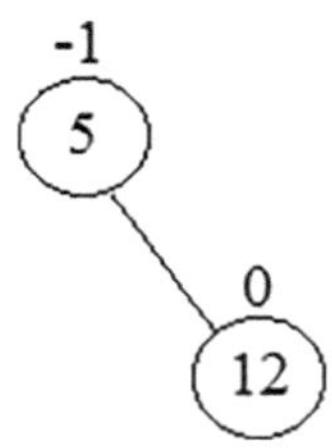

FIGURE 8.50 Sample AVL tree.

Let's insert a new node, 25, in the AVL tree as shown in Figure 8.51.

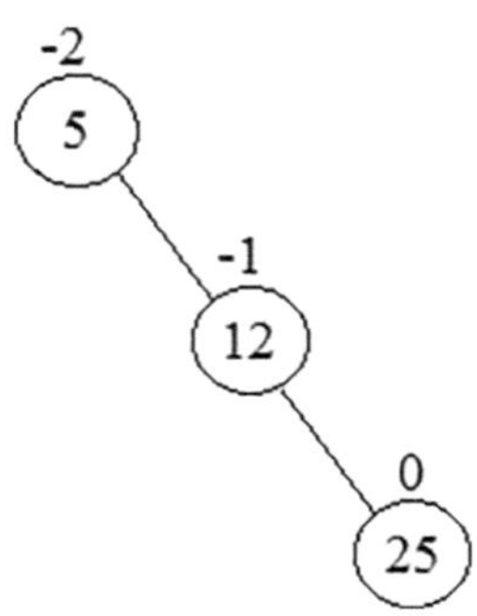

FIGURE 8.51 Insertion of key as right most child resulting to skewed tree.

After inserting 25, the balance factor of 5 is disturbed. Thus, 5 is the critical node. Hence, here we will apply an RR rotation to restore the balance factor of the tree. After rotation, node

12 becomes the root node, and nodes 5 and 25 become the left and the right child of the tree, respectively, as shown in Figure 8.52.

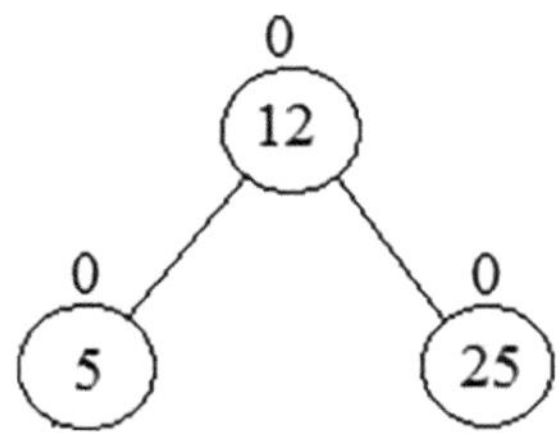

FIGURE 8.52 RR rotation in an AVL tree.

Therefore, the RR rotation is performed, and the balance factor of each node is also restored.

RL Rotation

RL rotation is also known as a right-left rotation, as the new node is inserted in the left subtree of the right subtree of the critical node. It is a double rotation. Let us take an example and perform an RL rotation in it.

Initially, the AVL tree is as shown in Figure 8.53:

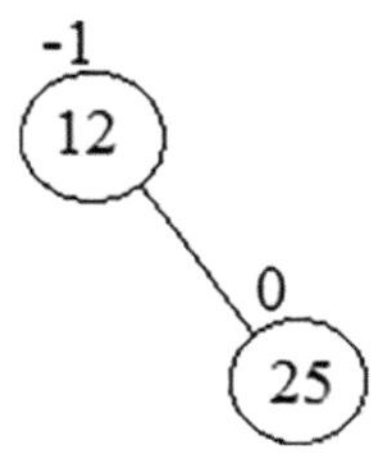

FIGURE 8.53 Sample AVL tree.

Let's insert a new node, 15, in the AVL tree, as shown in Figure 8.54.

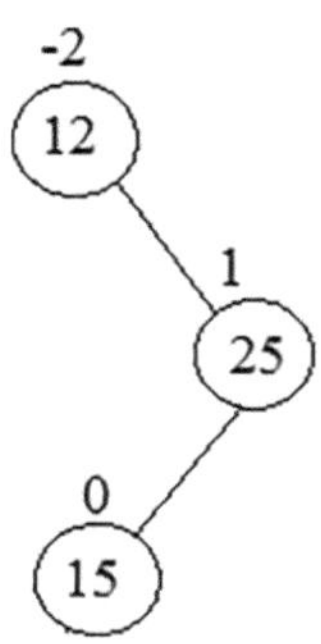

FIGURE 8.54 Insertion of key as left most child resulting to disbalanced tree.

After inserting 15, the balance factor of 12 is disturbed. Thus, 12 is the critical node. Hence, we will apply an RL rotation to restore the balance factor of the tree. After rotation, node 15 becomes the root node, and nodes 12 and 25 become the left and the right child of the tree, respectively, as shown in Figure 8.55.

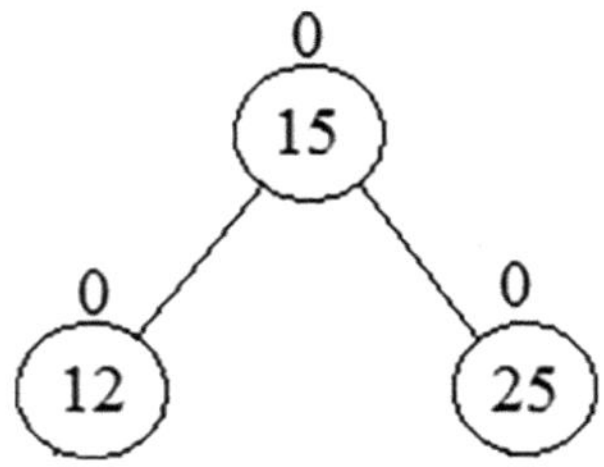

FIGURE 8.55 RL rotation in an AVL tree.

Therefore, the RL rotation is performed, and the balance factor of each node is also restored.

Frequently Asked Questions

Q2. Create an AVL tree by inserting the following elements:

60, 10, 20, 30, 19, 120, 100, 80, 19

Answer.

1. Insert 60:

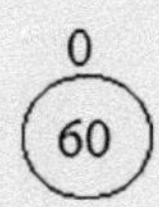

2. Insert 10. No rebalancing is required:

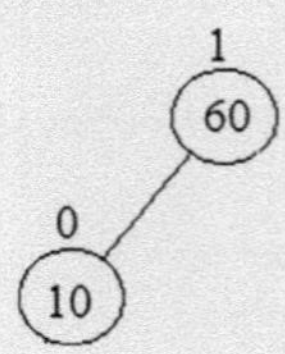

3. Insert 20. Now, rebalancing is required. We will perform LR rotation:

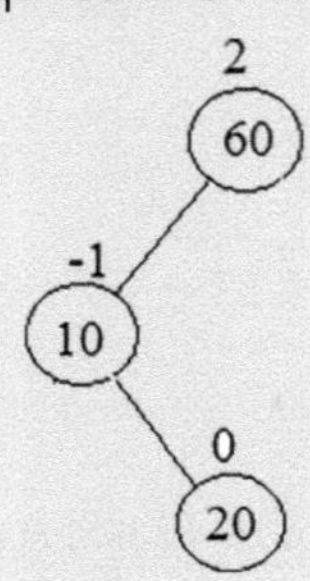

After performing LR rotation, the AVL tree is as follows:

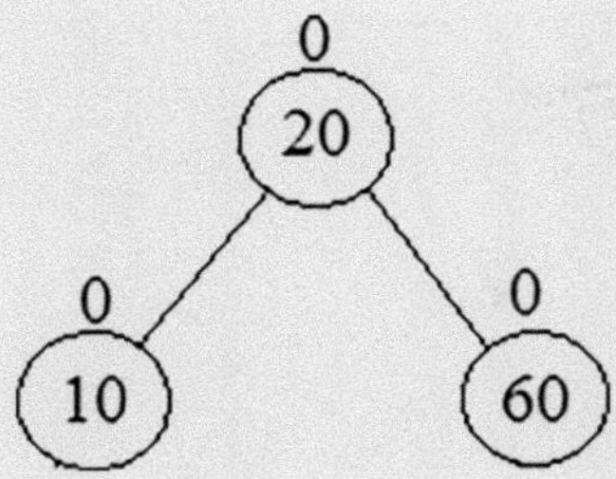

4. Insert 30. No rebalancing is required:

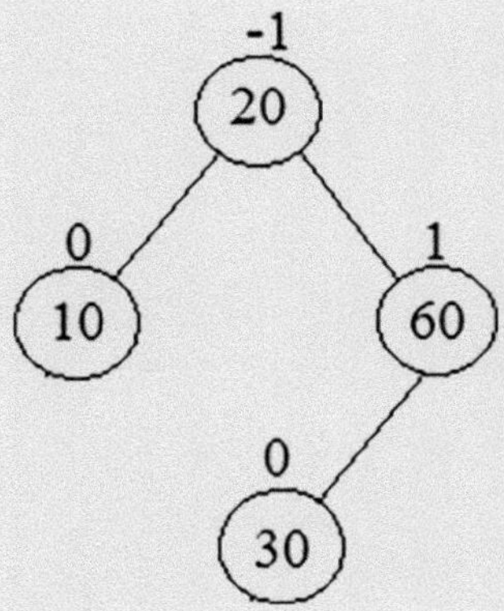

5. Insert 19. No rebalancing is required:

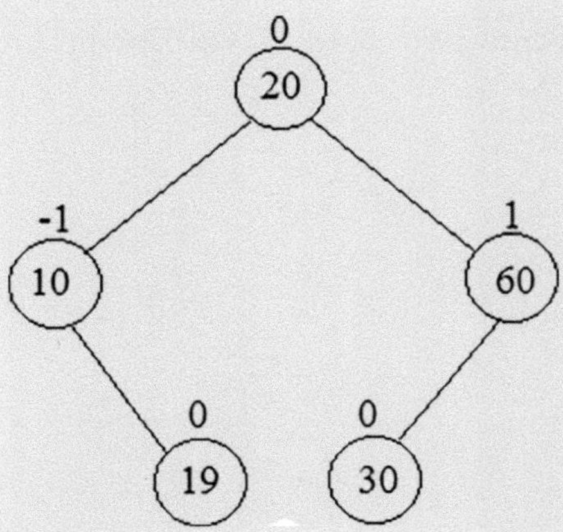

6. Insert 120. No rebalancing is required:

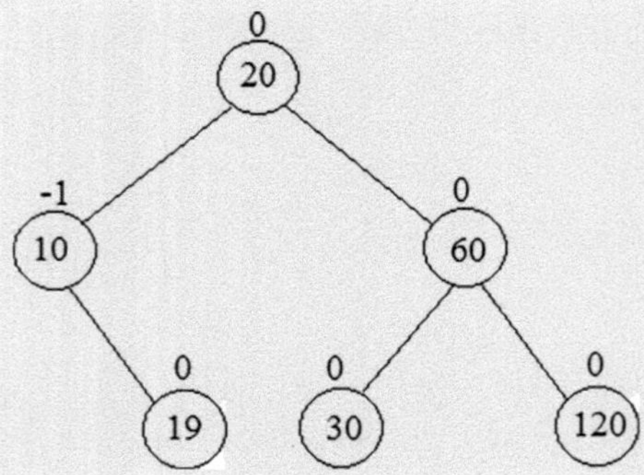

7. Insert 100. No rebalancing is required:

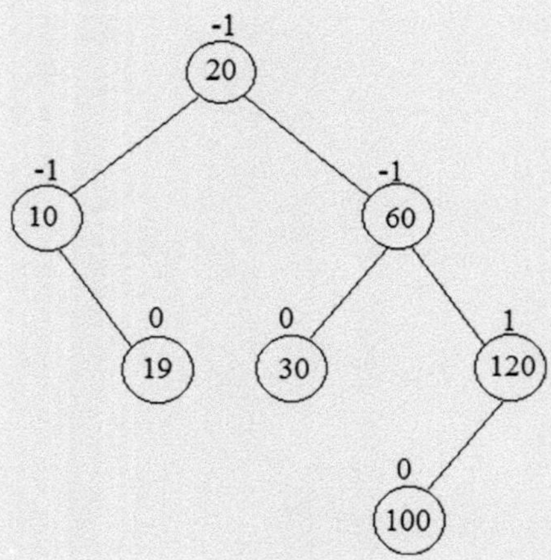

8. Insert 80. Now, rebalancing is required. We will perform LL rotation:

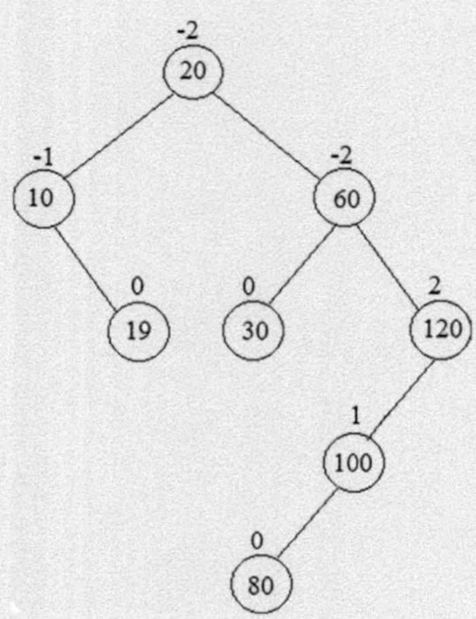

After performing LL rotation, the AVL tree is as follows:

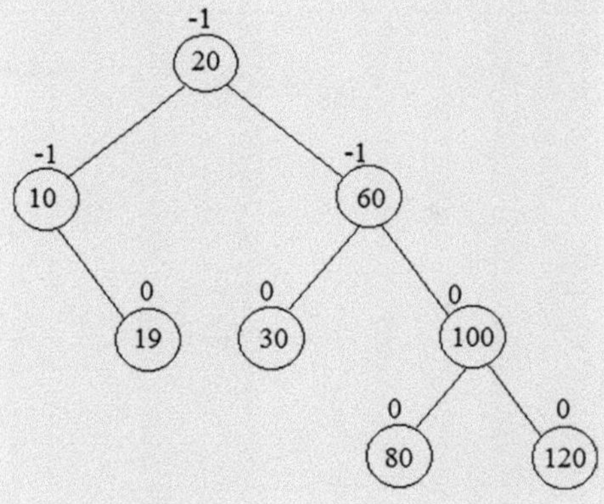

9. Insert 19. Now, rebalancing is required. We will perform RR rotation:

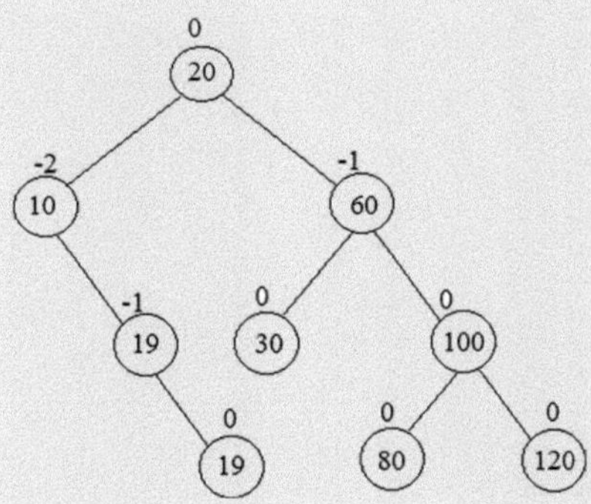

After performing RR rotation, the AVL tree is as follows:

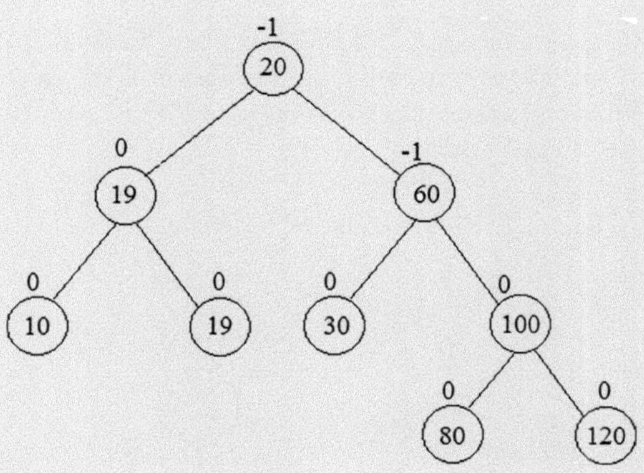

LR Rotation

LR rotation is also known as a left-right rotation, as the new node is inserted in the right subtree of the left subtree of the critical node. It is also a double rotation. Let us take an example and perform an LR rotation in it.

Initially, the AVL tree is as shown in Figure 8.56:

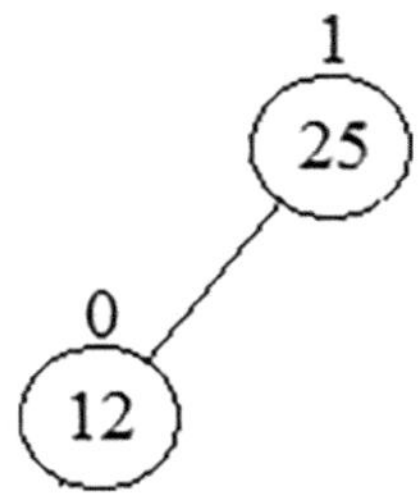

FIGURE 8.56 Sample AVL tree.

Let's insert a new node, 15, in the AVL tree, as shown in Figure 8.57.

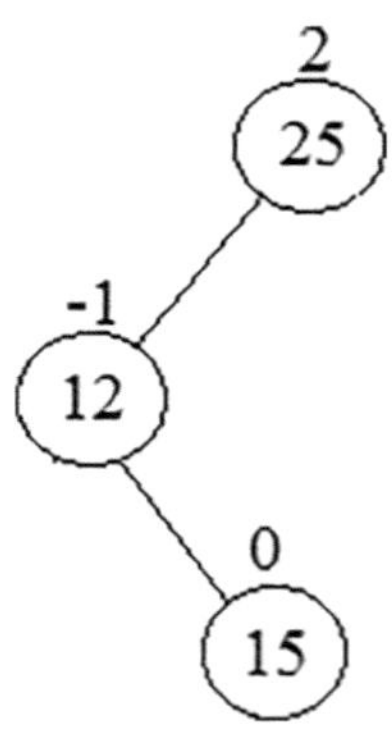

FIGURE 8.57 Insertion of key as right most child resulting to skewed tree.

After inserting 15, the balance factor of 25 is disturbed. Thus, 25 is the critical node. Hence, we will apply an LR rotation to restore the balance factor of the tree. After rotation, node 15 becomes the root node, and nodes 12 and 25 become the left and the right child of the tree, respectively, as shown in Figure 8.58.

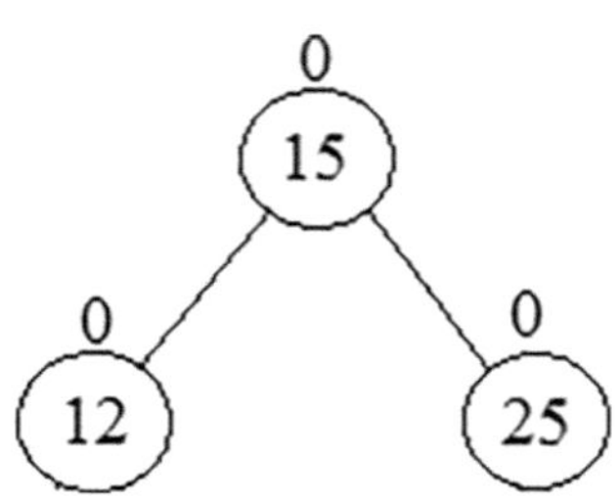

FIGURE 8.58 LR rotation in an AVL tree.

Therefore, an LR rotation is performed, and the balance factor of each node is also restored.

SUMMARY

- A tree is defined as a collection of one or more nodes, where one node is designated as a root node, and the remaining nodes can be partitioned into the left and the right subtrees. It is used to store hierarchical data.
- The root node is the topmost node of the tree. It does not have a parent node. If the root node is empty, then the tree is empty. A leaf node is one that does not have any child nodes.
- A path is a unique sequence of consecutive edges that is required to be followed to reach the destination from a given source.
- The degree of a node is equal to the number of children that a node has.
- A binary tree is a collection of nodes where each node contains three parts: a left pointer, a right pointer, and the data item. A binary tree can have at most two children; that is, a parent can have either zero, one, or at most two children.
- There are two types of binary trees: complete binary trees and extended binary trees.
- In a complete binary tree, every level except the last one must be completely filled. Also, all the nodes in the complete binary tree must appear to the left as much as possible.
- Extended binary trees are also known as 2T-trees. A binary tree is said to be an extended binary tree if and only if every node in the binary tree can have either zero or two children.
- Binary trees can be represented in memory in two ways: array representation and linked representation. Array representation, also known as sequential representation, is done using one-dimensional arrays. Linked representation is done using linked lists.
- A binary search tree is a variant of a binary tree in which all the nodes in the left subtree have a value less than that of the root node. Similarly, all the nodes in the right subtree have a value more than that of the root node. It is also known as an ordered binary tree.
- Searching is one of the most common operations performed in the binary search tree. This operation is performed to find whether a particular key exists in the tree or not.
- An insertion operation is performed to insert a new node with the given value in a binary search tree.
- The mirror image of the binary search tree means interchanging the right subtree with the left subtree at every node of the tree.
- Traversing is the process of visiting each node in the tree exactly once in a particular order. A tree can be traversed in various ways: pre-order traversal, in-order traversal, and post-order traversal.
- The term *pre* in *pre-order* determines that the root node is accessed before accessing any other node in the tree. Hence, it is also known as a DLR traversal, that is, *Data Left Right*.
- The term *in* in *in-order* determines that the root node is accessed between the left and the right subtrees. Hence, it is also known as an LDR traversal, that is, *Left Data Right*.
- The term *post* in *post-order* determines that the root node will be accessed last after the left and right subtrees. Hence, it is also known as an LRD traversal, that is, *Left Right Data*.
- A binary tree can be constructed if we are given at least two of the traversal results; one traversal should always be an in-order traversal, and the second can be either a pre-order traversal or a post-order traversal.
- An AVL tree is a self-balancing binary search tree. Every node in the AVL tree has a balance factor associated with it. The balance factor is calculated by subtracting the height of the right subtree from the height of the left subtree. Thus, a node with a balance factor of –1, 0, or 1 is said to be a height-balanced tree.

EXERCISES

Theory Questions

1. What is a tree? Discuss its various applications.
2. Differentiate between *height* and *level* in a tree.
3. Explain the concept of binary trees.
4. In what ways can a binary tree be represented in the computer's memory?
5. What do you understand by 2T trees? Explain.
6. What do we mean by a binary search tree?
7. List the various operations performed on binary search trees.
8. What are complete binary trees?
9. How can a node be deleted from a binary search tree? Discuss all the cases in detail with an example.
10. Create a binary search tree by inserting the following keys: 76, 12, 56, 31, 199, 17, 40, 76, 75. Also, find the height of the binary search tree.
11. Create a binary search tree by performing the following operations:
    a) Insert 50, 34, 23, 87, 100, 67, 43, 51, 18, and 95.
    b) Delete 100, 34, 95, and 50 from the binary search tree.
    c) Find the smallest value in the binary search tree.
12. How can we find the mirror image of a binary search tree?
13. List the various traversal methods of a binary tree.
14. What do you understand by an AVL tree?
15. Explain the concept of the balance factor in AVL trees.
16. List the advantages of an AVL tree.
17. Consider the following binary search tree and perform the following operations:
    a) Find the pre-order and post-order traversals of the tree.
    b) Insert 25, 32, 50, 75, and 87 in the tree.
    c) Find the largest value in the tree.
    d) Delete the root node.

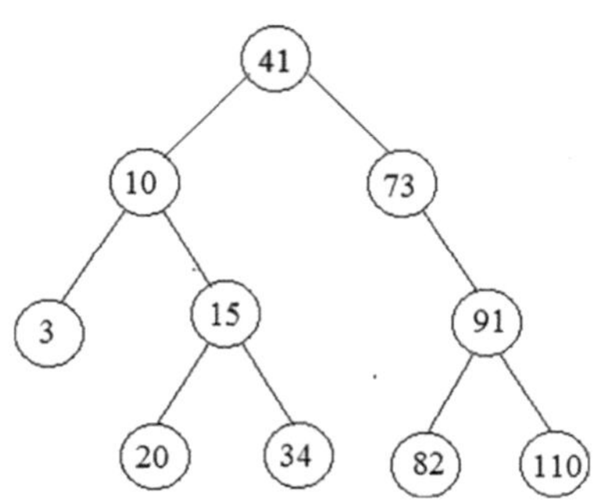

18. Give the linked representation of the binary search tree given above.

19. Construct a binary search tree of the word VIVEKANANDA. Find its pre-order, in-order, and post-order traversals.

20. Create an AVL tree by inserting the following keys: 50, 19, 59, 90, 100, 12, 10, and 150 into the tree.

21. Consider the following AVL search tree and perform these operations in it:

a) Insert 100, 58, 93, 40, and 7 into the tree.

b) Search for 93 in the AVL tree.

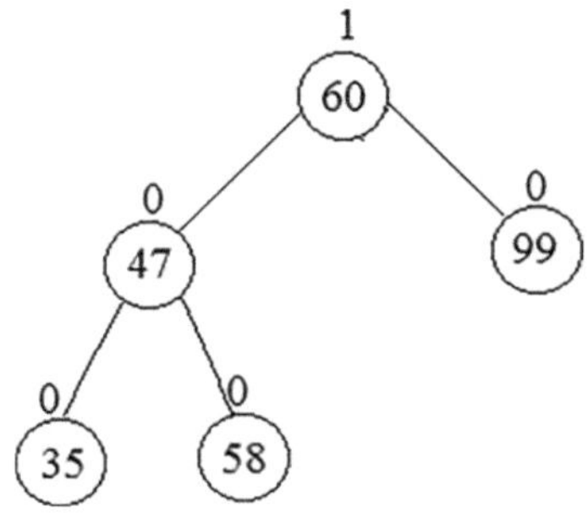

22. Discuss the various types of rotations performed in AVL trees.

23. Which are better, AVL trees or binary search trees? Why?

24. Consider the given tree and determine the following:

a) Height of the tree

b) Names of the leaf nodes

c) Siblings of C

d) Level number of node *J*

e) Root node of the tree

f) Left and right subtrees

g) Depth of the tree

h) Ancestors of *E*

i) Descendants of *H*

j) Path from node *A* to *F*

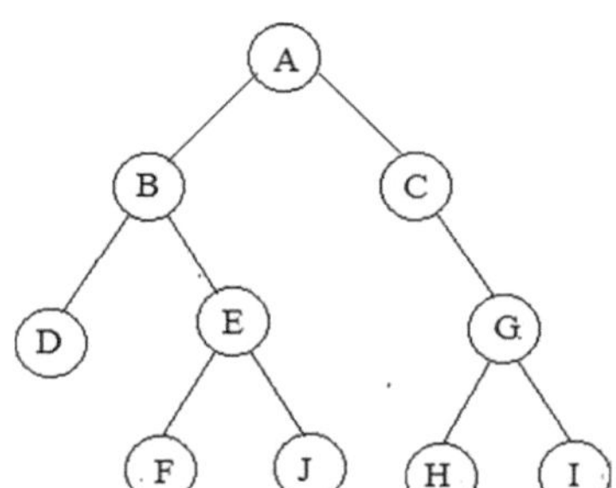

Programming Questions

1. Write a function to find the height of the binary search tree.
2. Write a program to insert and delete nodes from the binary search tree.
3. Write a program to show insertion in AVL trees.
4. Write a function to calculate the total number of nodes in the tree.
5. Write a program to traverse a binary search tree, showing all the traversal methods.
6. Write a function to find the largest value in the binary search tree.
7. Write an algorithm showing post-order traversal of a binary search tree.
8. Write an algorithm to find the total number of internal nodes in the binary search tree.
9. Write a function to search for a node in the binary search tree.

Multiple Choice Questions

1. The maximum height of a binary tree with n nodes is ______.
   - **A.** 0
   - **B.** n
   - **C.** $n+1$
   - **D.** $n-1$
2. The degree of a terminal node is always ______.
   - **A.** 1
   - **B.** 2
   - **C.** 0
   - **D.** 3
3. A binary tree is a tree in which ______.
   - **A.** Every node must have two children.
   - **B.** Every node must have at least two children.
   - **C.** No node can have more than two children.
   - **D.** All of these.
4. What is the post-order traversal of the binary search tree having pre-order traversal as DBAEFGCH and in-order traversal as BEAFDCHG?
   - **A.** EFBAHGCD
   - **B.** EFBAHCGD
   - **C.** EFABHGCD
   - **D.** EFABHCGD
5. How many rotations are required during the construction of an AVL tree if the following keys are to be added in the order given?

   36, 51, 39, 24, 29, 60, 79, 20, 28
   - **A.** 3 left rotations, 3 right rotations
   - **B.** 2 left rotations, 2 right rotations
   - **C.** 2 left rotations, 3 right rotations
   - **D.** 3 left rotations, 2 right rotations

6. A binary tree of height h has at least h nodes and at most ______ nodes.
   - A. 2
   - B. 2^h
   - C. $2^h - 1$
   - D. $2^h + 1$
7. How many distinct binary search trees can be created out of four distinct keys?
   - A. 5
   - B. 12
   - C. 14
   - D. 23
8. Nodes at the same level that also share the same parent are called ________.
   - A. Cousins
   - B. Siblings
   - C. Ancestors
   - D. Descendants
9. The balance factor of a node is calculated by _______.
   - A. $Height_{Left\ subtree} - Height_{Right\ subtree}$
   - B. $Height_{Right\ subtree} - Height_{Left\ subtree}$
   - C. $Height_{Left\ subtree} + Height_{Right\ subtree}$
   - D. $Height_{Right\ subtree} + Height_{Left\ subtree}$
10. The following sequence is inserted into an empty binary search tree:

    6 11 26 12 5 7 16 8 35

    What type of traversal is given by the following?

    6 5 11 7 26 8 12 35 16
   - A. Pre-order traversal
   - B. In-order traversal
   - C. Post-order traversal
   - D. None of these
11. In tree creation, which one will be the most suitable and effective data structure?
   - A. Stack
   - B. Linked list
   - C. Queue
   - D. Array
12. What can a binary tree be represented with?
   - A. Linked list
   - B. Array
   - C. Both of the above
   - D. None of the above
13. A binary tree of n nodes has exactly n+1 edges. True or false?
   - A. True
   - B. False

14. The in-order traversal of a tree will yield a sorted listing of the elements of the tree in __________.

A. Binary heaps
B. Binary trees
C. Binary search trees
D. All of these

15. Which is the nearest ancestor node on the path from the root node to the newly inserted node of the AVL tree having a balance factor of –1, 0, or 1?

A. Parent node
B. Child node
C. Root node
D. Critical node

CHAPTER 9

MULTI-WAY SEARCH TREES

INTRODUCTION

We have already studied binary search trees, in which we discussed how every node contains three parts: an information part and two pointers (left and right), which respectively point to the left and right subtrees. The same concept is used for *multi-way* (*M-way*) search trees. An M-way search tree is a tree that contains $(M - 1)$ values per node. It also has M subtrees. In an M-way search tree, M is called the degree of the node. For example, if the value of $M = 3$ in an M-way search tree, then the tree will contain two values per node, and it will have three subtrees. When an M-way search tree is not empty, it has the following properties:

- Each node in an M-way search tree is of the following structure, where P_0, P_1, P_2, ..., and P_n are the pointers to the node's subtrees, and K_0, K_1, K_2,..., and K_n are the key values stored in the node:

n	P_0	K_0	P_1	K_1	P_2	K_2 _ _ _ _ _ _	P_{n-1}	K_{n-1}	P_n

- The key values in a node are stored in ascending order, that is, $K_i < K_{i+1}$, where $i = 0, 1, 2, \ldots$, and $n-2$.
- All the key values stored in the left subtree are always less than the root node.
- All the key values stored in the right subtree are always greater than the root node.
- The subtrees pointed to by P_i for $i = 0, 1, 2, \ldots$, and n are also M-way search trees.

Figure 9.1 shows an example of an M-way search tree:

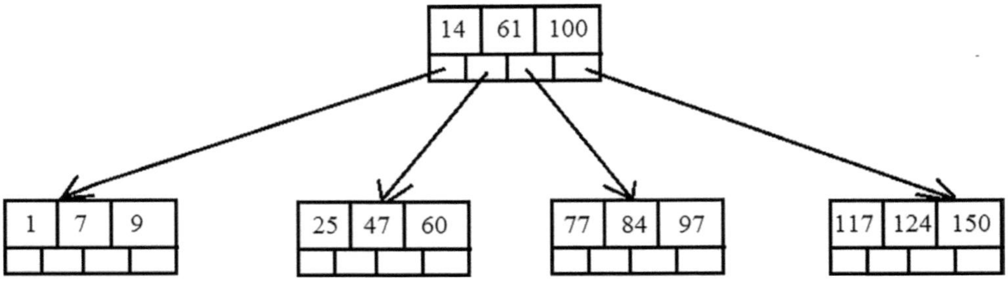

FIGURE 9.1 M-way search tree of order 4.

B-TREES

A *B-tree* is a specialized M-way tree that is widely used for disk access. The B-tree was developed in 1970 by Rudolf Bayer and Ed McCreight. In a B-tree, each node may contain a large number of keys. A B-tree is designed to store a large number of keys in a single node so that the height remains relatively small. A B-tree of order m has all the properties of an M-way search tree. In addition, it has the following properties:

- All leaf nodes are at the bottom level or at the same level.
- Every node in a B-tree can have at most m children.
- The root node can have at least two children if it is not a leaf node, and it can have no children if it is a leaf node.
- Each node in a B-tree can have at least $(m/2)$ children, except the root node and the leaf node.
- Each leaf node must contain at least *ceil* $[(m/2) - 1]$ keys.

Let's look at an example, as shown in Figure 9.2. A B-tree of order 4 can have at least *ceil* $[4/2] = 2$ children and *ceil* $[(4/2) - 1] = 1$ key. The maximum number of children a node can have is four . Each leaf node must contain at least one key.

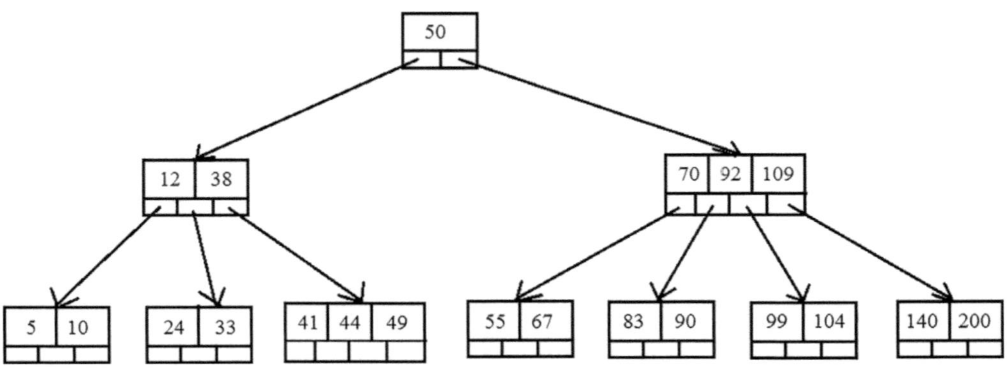

FIGURE 9.2 B-tree of order 4.

Practical Application:

In database programs, the data is too large to fit in memory; therefore, it is stored in secondary storage, that is, Hard Disk Drive (HDD) or Solid State Drive (SSD).

OPERATIONS ON A B-TREE

A B-tree stores sorted data, and we can perform various operations on it:

- Inserting a new element in a B-tree
- Deleting an element from a B-tree

So, let's discuss both of these operations in detail.

Insertion in a B-Tree

First of all, insertions in a B-tree are done at the leaf-node level. The following are the steps to insert an element in a B-tree:

1. First, we will search the B-tree to find the leaf node where the new value or key is to be inserted.
2. If the leaf node is full, that is, it already contains $(m-1)$ keys, then follow these steps:
   a) Insert the new key into the existing set of keys in order.
   b) Now, the node is split into two halves.
   c) Finally, push the middle (median) element upward to its parent node. Also, if the parent node is full, then split the parent node by following these steps.
3. If the leaf node is not full, that is, it contains less than $(m-1)$ keys, then insert the new key into the node, keeping the elements of the node in alphabetical or numerical order.

Frequently Asked Questions

Q1. Construct a B-tree of order 5 and insert the following values into it:

B, N, G, A, H, E, J, Q, M, D, V, L, T, Z

Answer.

1. Since *order* = 5, we can store at least 3 values and at most 4 values in a single node. Hence, we will insert *B*, *N*, *G*, and *A* into the B-tree in sorted order:

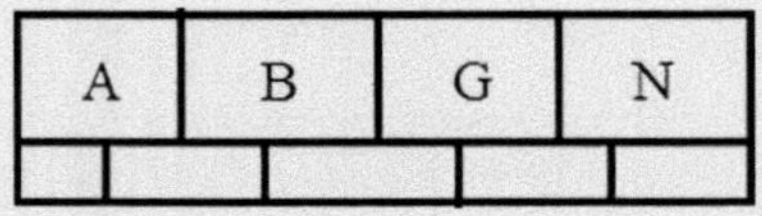

2. Now, *H* is to be inserted between *G* and *N*, so now the order will be *A B G H N*, which is not possible, as at most four values can be accommodated in a single node. So, now, we will split the node, and the middle element *G* will become the root node:

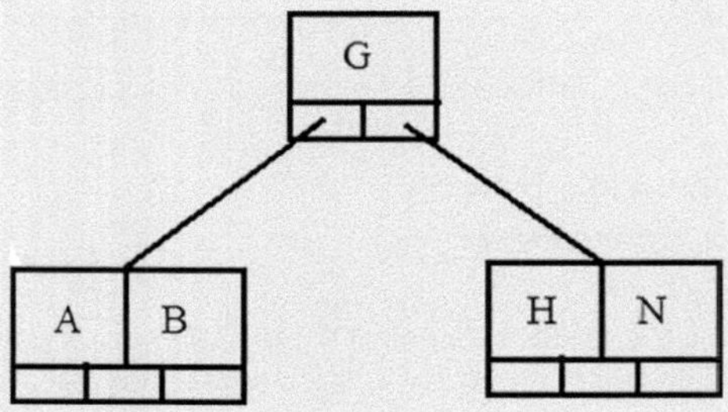

3. Now, we will insert *E*, *J*, and *Q* into the B-tree:

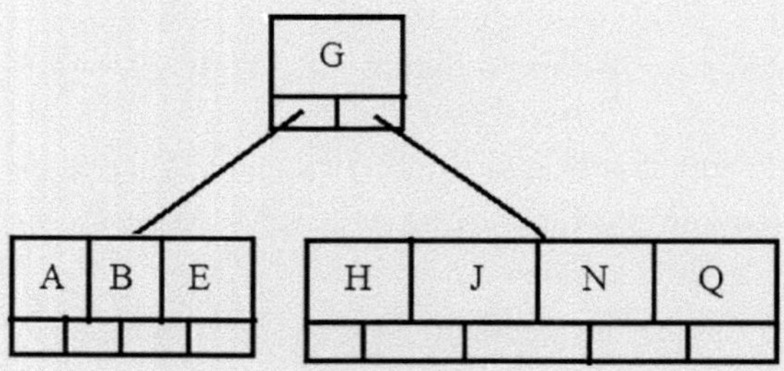

4. *M* is to be inserted into the right subtree, but at most four values can be stored in the node, so now we will push the middle element, *M*, into the root node. Thus, the node is split into two halves:

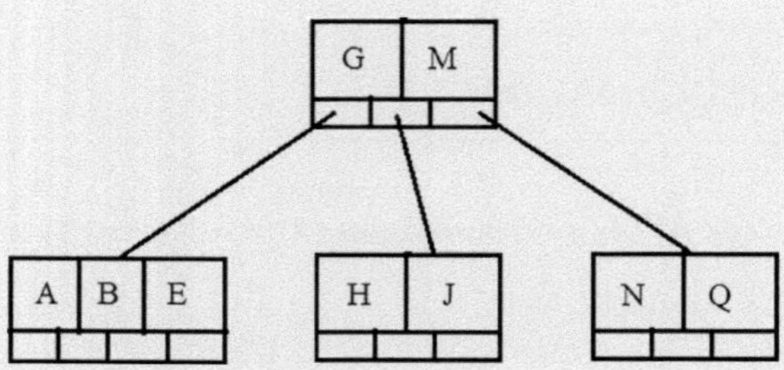

5. Now, we will insert *D*, *V*, *L*, and *T* into the tree:

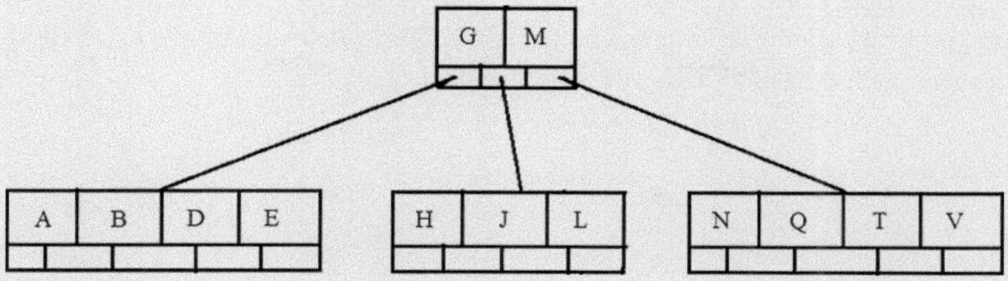

6. Finally, *Z* is to be inserted. It will be inserted into the right subtree. Hence, the last node will split into two halves, and the middle element, *T*, will push up to the root node:

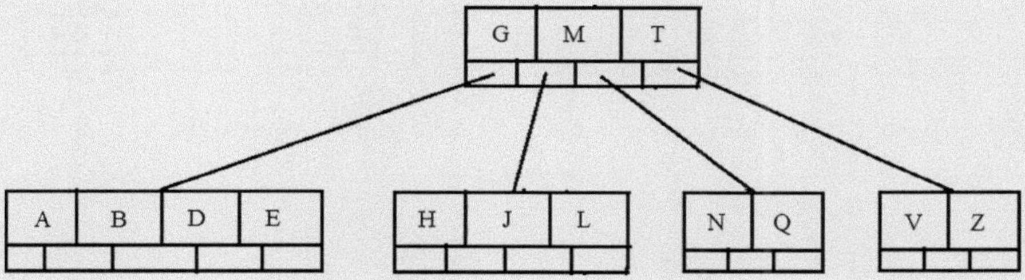

Deletion from a B-Tree

Deletion of keys in a B-tree also first requires traversal in the B-tree. After reaching a particular node, we can come across two cases:

- The node is a leaf node.
- The node is not a leaf node.

Now, let us discuss both these cases in detail.

Node Is a Leaf Node

In this case, if the node has more than the minimum number of keys, then deletion can be done very easily. If the node has a minimum number of keys, then, first, we will check the number of keys in the adjacent leaf node. If the number of keys in the adjacent node is more than the minimum number of keys, then the first key of the adjacent leaf node will go to the parent node, and the key present in the parent node will be combined together in a single node. Now, if the parent node also has less than the minimum number of keys, then the same steps will be repeated until we get a node that has more than the minimum number of keys present in it.

Node Is Not a Leaf Node

In this case, the key from the node is deleted, and its place will be occupied by either its successor or predecessor key. If both the predecessor and successor nodes have keys less than the minimum number, then the keys of the successor and predecessor are combined.

As an example, let's consider a B-tree of order 5:

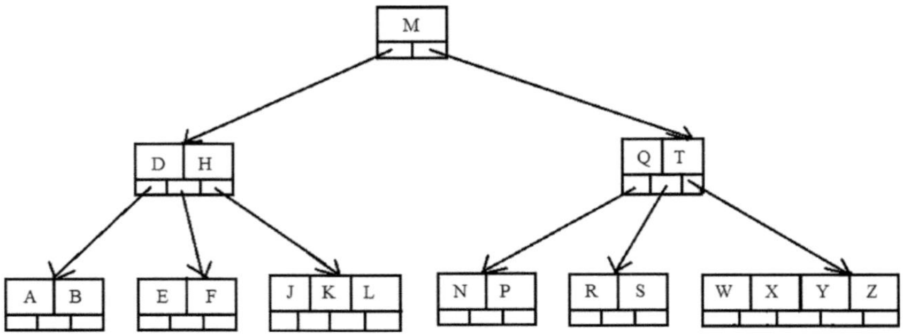

First, delete *J* from the tree. *J* is in the leaf node, so it is simply deleted from the B-tree:

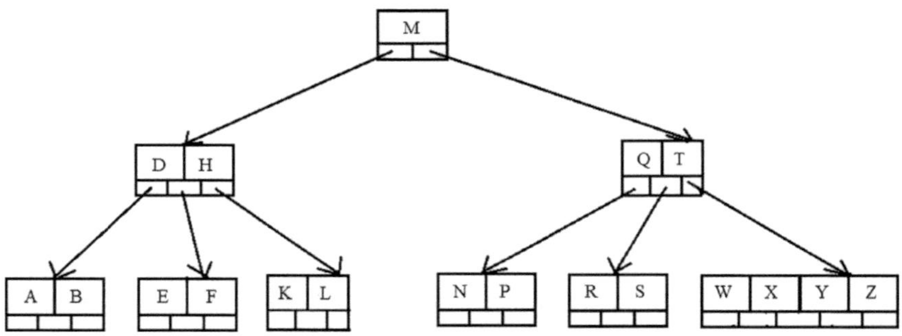

Now *T* is to be deleted, but it is not the leaf node, so we will replace *T* with its successor, *W*. Hence, *T* is deleted:

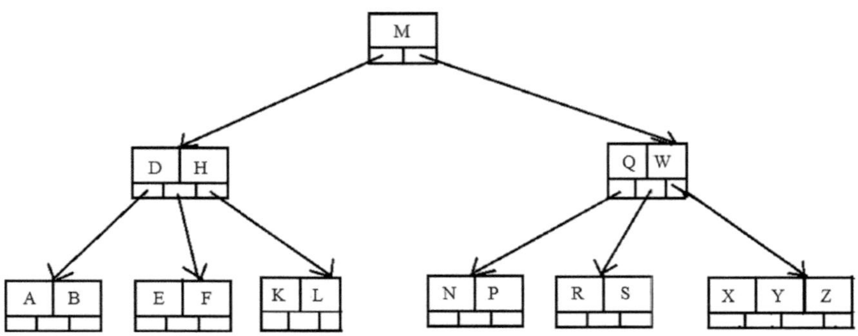

Now, delete *R*; in this case, we will borrow keys from the adjacent leaf node:

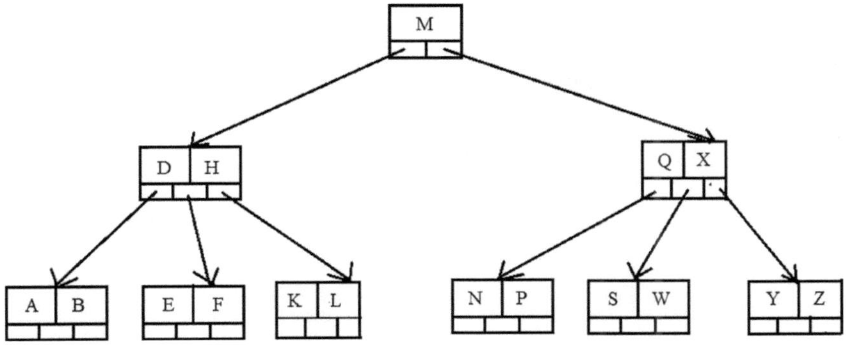

Now we want to delete *E*. We will also borrow keys from an adjacent node here. We can see that there are no free keys in an adjacent node, so the leaf node has to be combined with one of the two siblings. This includes moving down the parent's key that was between those two leaves:

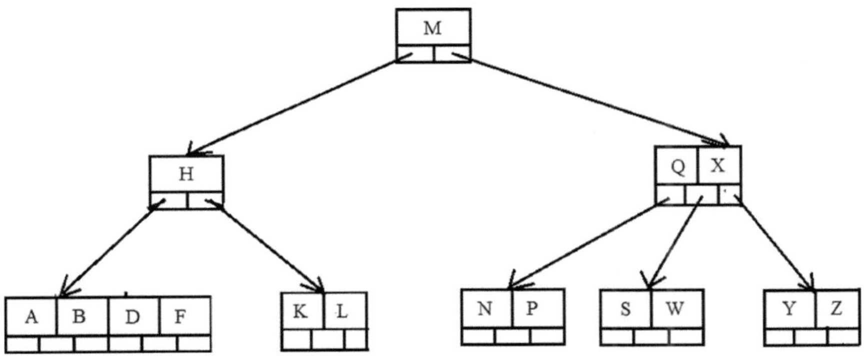

We can see that *H* is still unstable according to the definition. Therefore, the final tree after all deletions is shown as follows:

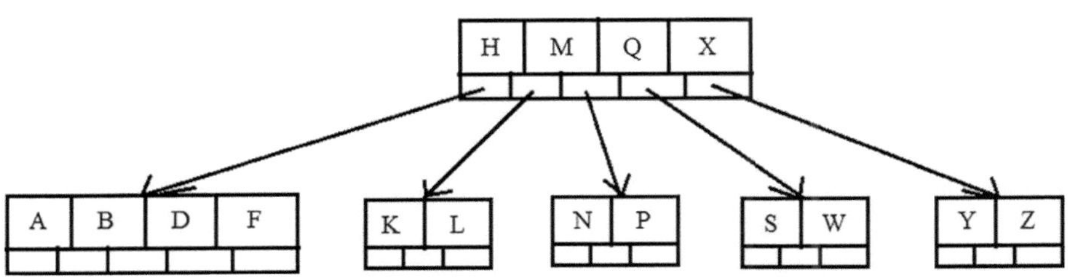

Frequently Asked Questions

Q2. Consider the following B-tree of order 5 and insert 81, 7, 49, 61, and 30 into it:

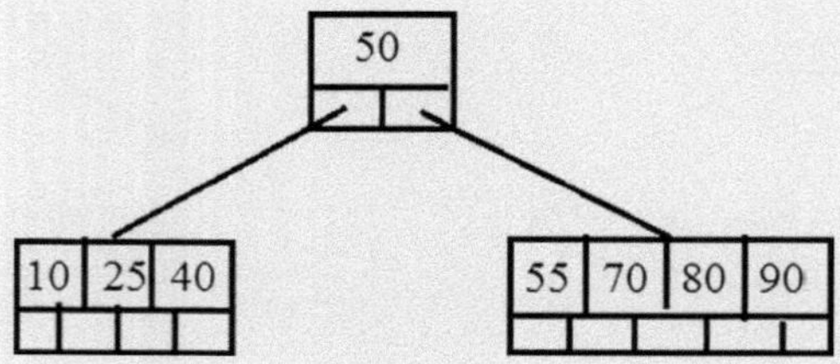

Answer.

1. Insert 81:

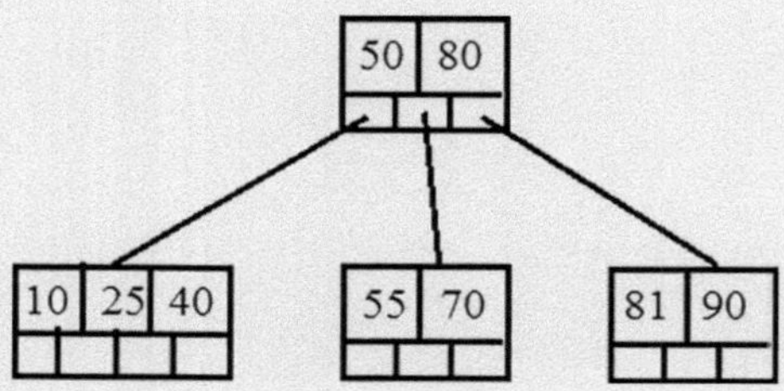

2. Insert 7 and 49:

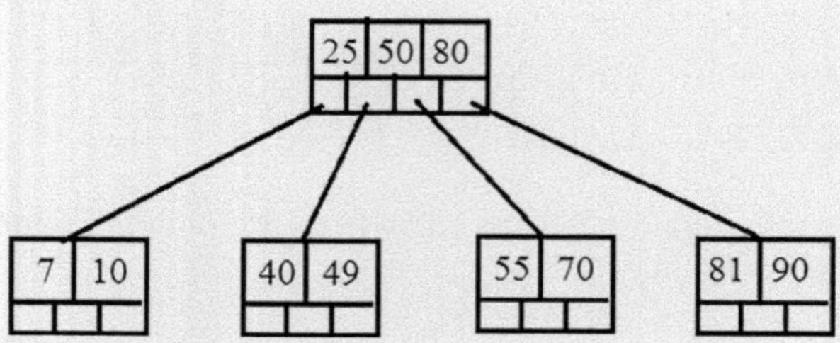

3. Insert 61 and 30:

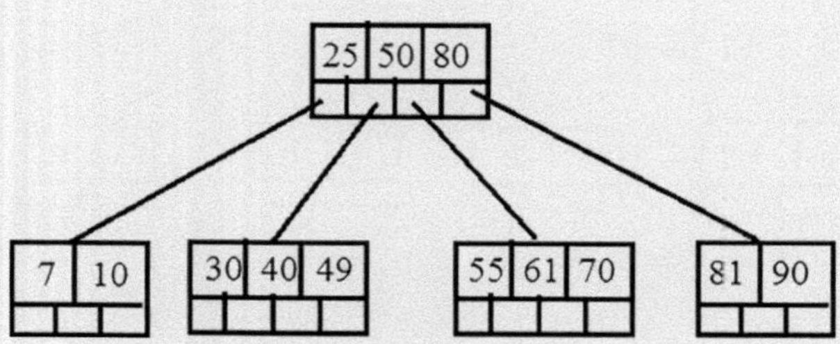

Frequently Asked Questions

Q3. Consider the following B-tree of order 5 and delete the values 95, 200, 176, and 70 from it:

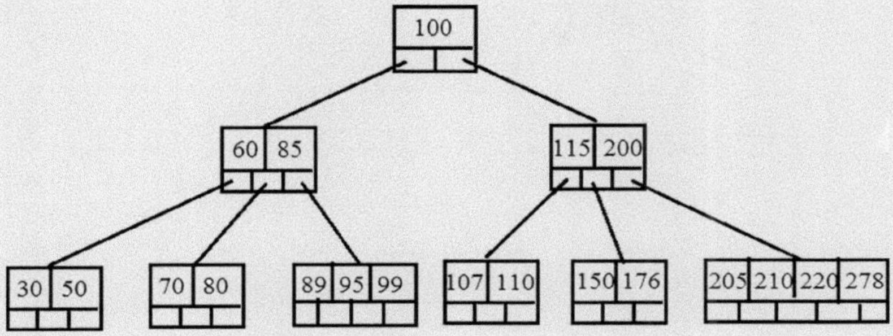

Answer.

1. Delete 95:

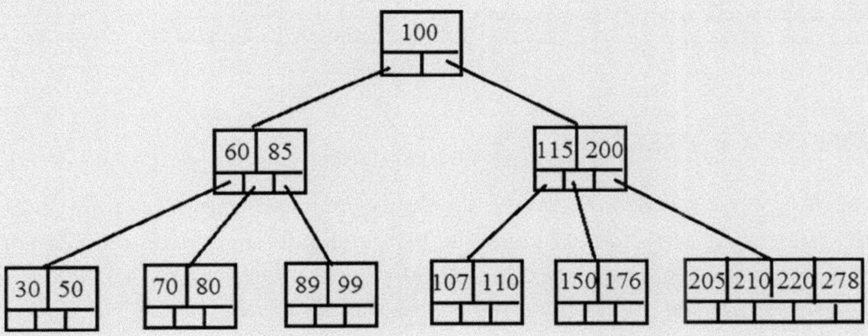

2. Delete 200:

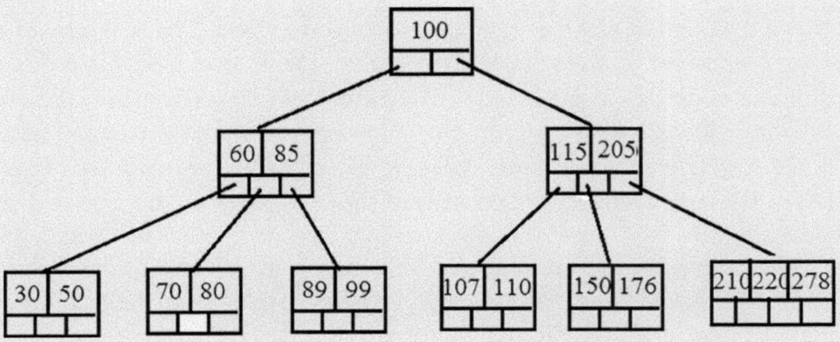

3. Delete 176:

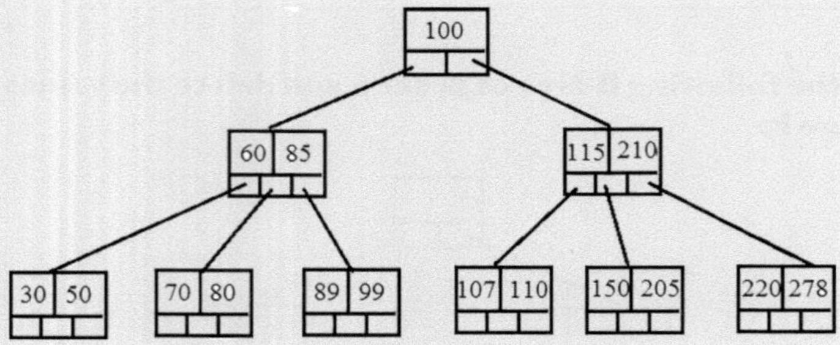

4. Delete 70:

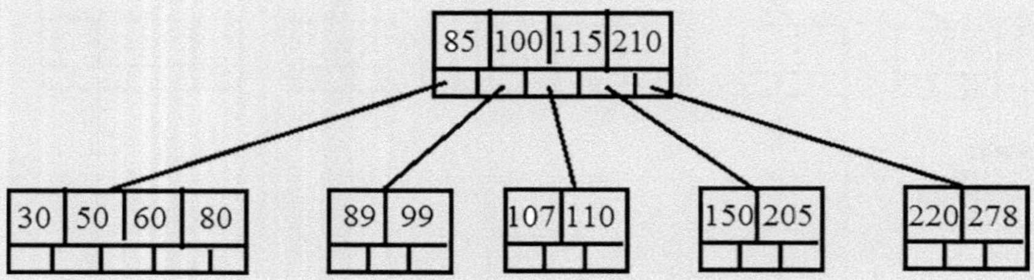

APPLICATION OF A B-TREE

The main application of a B-tree is the organization of a large amount of data or a huge collection of records into a file structure. A B-tree should search the records very efficiently, and all the operations (such as insertion, deletion, searching, and so on) should be done very efficiently; therefore, the organization of records should be very good.

B+ TREES

A *B+ tree* is a variant of a B-tree that also stores sorted data. The structure of a B-tree is the standard organization for indexes in database systems. Multilevel indexing is done in a B+ tree; that is, leaf nodes constitute a dense index while non-leaf nodes constitute a sparse index. A B+ tree has a slightly different data structure that allows sequential processing of data and stores all the data in the lowest level of the tree. A B-tree can store both records and keys in its interior nodes, while a B+ tree stores all the records in the leaf nodes of the tree and the keys in the interior nodes. In a B+ tree, the leaf nodes are linked to one another like a linked list. It helps in making queries simpler and more efficient. A B+ tree is usually used to store large amounts of data that cannot be stored in the main memory. Hence, in a B+ tree, the leaf nodes are stored in the secondary storage while the internal nodes are stored in the main memory.

In a B+ tree, all the internal nodes are called *index nodes* because they store the index values. Similarly, all the external nodes are called *data nodes* because they store the keys. A B+ tree is always balanced and is very efficient for searching data, as all the data is stored in the leaf nodes. The advantages of a B+ tree are as follows:

- A B+ tree is always balanced, and the height of the tree always remains less than equivalent Binary Search Tree and hence facilitating fast search operation
- All the leaf nodes are linked to one another, which makes it very efficient.
- The leaf nodes are also linked to the nodes at an upper level; thus, it can be easily used for a wide range of search queries.
- The records can be fetched with an equal number of disk accesses.
- The records can be accessed either sequentially or randomly.
- Searching data becomes very easy as all the data is stored only in the leaf nodes.
- Similarly, deletion is also very simple as it will only take place in the leaf nodes.

Figure 9.3 shows an example of a B+ tree:

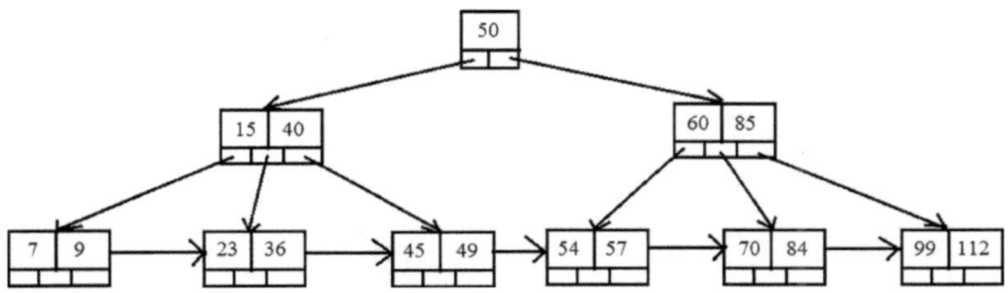

FIGURE 9.3 B+ tree of order 3.

SUMMARY

- An M-way search tree has $M - 1$ values per node and M subtrees. M is called the order of the node.
- A B-tree is a specialized multi-way tree that is widely used for disk access. B-trees were developed in 1970 by Rudolf Bayer and Ed McCreight.
- A B-tree of order M has all the properties of a multi-way search tree.
- The main application of a B-tree is the organization of a large amount of data or a huge collection of records into a file structure.
- A B+ tree is a variant of a B-tree that also stores sorted data. The structure of a B-tree is the standard organization for indexes in database systems. A B+ tree has a slightly different data structure that allows the sequential processing of data and stores all the data in the lowest level of the tree.

EXERCISES

Review Questions

1. Define the following:
   a) M-way search tree
   b) B-tree
   c) B+ tree
2. Describe the difference between B-trees and B+ trees.
3. Give some important applications of a B-tree.

4. Write down some advantages of a B+ tree over a B-tree.
5. Create a B-tree of order 5 on inserting the keys 10, 20, 50, 60, 40, 80, 100, 70, 130, 90, 30, 120, 140, 25, 35, 160, and 180 in a left-to-right sequence. Show the trees after deleting 190 and 60.
6. Explain the insertion of a node in a B-tree.
7. Explain B+ tree indexing with the help of an example.
8. What do you know about B-trees? Write the steps to create a B-tree. Construct an M-way search tree of order 4 and insert the values 34, 45, 98, 1, 23, 41, 78, 100, 234, 122, 199, 10, and 40.
9. Why do we always prefer a higher value of m in a B-tree? Explain.
10. Are B-trees of order 2 full binary trees? Explain.

Multiple Choice Questions

1. Why are B+ trees preferred to binary trees in databases?
   - A. Disk capacities are greater than memory capacities.
   - B. Disk access is much slower than memory access.
   - C. Disk data transfer rates are much less than the memory data transfer rates.
   - D. Disks are more reliable than memory.
2. In an M-way search tree, M stands for ________.
   - A. Degree of the node
   - B. External nodes
   - C. Internal nodes
   - D. None of these
3. A B-tree of order 4 is built. What is the maximum number of keys that a node may accommodate before splitting operations take place?
   - A. 5
   - B. 2
   - C. 4
   - D. 3
4. In a B-tree of order M, every node has at most ________ children.
   - A. $M + 1$
   - B. M − 1
   - C. $M/2$
   - D. M
5. What is the best data structure to search the keys in less time?
   - A. B-tree
   - B. M-way search tree
   - C. B+ tree
   - D. Binary search tree

6. What is the best case of searching for a value in a binary search tree?
   A. $O(n^2)$
   B. $O(log\ n)$
   C. $O(n)$
   D. $O(n\ log\ n)$
7. External nodes are also called ________.
   A. Index nodes
   B. Data nodes
   C. Value nodes
   D. None of the above
8. A B-tree of order 5 can store, at most, how many keys?
   A. 3
   B. 4
   C. 5
   D. 6
9. Does a B+ tree store redundant keys?
   A. No
   B. Yes
10. A B-tree of order 3 can store at least how many keys?
   A. 0
   B. 1
   C. 2
   D. 3

CHAPTER 10

HASHING

INTRODUCTION

In previous chapters, we have discussed three types of searching techniques: linear search, binary search, and interpolation search. Linear search has a running time complexity of $O(n)$, whereas binary search has a running time proportional to $O(log\ n)$, where n is equal to the number of elements in the array. Although the searching algorithms discussed in Chapter 6 are efficient, their search time is dependent on the number of elements in the array, and none of them can search for an element within a constant time equal to $O(1)$. It is very difficult to achieve in all the searching algorithms, such as linear search, binary search, and so on, as all these algorithms are dependent on the number of elements present in the array. Also, there are many comparisons involved while searching for an element using the previous searching algorithms. Therefore, our primary need is to search for the element in a constant time, along with fewer key comparisons. Now, let us take an example. Suppose there is an array of size N, and all the keys to be stored in the array are unique and are in the range 0 to $N-1$. Now, we will store all the records in the array based on the key, where the array index and the keys are the same. Thus, we can access the records in a constant time, with no key comparisons involved. This can be explained by the example shown in Figure 10.1:

arr[0]	arr[1]	arr[2]	arr[3]	arr[4]	arr[5]	arr[6]	arr[7]	arr[8]	arr[9]
	1		3	4		6		8	

FIGURE 10.1 An array.

In the previous figure, there is an array containing five elements. Note that the keys and the array index are the same; that is, the record with the key value `3` can be directly accessed by array index `arr[3]`. Similarly, all the records can be accessed through key values and the array index. This can be done by *hashing*, where we will convert the key into an array index and store the records in the array, as shown in Figure 10.2:

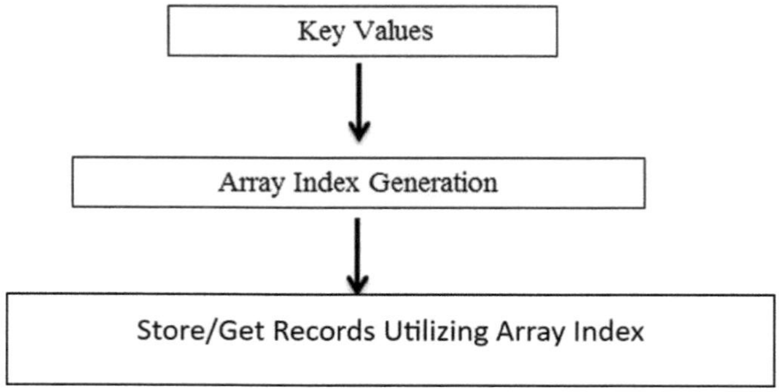

FIGURE 10.2 Array index generation using hashing.

The process of array index generation uses a hash function, which is used to convert the keys into an array index. The array in which such records are stored is known as a *hash table*.

Practical Application:

A simple real-life example is when we search for a word in the dictionary; then, we find the definition or meaning with the help of a key and its index.

Driver's license numbers and insurance card numbers are created using hashing from data items that never change, that is, date of birth, name, and so on.

Frequently Asked Questions

Q1. Explain the term hashing.

Answer.

Hashing is the process of mapping keys to their appropriate locations in the hash table. It is the most effective technique for searching the values in an array or in a hash table.

Difference Between Hashing and Direct Addressing

In *direct addressing*, we store the key at the same address as the value of the key, as shown in Figure 10.3. In hashing, however, as shown in Figure 10.4, the address of the key is determined by using a mathematical function known as a *hash function*. The hash function will operate on the key to determine the address of the key. Direct addressing may result in a more random distribution of the key throughout the memory, and hence sometimes leads to more wastage of space when compared with hashing.

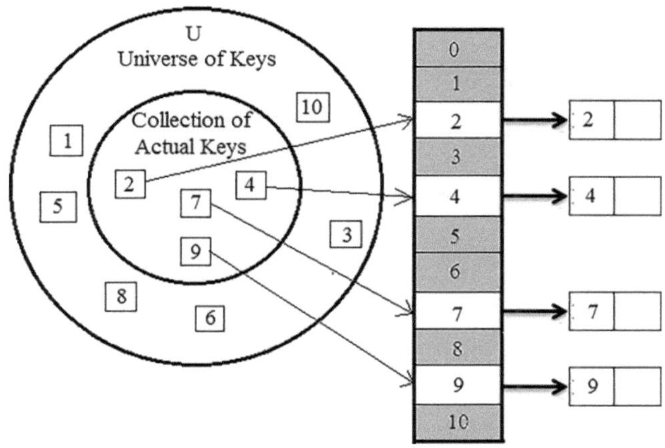

FIGURE 10.3 Mapping of keys using a direct addressing method.

Hash Tables

A hash table is a data structure that supports one of the most efficient searching techniques, that is, hashing. A hash table is an array in which the data is accessed through a special index called a key. In a hash table, keys are mapped to the array positions by a hash function. A hash function is a function or a mathematical formula that, when applied to a key, produces an integer that is used as an index to find a key in the hash table. Thus, a value stored in a hash table can be searched in $O(1)$ time with the help of a hash function. The main idea behind a hash table is to establish a direct mapping between the keys and the indices of the array.

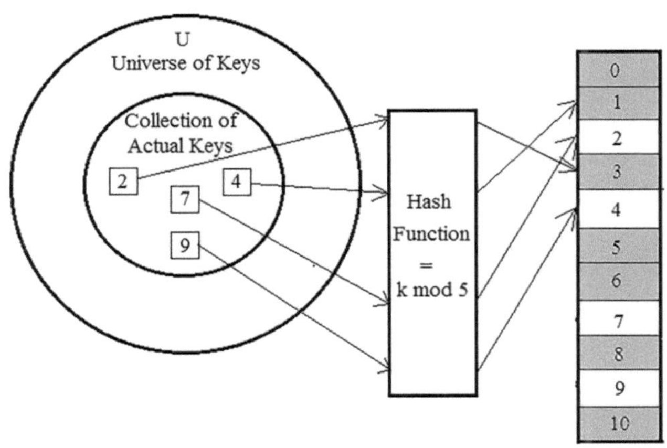

FIGURE 10.4 Mapping of keys to the hash table using hashing.

Hash Functions

A hash function is a mathematical formula used to generate address index within hash table using key to be stored.

There are four main characteristics of hash functions:

- They use all the input data.
- They must generate different hash values.
- They are fully determined by the data being hashed.
- They must distribute the keys uniformly across the entire hash table.

Different Types of Hash Functions

In this section, we will discuss some of the most common hash functions:

- *Division method* – In the division method, a key k is mapped into one of the m slots by taking the remainder of k divided by m. In simple terms, we can say that this method divides an integer, say x, by m and then uses the remainder obtained. It is the simplest method of hashing. The hash function is given by:

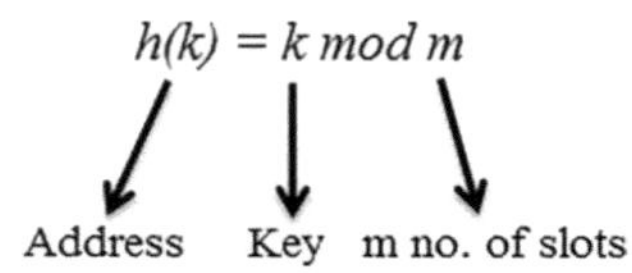

For example, if $m = 5$ and the key $k = 10$, then $h(k) = 2$. Thus, the division method works very fast as it requires only a single division operation. Although this method is good for any value of m, consider that if m is an even number, then $h(k)$ is even when the value of k is even, and similarly, $h(k)$ is odd when the value of k is odd. Therefore, if the even and odd keys are almost equal, then there will be no problem. If there is a larger number of even keys, then the division method is not good as it will not distribute the keys uniformly in the hash table. Also, we avoid certain values of m; that is, m should not be a power of 2, because if $h(k) = k \ mod \ 2^x$, then $h(k)$ will extract the lowest x bits of k. The main drawback of the division method is that many consecutive keys map to consecutive hash values respectively, which means that consecutive array locations will be occupied, and hence there will be an effect on the performance.

Frequently Asked Questions

Q2. Given a hash table of 50 memory locations, calculate the hash values of keys 20 and 75 using the division method.

Answer.

$m = 50$, $k1 = 10$, $k2 = 75$ hash values are calculated as:

$h(10) = 10 \% 50 = 10$

$h(75) = 75 \% 50 = 25$

- *Mid-square method* – In the mid-square method, we will calculate the square of the given key. After getting the number, we will extract some digits from the middle of that number as an address.

For example, if key $k = 5025$, then $k^2 = 25250625$. Thus, $h(5025) = 50$.

This method works very well as all the digits of the key contribute to the output, that is, all the digits contribute in producing the middle digits. Also in this method, the same r digits must be chosen from all the keys.

Frequently Asked Questions

Q3. Given a hash table of 100 memory locations, calculate the hash values of keys 2045 and 1357 using the mid-square method.

Answer.

Now, there are 100 memory locations where indices will be from 0 to 99. Hence, only two digits will be taken to map the keys. So, the value of *r* is equal to 2.

$k1 = 2045$, $k2 = 4182025$, $h(2045) = 20$

$k2 = 1357$, $k2 = 1841449$, $h(1357) = 14$

Note: The 3rd and 4th digits are chosen to start from the right.

- *Folding method* – In the folding method, we will break the key into pieces such that each piece has the same number of digits except the last one, which may have fewer digits as compared to other pieces. Now, these individual pieces are added. We will ignore the carry if it exists. Hence, the hash value is formed.

For example, if $m = 100$ and the key $k = 12345678$, then the indices will vary from 0 to 99 and thus, each piece of the key must have two digits. Therefore, the given key will be broken into four pieces, that is, 12, 34, 56, and 78. Now, we will add all these, that is, $12 + 34 + 56 + 78 = 180$. Thus, the hash value will be 80 (ignore the last carry).

Frequently Asked Questions

Q4. Given a hash table of 100 memory locations, calculate the hash values of keys 2486 and 179 using the folding method.

Answer.

Now, there are 100 memory locations where indices will be from 0 to 99. Hence, each piece of the key must have two digits.

$h(2486) = 24 + 86 = 110$

$h(2486) = 10$ (ignore the last carry)

$h(179) = 17 + 9 = 26$

$h(179) = 26$

Collision

A *collision* is a situation that occurs when a hash function maps two different keys to a single/same location in the hash table. Suppose we want to store a record at one location. Now, another record cannot be stored at the same location as it is obvious that two records cannot be stored at the same location. Thus, there are methods to solve this problem, which are called collision resolution techniques.

Collision Resolution Techniques

As already discussed, *collision resolution techniques* are used to overcome the problem of collision in hashing. There are two popular methods that are used for resolving collisions:

- Chaining method
- Open addressing method

Now, we will discuss these methods in detail.

Chaining Method

In the *chaining method*, a chain of elements is maintained that have the same hash address. The hash table here behaves like an array of pointers. Each location in the hash table stores a pointer to the linked list, which contains all the key elements that were hashed to that location. For example, location 5 in the hash table points to the key values that were hashed to location 5. If no key value hashes to location 5, then location 5 will contain *NULL*. Figure 10.5 shows how the key values are mapped to the hash table and how they are stored in the linked list:

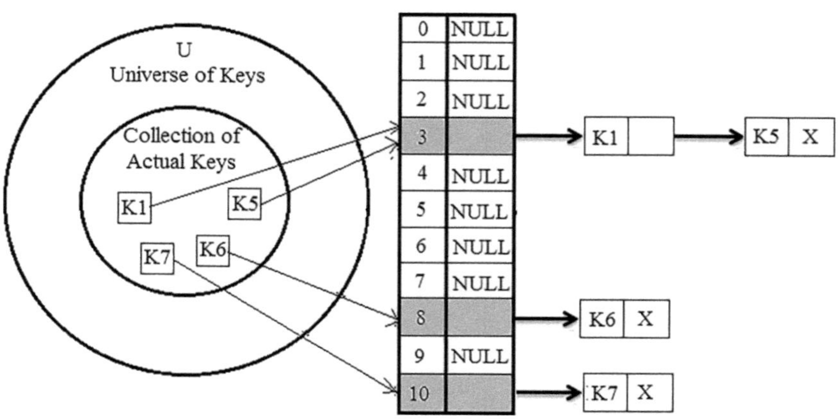

FIGURE 10.5 Keys hashed by the chaining method.

The following operations can be performed on a chained hash table:

- *Insertion into a chained hash table* – The process of inserting an element is quite simple. First, we get the hash value from the hash function that will map to the hash table. After mapping, the element is inserted into the linked list. The running time complexity of inserting an element in a chained hash table is $O(1)$.
- *Deletion from a chained hash table* – The process of deleting an element from the chained hash table is the same as we did in the singly linked list. First, we will perform a search operation, and then the delete operation, as in the case of the singly linked list, is performed. The running time complexity of deleting an element from a chained hash table is $O(m)$, where m is the number of elements present in the linked list at that location.
- *Searching in a chained hash table* – The process of searching for an element in a chained hash table is also very simple. First, we will get the hash value of the key by the hash function in the hash table. Then, we will search for the element in the linked list. The running time complexity of searching for an element in a chained hash table is $O(m)$, where m is the number of elements present in the linked list at that location.

Frequently Asked Questions

Q5. Insert the keys 4, 9, 20, 35, and 49 in a chained hash table of 10 memory locations. Use hash function h(k) = k mod m.

Answer.

Initially, the hash table is given as:

0	NULL
1	NULL
2	NULL
3	NULL
4	NULL
5	NULL
6	NULL
7	NULL
8	NULL
9	NULL

Now, we will insert 4 in the hash table:

Key to be inserted = 4

$h(4) = 4 \; mod \; 10$

$h(4) = 4$

Now, we will create a linked list for location 4, and the key element 4 is stored in it:

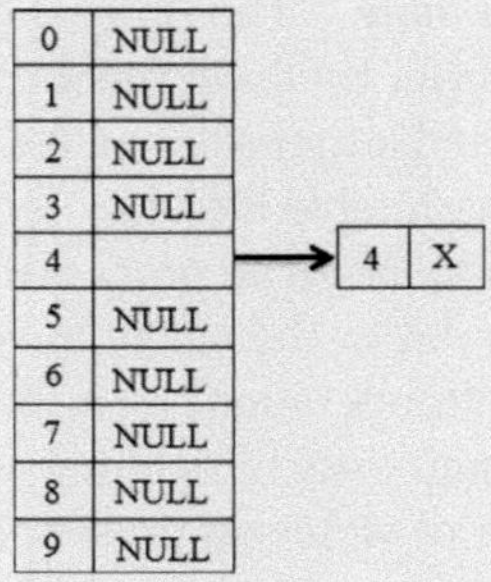

Key to be inserted = 9

$h(9) = 9 \textit{ mod } 10$

$h(9) = 9$

Now, we will create a linked list for location 9, and the key element 9 is stored in it:

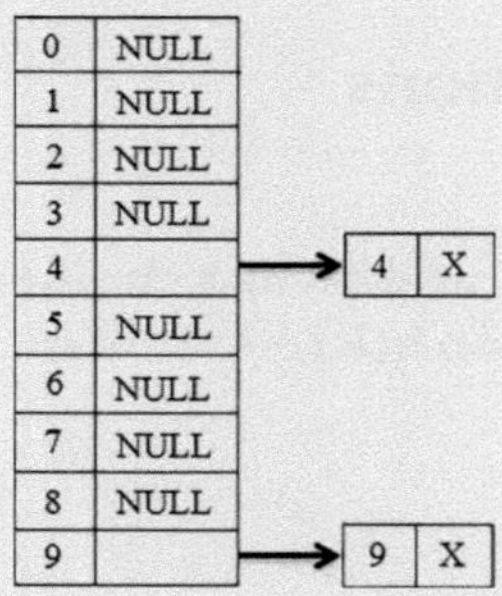

Key to be inserted = 20

$h(20) = 20 \textit{ mod } 10$

$h(20) = 2$

Now, we will create a linked list for location 2, and the key element 20 is stored in it:

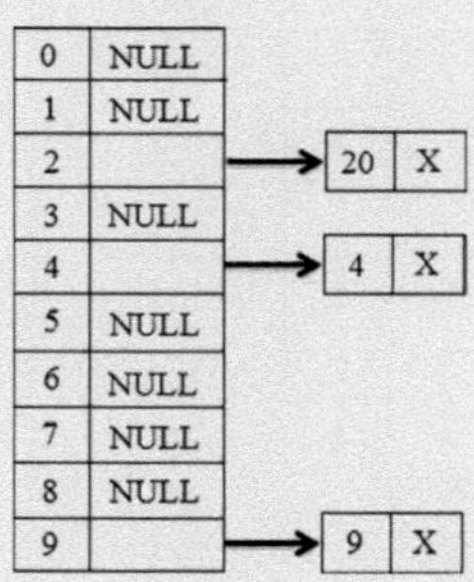

Key to be inserted = 35

$h(35) = 35 \textit{ mod } 10$

$h(35) = 5$

Now, we will create a linked list for location 5, and the key element 35 is stored in it:

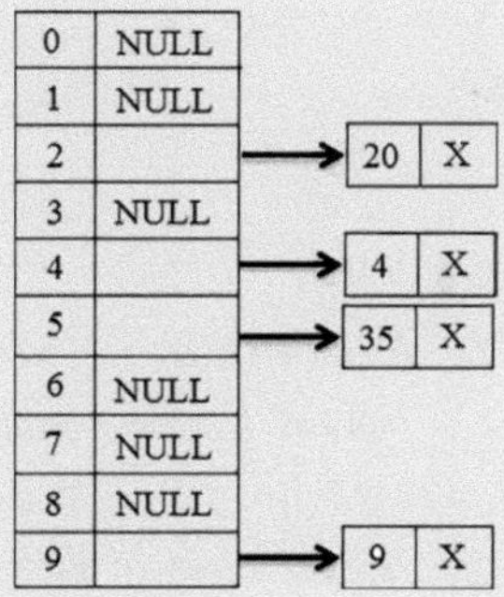

Key to be inserted = 49

h(49) = 49 *mod* 10

h(49) = 9

Now, we will insert 49 at the end of the linked list of location 9:

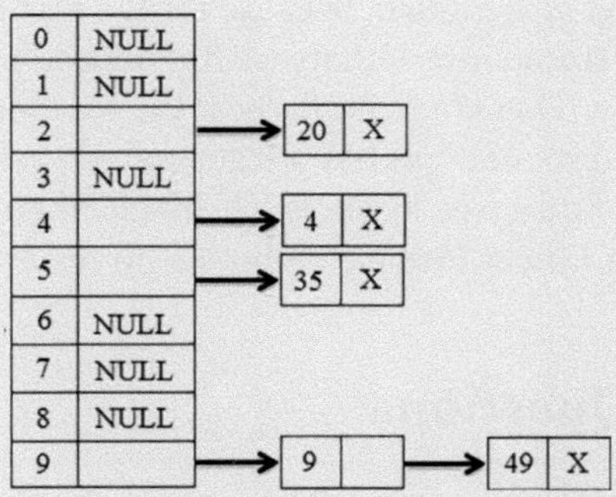

The main advantage of this method is that it completely resolves the problem of collision. It remains effective even when the key elements to be stored in the hash table are higher than the number of locations in the hash table. It is quite obvious that with an increase in the number of key elements, the performance of this method will decrease.

The disadvantage of this method is the waste of storage space as the key elements are stored in the linked list; in addition, pointers are required for each element to get accessed, which, in turn, consumes more space.

Open Addressing Method

In the open addressing method, all the elements are stored in the hash table itself. There is no need to provide the pointers, which is the biggest advantage of this method. Once a collision takes place, open addressing computes new locations using the probe sequence, and the next element or record is stored at that location. *Probing* is the process of examining the memory locations in the hash table. When we perform the insertion operation in the open addressing method, we first successively probe/examine the hash table until we find an empty slot in which the new key can be inserted. The open addressing method can be implemented using the following:

- Linear probing
- Quadratic probing
- Double hashing

Now, let us discuss all of them in detail.

$$h'(k) + (h(k) + i) \bmod m$$
$$\text{where } h(k) = k \bmod m$$
$$i = Probe\ no. = 0, 1, 2, 3, \cdots (m-1)$$
$$m = no.\ of\ slots$$

Linear Probing

Linear probing is the simplest approach to resolving the problem of collision in hashing. In this method, if a key is already stored at a location generated by the hash function $h(k)$, then the situation can be resolved with the following hash function:

Now, let us understand the working of this technique. For a given key k, first, the location generated by $(h(k) + 0) \bmod m$ is probed, because for the first time, $i = 0$. If the location generated is free, then the key is stored in it. Otherwise, the second probe is generated for $i = 1$, given by the hash function $(h(k) + 1) \bmod m$. Similarly, if the location generated is free, then the key is stored in it; otherwise, subsequent probes are generated such as $(h(k) + 2) \bmod m$, $(h(k) + 3) \bmod m$, and so on, until we find a free location. The detailed numerical methodology and practical implementation code of Linear Probing technique to resolve collision is discussed below:

Frequently Asked Questions

Q6. Given keys 13, 25, 14, and 35, map these keys into a hash table of size m = 5 using linear probing.

Answer.

Initially, the hash table is given as:

0	1	2	3	4
NULL	NULL	NULL	NULL	NULL

Step 1:

$i = 0$

Key to be inserted = 13

$h'(k) = (k \bmod m + i) \bmod m$

$h'(13) = (13 \% 5 + 0) \% 5$

$h'(13) = (3 + 0) \% 5$

$h'(13) = 3 \% 5 = 3$

Now, since location *T*[3] is free, 13 is inserted at location *T*[3]:

0	1	2	3	4
NULL	NULL	NULL	13	NULL

Step 2:

$i = 0$

Key to be inserted = 25

$h'(25) = (25 \% 5 + 0) \% 5$

$h'(25) = (0 + 0) \% 5$

$h'(25) = 0 \% 5 = 0$

Now, since location *T*[0] is free, 25 is inserted at location *T*[0]:

0	1	2	3	4
25	NULL	NULL	13	NULL

Step 3:

$i = 0$

Key to be inserted = 14

$h'(14) = (14 \% 5 + 0) \% 5$

$h'(14) = (4 + 0) \% 5$

$h'(14) = 4 \% 5 = 4$

Now, since location *T*[4] is free, 14 is inserted at location *T*[4]:

0	1	2	3	4
25	NULL	NULL	13	14

Step 4:

$i = 0$

Key to be inserted = 35

$h'(35) = (35 \% 5 + 0) \% 5$

$h'(35) = (0 + 0) \% 5$

$h'(35) = 0 \% 5 = 0$

Now, since location $T[0]$ is not free, the next probe sequence, that is, $i = 1$, is computed as:

$i = 1$

$h'(35) = (35 \% 5 + 1) \% 5$

$h'(35) = (0 + 1) \% 5$

$h'(35) = 1 \% 5 = 1$

Now, since location $T[1]$ is free, 35 is inserted at location $T[1]$.

Thus, the final hash table is shown as:

0	1	2	3	4
25	35	NULL	13	14

Write a program to show the linear probing technique of the collision resolution method:

```
# include<stdio.h>
# include<conio.h>
# define SIZE 50

intarr[SIZE];
void linear( int,int[]);
void lprob(int k, intarr[SIZE]);
void disp(intarr[SIZE);
void main( )
{
int i, k , choice ;
clrscr();
  for (i=0; i<SIZE; i++)
{
   arr[i]= '\0';
do
{
  printf("\n\t  MENU");
  printf("\n 1. Insert Keys");
  printf("\n 2. Search Keys");
  printf("\n3. Display Keys");
  printf("\n 4. Exit ");
  printf("\n Select Operation ");
  scanf("%d" ,&choice);
  switch(choice)
   {
     case1: do
     {
       printf("\n Enter values of key ");
```

```
        scanf( "%d", &k);
        if(k!=-1)
        {
          linear(k,arr);
        }
        while(k!=1);
        disp(arr);
        break;

     case2:
      printf(" \n enter key value to search ");
      scanf("%d",&k);
      lprob(k, arr);
      break;

     case3:
     disp(arr);
   break;
}
}
while(choice!=4);
}

void linear(int k, intarr[SIZE])
{
int position ;
position = k%SIZE;
while(arr[position]!= '\0')
{
  position = ++ position %SIZE;
  arr[position] = k;
}

void lprob(int k, intarr[SIZE])
{
  int position ;
  position = k%SIZE;
  while ((arr[position] != k)&& (arr[position] != '\0'))
  position = ++ position %SIZE;
if(arr[position]!= '\0')
  printf("\n successfully searched at %d", position);
else
  printf(" \n unsuccessfull search");
}

void disp(intarr[SIZE])
{
  int i ;
  printf(" \n List of keys :\n);
  for(i=0; i<SIZE; i++)
  printf("%d" , arr[i]);
}
```

Linear probing is a very good technique, as the algorithm provides good memory caching through good locality of address. The main disadvantage of this method is that it results in clustering, which means there is a higher risk of collisions taking place. Also, the time required for searching increases with the size of the clusters. We can say that the higher the number of collisions, the higher the number of probes required to find a vacant location, and the performance decreases. This is known as *primary clustering*. We can avoid this clustering by using other techniques such as quadratic probing and double hashing.

Quadratic Probing

Quadratic probing is another approach to resolving the problem of collision in hashing. In this method, if a key is already stored at a location generated by the hash function $h(k)$, then the situation can be resolved with the following hash function:

$$h'(k) = (h(k) + c_1 i + c_2 i^2) \ mod \ m$$
$$\text{where } h(k) = k \ mod \ m$$
$$i = Probe\ no. = 0, 1, 2, 3\ldots(m-1)$$
$$C_1, c_2 = consonants$$
$$(c_1, c_2 \ should\ not\ be\ equal\ to\ zero)$$

The quadratic probing method is better than linear probing, as it terminates the phenomenon of primary clustering because of its searching speed; that is, it uses a quadratic search. For a given key k, first, the location generated by $(h(k) + 0 + 0) \ mod \ m$ is probed because, for the first time, $i = 0$. If the location generated is free, then the key is stored in it. Otherwise, subsequent positions probed are offset by the amounts/factors that depend in a quadratic manner on the probe number i. The quadratic probing method works better than linear probing, but to maximize the use of the hash table, the values of m, c_1, and c_2 are constrained.

Frequently Asked Questions

Q7. Given keys 25, 13, 14, and 35, map these keys into a hash table of size m = 5 using quadratic probing with c1 = 1 and c2 = 3.

Answer.

Initially, the hash table is given as follows:

0	1	2	3	4
NULL	NULL	NULL	NULL	NULL

Step 1:

$i = 0$

$c_1 = 1, c_2 = 3$

Key to be inserted = 25

$h'(k) = (k \bmod m + c_1 i + c_2 i^2) \bmod m$

$h'(25) = (25 \% 5 + 1 X 0 + 3 X (0)^2) \% 5$

$h'(25) = (0 + 0) \% 5$

$h'(13) = 0 \% 5 = 0$

Now, since location $T[0]$ is free, 25 is inserted at location $T[0]$:

0	1	2	3	4
25	NULL	NULL	NULL	NULL

Step 2:

$i = 0$

$c_1 = 1, c_2 = 3$

Key to be vinserted = 13

$h'(13) = (13 \% 5 + 1 X 0 + 3 X (0)^2) \% 5$

$h'(13) = (3 + 0) \% 5$

$h'(13) = 3 \% 5 = 3$

Now, since location $T[3]$ is free, 13 is inserted at location $T[3]$:

0	1	2	3	4
25	NULL	NULL	13	NULL

Step 3:

$i = 0$

$c_1 = 1, c_2 = 3$

Key to be inserted = 14

$h'(14) = (14 \% 5 + 1 X 0 + 3 X (0)^2) \% 5$

$h'(14) = (4 + 0) \% 5$

$h'(14) = 4 \% 5 = 4$

Now, since location $T[4]$ is free, 14 is inserted at location $T[4]$:

0	1	2	3	4
25	NULL	NULL	13	14

Step 4:

$i = 0$

$c_1 = 1, c_2 = 3$

Key to be inserted = 35

$h'(35) = (35 \% 5 + 1 X 0 + 3 X (0)^2) \% 5$

$h'(35) = (0 + 0) \% 5$

$h'(35) = 0 \% 5 = 0$

Now, since location *T*[0] is not free, the next probe sequence, that is, $i = 1$, is computed as:

$i = 1$

$h'(35) = (35 \% 5 + 1 X 1 + 3 X (1)^2) \% 5$

$h'(35) = (0 + 1 + 3) \% 5$

$h'(35) = 4 \% 5 = 4$

Again, since location *T*[4] is not free, the next probe *sequence, that is,* $i = 2$, is computed as:

$i = 2$

$h'(35) = (35 \% 5 + 1 X 2 + 3 X (2)^2) \% 5$

$h'(35) = (0 + 2 + 12) \% 5$

$h'(35) = 14 \% 5 = 4$

Again, since location *T*[4] is not free, the next probe sequence, that is, $i = 3$, is computed as:

$i = 3$

$h'(35) = (35 \% 5 + 1 X 3 + 3 X (3)^2) \% 5$

$h'(35) = (0 + 3 + 27) \% 5$

$h'(35) = 30 \% 5 = 0$

Again, since location *T*[0] is not free, the next probe sequence, that is, $i = 4$, is computed as:

$i = 4$

$h'(35) = (35 \% 5 + 1 X 4 + 3 X (4)^2) \% 5$

$h'(35) = (0 + 4 + 48) \% 5$

$h'(35) = 52 \% 5 = 2$

Now, since location *T*[2] is free, 35 is inserted at location *T*[2].

Thus, the final hash table is shown as:

0	1	2	3	4
25	NULL	35	13	14

Write a program to show the quadratic probing technique of the collision resolution method:

```
# include<stdio.h>
# include<conio.h>
# define SIZE 50

intarr[SIZE];
void quad( int ,int[]);
void qprob(int k, intarr[SIZE]);
void disp(intarr[SIZE);
void main( )
{
int i, k , choice ;
clrscr();
for (i=0; i<SIZE; i++)
{
arr[i]= '\0';
do
{
printf("\n\t MENU");
printf("\n 1. Insert Keys");
printf("\n 2. Search Keys");
printf("\n3. Display Keys");
printf("\n 4. Exit ");
printf("\n Select Operation ");
scanf("%d" ,&choice);
switch(choice)
{
     Case1:
do
{
printf("\n Enter values of key ");
scanf( "%d", &k);
if(k!=-1)
{
quad(k,arr);
}
while(k!=1);
```

```
disp(arr);
break;

case2:
printf(" \n enter value of search key ");
scanf("%d",&k);
qprob(k, arr);
break;

case3:
disp(arr);
break;
}
}
while(choice!=4);
}

void quad(int k, int arr[SIZE])
{
int position, i=1;
position = k%SIZE;
while(arr[position]!= '\0')
{
position = (k%SIZE + i*i) % SIZE;
   i++;
}
arr[position] = k;
}

void qprob(int k, intarr[SIZE])
{
int position ;
  position = k%SIZE;
  while ((arr[position] != k)&& (arr[position] !=-1))
  position = ++ position %SIZE;
if(arr[position]!= '\0')
  printf("\n successfully searched at %d", position);
else
  printf(" \n unsuccessfull search");
}

void disp(intarr[SIZE])
{
int i ;
printf(" \n List of keys :\n);
  for(i=0; i<SIZE; i++)
  printf("%d" , arr[i]);
}
```

As discussed previously, one of the biggest advantages of quadratic probing is that it eliminates the phenomenon of primary clustering. On the contrary, one of the major disadvantages of this method is that a sequence of successive probes may only cover some portion of the hash table, and this portion may be quite small. Therefore, if such a situation occurs, it will be difficult for us to find an empty location in the hash table, despite the fact that the table is not full. Hence, quadratic probing encounters a problem known as *secondary clustering*. In this method, the chance of multiple collisions increases as the hash table becomes full. This type of situation can be overcome by double hashing.

Double Hashing

Double hashing is one of the best methods available for open addressing. As the name suggests, this method uses two hash functions to operate rather than a single hash function. The hash function is given as follows:

$$h'(k) = (h_1(k) + ih_2(k)) \; mod \; m$$

Here, $h_1(k) = k \; mod \; m$ and $h_2(k) = k \; mod \; m'$ are the two hash functions, m is the size of the hash table, m' is less than m (can be $(m - 1)$ or $(m - 2)$), and i is the probe number that varies from 0 to $(m - 1)$.

Now, let us understand the working of this technique. For a given key, k, first, the location generated by $(h_1(k) \; mod \; m)$ is probed because, for the first time, $i = 0$. If the location generated is free, then the key is stored in it. Otherwise, subsequent probes generate locations that are at an offset of $(h_2(k) \; mod \; m)$ from the previous location. Also, the offset may vary with every probe depending upon the value generated by the second hash function, that is, $(h_2(k) \; mod \; m)$. As a result, the performance of double hashing is very close to the performance of the *ideal* scheme of uniform hashing.

The double hashing method is free from all the problems of primary clustering and secondary clustering. It also minimizes repeated collisions.

Frequently Asked Questions

Q8. Given keys k = 71, 29, 38, 61, and 100, map these keys into a hash table of size m = 5 using double hashing. Take h1 = (k mod 5) and h2 = (k mod 4).

Answer.

nitially, the hash table is given as:

0	1	2	3	4
NULL	NULL	NULL	NULL	NULL

Step 1:

$i = 0$

Key to be inserted = 71

$h'(k) = (h_1(k) + ih_2(k)) \; mod \; m$

$h'(k) = (k \bmod m + (i\, k \bmod m')) \bmod m$

$h'(71) = (71 \% 5 + (0 X 71 \% 4)) \% 5$

$h'(71) = (1 + (0 X 3)) \% 5$

$h'(71) = 1 \% 5 = 1$

Now, since location *T*[1] is free, 71 is inserted at location *T*[1].

0	1	2	3	4
NULL	71	NULL	NULL	NULL

Step 2:

$i = 0$

Key to be inserted = 29

$h'(k) = (k \bmod m + (i\, k \bmod m')) \bmod m$

I$h'(29) = (29 \% 5 + (0 X 29 \% 4)) \% 5$

$h'(29) = (4 + (0 X 1)) \% 5$

$h'(29) = 4 \% 5 = 4$

Now, since location *T*[4] is free, 29 is inserted at location *T*[4].

0	1	2	3	4
NULL	71	NULL	NULL	29

Step 3:

$i = 0$

Key to be inserted = 38

$h'(k) = (k \bmod m + (i\, k \bmod m')) \bmod m$

$h'(38) = (38 \% 5 + (0 X 38 \% 4)) \% 5$

$h'(38) = (3 + (0 X 2)) \% 5$

$h'(38) = 3 \% 5 = 3$

Now, since location *T*[3] is free, 38 is inserted at location *T*[3].

0	1	2	3	4
NULL	71	NULL	38	29

Step 4:

$i = 0$

Key to be inserted = 61

$h'(k) = (k \bmod m + (i\, k \bmod m')) \bmod m$

$h'(61) = (61 \% 5 + (0 X 61 \% 4)) \% 5$

$h'(61) = (1 + (0 X 1)) \% 5$

$h'(61) = 1 \% 5 = 1$

Now, since location $T[1]$ is not free, the next probe sequence, that is, $i = 1$, is computed as:

$i = 1$

$h'(61) = (61 \% 5 + (1 X 61 \% 4)) \% 5$

$h'(61) = (1 + (1 X 1)) \% 5$

$h'(61) = (1 + 1) \% 5$

$h'(61) = 2\% 5 = 2$

Now, since location $T[2]$ is free, 61 is inserted at location $T[2]$.

0	1	2	3	4
NULL	71	61	38	29

Step 5:

$i = 0$

Key to be inserted = 100

$h'(k) = (k \bmod m + (i\, k \bmod m')) \bmod m$

$h'(100) = (100 \% 5 + (0 X 100 \% 4)) \% 5$

$h'(100) = (0 + (0 X 0)) \% 5$

$h'(100) = 0 \% 5 = 0$

Now, since location $T[0]$ is free, 100 is inserted at location $T[0]$.

Thus, the final hash table is shown as:

0	1	2	3	4
100	71	61	38	29

SUMMARY

- A hash table is an array in which the data is accessed through a special index called a key. In a hash table, keys are mapped to the array positions by a hash function.
- A hash function is a mathematical formula that, when applied to a key, produces an integer that is used as an index to find a key in the hash table.
- There are different types of hash functions that use numeric keys. Popular methods are the division method, the mid-square method, and the folding method.
- In the division method, a key k is mapped into one of the m slots by taking the remainder of k divided by m. The main drawback of the division method is that many consecutive keys map to consecutive hash values, respectively, which means that consecutive array locations will be occupied, and hence, there will be an effect on the performance.
- In the mid-square method, we calculate the square of the given key. After getting the number, we extract some digits from the middle of that number as an address.
- In the folding method, we break the key into pieces such that each piece has the same number of digits except the last one, which may have fewer digits as compared to other pieces. Now, these individual pieces are added. Hence, the hash value is formed.
- A collision is a situation that occurs when a hash function maps two different keys to a single/same location in the hash table.
- Collision resolution techniques are used to overcome the problem of collision in hashing. There are two popular methods that are used for resolving collisions, which are the chaining method and the open addressing method.
- In the chaining method, a chain of elements is maintained that have the same hash address. Hash tables here behave like an array of pointers. In this, each location in the hash table stores a pointer to the linked list, which contains all the key elements that were hashed to that location. The disadvantage of this method is the wastage of storage space as the key elements are stored in the linked list, and for each element to be accessed, pointers are required, which, in turn, consume more space.
- In the open addressing method, all the elements are stored in the hash table itself. There is no need to provide the pointers in this method, which is the biggest advantage. Once a collision takes place, open addressing computes new locations using the probe sequence, and the next element or next record is stored in that location.
- Probing is the process of examining the memory locations in the hash table.
- Linear probing is the simplest approach to resolving the problem of collision in hashing. In this method, if a key is already stored at a location generated by the hash function $h(k)$, then the situation can be resolved with the following hash function:

$$h'(k) = (h(k) + i) \ mod \ m$$

- Quadratic probing is another approach to resolving the problem of collision in hashing. In this method, if a key is already stored at a location generated by the hash function $h(k)$, then the situation can be resolved with the following hash function:

$$h'(k) = (h(k) + c_1 i + c_2 i^2) \ mod \ m$$

- Double hashing is one of the best methods available for open addressing. As the name suggests, this method uses two hash functions to operate rather than a single hash function. The hash function is given as:

$$h'(k) = (h_1(k) + ih_2(k))\ mod\ m$$

EXERCISES

Review Questions

1. What are hash tables?
2. What is hashing? Give some of its practical applications.
3. Define the hash function and explain the various characteristics of a hash function.
4. What is a collision in hashing, and how can it be resolved?
5. Explain the different types of hash functions along with an example.
6. Discuss the collision resolution techniques in hashing.
7. What is clustering in hashing? What are the two types of clustering?
8. What do you understand by double hashing?
9. Define the following terms:
   - **a)** Quadratic probing
   - **b)** Linear probing
10. What is the chaining method in hashing, and how can it help in resolving collisions?
11. Consider a hash table of size 10. Using linear probing, insert the keys 12, 45, 67, 122, 78, and 34 into it.
12. Consider a hash table of size 9. Using double hashing, insert the keys 4, 17, 30, 55, 90, 11, 54, and 77 into it. Take $h_1 = k\ mod\ 9$ and $h_2 = k\ mod\ 6$.
13. Consider a hash table of size 11. Using quadratic probing, insert the keys 10, 45, 56, 97, 123, and 1 into it.
14. How can the open addressing method be used in resolving collisions?
15. Write a C function to retrieve an item from the hash table using linear probing and quadratic probing.

Multiple Choice Questions

1. Which of the following collision resolution techniques is free from the clustering phenomenon?
   - **A.** Linear probing
   - **B.** Quadratic probing
   - **C.** Double hashing
   - **D.** None of these
2. The process of examining a memory location is called ___________.
   - **A.** Probing
   - **B.** Hashing
   - **C.** Chaining
   - **D.** Addressing

3. What does a hash table with chaining as a collision resolution technique degenerate to?
   A. Tree
   B. Graph
   C. Array
   D. Linked list
4. Which of the probing techniques suffers from the problem of primary clustering?
   A. Quadratic probing
   B. Linear probing
   C. Double hashing
   D. All of these
5. Given the hash function $h(k) = k\ mod\ 6$, what is the number of collisions to store the following sequence of keys using open addressing?
   A. 1
   B. 3
   C. 2
   D. 5
6. In a hash table, an element with the key k is stored at ___________.
   A. k
   B. $h(k^2)$
   C. $h(k)$
   D. $log\ h(k)$
7. A good hash function eliminates the problem of collision. True or false?
   A. True
   B. False
8. Given the hash function of size 7 and hash function $h(k) = k\ mod\ 7$, what is the number of collisions with linear probing for insertion of the following keys: 29, 36, 16, and 30?
   A. 1
   B. 2
   C. 3
   D. 4
9. ___________ is the process of mapping keys to appropriate locations in the hash table.
   A. Probing
   B. Hashing
   C. Collision
   D. Addressing
10. When there is no free location in the hash table, then ___________ occurs.
    A. Underflow
    B. Overflow
    C. Collision
    D. None of the above

CHAPTER 11

FILES

INTRODUCTION

We all know that in most organizations nowadays, a large amount of data is collected in one form or another. The organizations use various types of data collection applications to collect the data. When we talk about an *organization*, we are not just talking about big ones, such as schools, colleges, and companies, but also small ones, such as the bakery shop at the corner of the street; it can be observed that the collection and exchange of data take place everywhere. For example, when we get admitted into a school, a lot of data is collected by the school, such as name, age, address, parents' names, blood type, and so on. In older times, data was collected in the form of paper documents, which were very difficult to handle and store. Therefore, to efficiently and effectively analyze the collected data, computers are used to store the data in the form of files. A *file* in computer terminology is defined as a block of useful data in a persistent storage medium; that is, the file is available for future use. The data is organized in a hierarchical order in the files. The hierarchical order includes items such as records, fields, and so forth, which are all defined as follows.

TERMINOLOGY

Let's look at the definitions of the terms used when discussing files:

- *Data field* – A data field is a unit that stores a unary fact. It is usually characterized by its type and size. For example, "employee name" is a data field that stores the names of employees.
- *Record* – The collection of related data fields is called a record. For example, an employee's record may contain various data fields such as name, ID, address, contact number, and so on.
- *File* – The collection of related records is called a file. An example is a file of the employees working in an organization.
- *Directory* – The collection of related files is called a directory. Every file in a computer system is stored in a directory.

- *File name* – This is the name of a file is a string of characters.
- *Read-only* – A file named read-only cannot be modified or deleted. If we try to delete the file, then a particular message is displayed.
- *Hidden* – A file marked as hidden is not displayed in the directory.

FILE OPERATIONS

There are various operations that can be performed on files, as shown in Figure 11.1:

1. *File creation* – This is the first operation to be performed if the file is not created. A file is created by specifying its name and mode. Records are inserted into the file by opening the file in write mode. Once all the records are inserted into the file, the file can be used for future read and write operations. For example, we could create a new file named `EMPLOYEE`.
2. *Updating a file* – This means changing the contents of a file. It is usually done in the following ways:
   a) *Inserting into a file* – The new record is inserted into the file. For example, if a new employee joins an organization, their record is inserted into the `EMPLOYEE` file.
   b) *Modifying a file* – The existing records are modified in the file. For example, if the address of an employee is changed, then the new address must be modified in the `EMPLOYEE` file.
   c) *Deleting from a file* – The existing record is deleted from the file. For example, if an employee quits a job, then their record is deleted from the `EMPLOYEE` file.
3. *Retrieving from a file* – This refers to the process of extracting some useful data from a file. It is usually done in the following ways:
   a) *Inquiring* – This retrieves a small amount of data from the file.
   b) *Generating a report* – This retrieves a huge amount of data from the file.

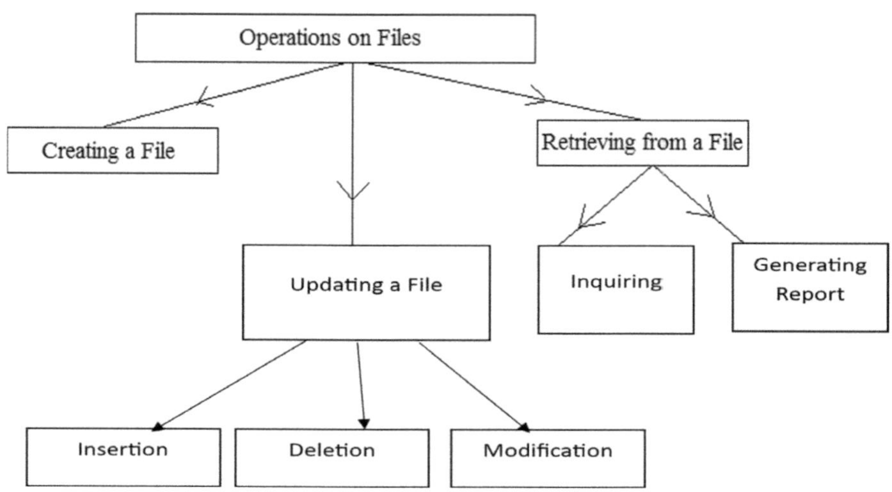

FIGURE 11.1 Operations on files.

FILE CLASSIFICATION

A file is classified into two types, as follows:

- *Text files* – A text file, often called a flat file, is a file that stores all the numeric or non-numeric data using its corresponding ASCII values. The data can be a string of letters, numbers, or special symbols. Therefore, it is also known as an ASCII file. Usually, a text file has a special marker known as the end-of-file marker, which denotes the end of the file.
- *Binary files* – A binary file is a file that contains all the data in the binary form of 1s and 0s. It stores the data in the same form as that of primary memory. Thus, a binary file is not readable by human beings. Binary files are read by computer programs, and they decode the binary files into something meaningful. Data is efficiently stored in binary files.

FILE ORGANIZATION

File organization refers to how records are physically arranged on a storage device. Further, there may be a single key or multiple keys associated with it. Based on its physical storage and the keys used to access the records, files are classified as sequential files, relative files, indexed sequential files, and inverted files. Various factors should be taken into consideration while choosing a particular type of file organization, which are:

- Ease of retrieval of the records
- Economy of storage
- Reliability, that is, whether a file organization is reliable or not
- Security, that is, whether a file organization is secured or not

Now, we will discuss some of the techniques that are commonly used for file organization.

SEQUENCE FILE ORGANIZATION

Sequence file organization is the most basic way to organize a collection of records in a file. This is when the file is created, when the records are written, one after the other in order, and can be accessed only in that order in which they are written when the file is used for input. All the records are numbered from zero onward. Thus, if there are *N* records in a file, then the first record is numbered as `0`, and the last record will be numbered as `N-1`. In some cases, records of sequential files are sorted by the value of some field in each record. The field whose value is used to sort the records is known as a sort key. If a file is sorted by the value of a field named `key field`, then the record `i` proceeds record `j` if and only if the value of `key field` in record `i` is less than or equal to the value of `key field` in record `j`. Also, a file can be sorted in either ascending or descending order by a sort key comprising one or more fields. As the records in a sequential file can only be accessed sequentially, these files are used more commonly in batch processing than in interactive processing. For example,. Optical Discs (Blu-ray Archival Systems) offers sequential access of data via Laser reading Figure 11.2 shows the sequential storage structure.

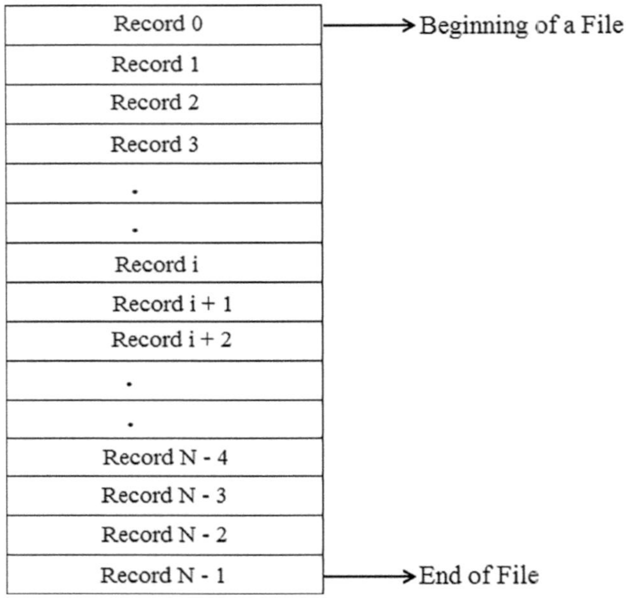

FIGURE 11.2 Structure of sequence file organization.

The advantages of a sequence file organization are as follows:

- It is easy to handle.
- It does not involve extra overheads/problems.
- Records can be of varying lengths in this organization.
- It can be stored on magnetic disks.

These are the disadvantages of sequence file organization:

- Records can be accessed only in sequence.
- It does not support the update operation between the files.
- It does not support interactive applications.

INDEXED SEQUENCE FILE ORGANIZATION

An indexed sequential file organization is an efficient way of organizing the records when there is a need to access both sequentially by some key values and also to access the records individually by the same key value. It provides the combination of access types that are supported by a sequential file or a relative file. The index is structured as a binary search tree. This index is used to serve as a request for access to a particular record, and the sequential data file alone is used to support sequential access to the entire collection of records. Because of its capability to support both sequential and direct access, indexed sequence file organization is used to support applications that require both batch and interactive processing. Figure 11.3 shows how this organization is used.

The advantages of indexed sequence file organization are as follows:

- Records can be accessed sequentially and randomly.
- It supports batch as well as interactive-oriented applications.
- It supports the update operation between records in the file.

These are the disadvantages of indexed sequence file organization:

- In this organization, files can only be stored on magnetic disks.
- It involves extra overhead in the form of maintenance.
- Records can only be of a fixed length, as we maintain the structure of each node like a linked list.

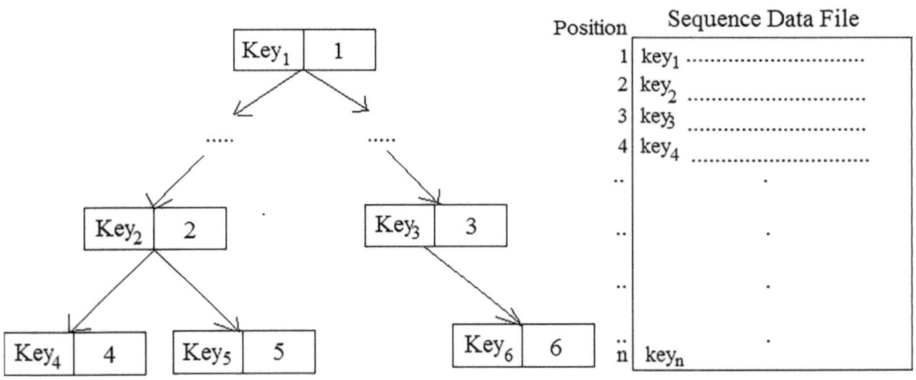

FIGURE 11.3 Use of BST and sequential files to provide indexed sequential access.

RELATIVE FILE ORGANIZATION

Relative file organization provides an effective way of accessing individual records directly. In a relative file organization, there is a predictable relationship between the key and the record's location in the file. The records do not necessarily appear physically in sorted order by their keys. How is a given record found? The relationship that will be used to translate between the key value and the physical address is designated using a mapping function, for example, *R(Key value, address)*. When a record is to be written into a relative file, the mapping function *R* is used to translate the record's key to an address, which indicates where the record is to be stored. The fundamental techniques that are used for mapping function *R* are directory lookup and address calculation (hashing):

- *Directory lookup technique* – This is the simplest technique for implementing a mapping function *R*. The basic idea of this technique is to keep a directory of key values: address pairs. To find a record in a relative file, one locates its key value in the directory, and then the indicated address is used to find the record on the storage device. The directory can be organized as a binary search tree.

- *Address calculation technique* – Another common technique for implementing a mapping function R is to perform a calculation on the key value (hashing) such that the result is a relative address.

The advantages of relative file organization are as follows:

- Records can be accessed out of sequence.
- It is well suited for interactive applications.
- It supports an update operation between the files.

These are the disadvantages of relative file organization:

- It can only be stored on magnetic disks.
- It also involves extra overhead in the form of the maintenance of indexes.

INVERTED FILE ORGANIZATION

One fundamental approach for providing a linkage between an index and a file is called *inversion*. A key's inversion index contains all the values that the key presently has in records of the file. Each key-value entry in the inversion index points to all the data records that have the corresponding value. Then, the file is said to be inverted on that key. The inversion approach for providing multi-key access has been used as the basis for a physical data structure in commercially available relational DBMSs such as Oracle, DB2, and so on. These systems were designed to provide rapid access to the records via as many inversion keys as the designer cares to identify. They have user-friendly natural language-like query languages to assist the user in formulating inquiries. A complete inverted file has an inversion index for every data field. If a file is not completely inverted but has at least one inversion index, then it is said to be a partially inverted file.

The advantages of inverted file organization are as follows:

- The Boolean query requires only one access per record satisfying the query, plus some access to process the indexes.
- Records can be stored in any way: for example, sequentially ordered by primary key, randomly linked ordered by primary key, and so forth.
- It also results in space saving as compared with the other file structures.

There is one main disadvantage of inverted file organization: since the index entries are of variable lengths, index maintenance becomes more complex.

SUMMARY

- A file is a collection of records. It is usually stored on a secondary storage device.
- The data is organized in a hierarchical order in the files. The hierarchical order includes items such as records, fields, and so on.

- File creation is the first operation to be performed if the file does not exist. A file is created by specifying its name and mode.
- A file is classified into two types, which are text files and binary files.
- A text file, often called a flat file, is a file that stores all the numeric or non-numeric data using its corresponding ASCII values. The data can be a string of letters, numbers, or special symbols.
- A binary file is a file that contains all the data in the binary form of 1s and 0s. It stores the data in the same form as that of primary memory.
- A file organization refers to the way in which records are physically arranged on a storage device.
- Sequence file organization is the most basic way to organize a collection of records in a file. In sequence file organization, the file is created when the records are written, one after the other in order, and can be accessed only in the order in which they are written when the file is used for input. All the records are numbered from zero onward.
- An indexed sequential file organization is an effective way of organizing the records when there is a need to access both sequentially by some key values and also to access the records individually by the same key value. It provides the combination of access types that are supported by a sequential file or a relative file.
- Relative file organization provides an effective way of accessing individual records directly. In a relative file organization, there is a predictable relationship between the key and the record's location in the file.
- One fundamental approach for providing a link between an index and a file is called inversion. The inversion approach for providing multi-key access has been used as the basis for the physical data structure.

EXERCISES

Review Questions

1. What is a file?
2. Why is there a need to store the data in files? Explain.
3. What do you understand by the terms *record* and *field*?
4. Discuss various operations that can be performed on files.
5. Differentiate between a text file and a binary file.
6. Write a short note on file attributes.
7. What do you understand by *file organization*? Discuss in detail.
8. Explain sequential file organization.
9. What are inverted files? Discuss.
10. Explain indexed sequential file organization.
11. Give the advantages and disadvantages of indexed sequential file organization.
12. What is relative file organization? Discuss the advantages and disadvantages of relative file organization.

Multiple Choice Questions

1. What is a collection of related fields called?
   - A. Data
   - B. Record
   - C. Field
   - D. File
2. A file marked as ________ can't be modified or deleted.
   - A. Hidden
   - B. Read-only
   - C. Archive
   - D. None of these
3. Which of the following is often known as a flat file?
   - A. Binary file
   - B. Text file
   - C. String file
   - D. None of these
4. A ________ is a collection of data organized in a fashion that facilitates various operations such as updating, retrieving, and so forth.
   - A. Record
   - B. Data word
   - C. Field
   - D. File
5. Can relative files be used both for random as well as sequential access?
   - A. Yes
   - B. No
   - C. Not enough information
6. A file marked as ________ is not displayed in the directory.
   - A. Read-only
   - B. Archive
   - C. Hidden
   - D. None of these
7. What is a data field characterized by?
   - A. Type
   - B. Size
   - C. Mode
   - D. Both (a) and (b)
8. What is used to store a collection of files?
   - A. Record
   - B. Dictionary
   - C. Directory
   - D. System

CHAPTER 12

GRAPHS

INTRODUCTION

So far, we have studied various types of linear data structures, which are widely used in various applications, but the only non-linear data structure we have studied thus far is trees. When we talked about trees, we discussed the parent-child relationship in which one parent can have many children. In graphs, however, this parent-child relationship is less restricted; that is, any complex relationship can exist. Thus, a tree can be generalized as a special type of graph. A *graph* is a non-linear data structure that has a wide range of real-life applications. A graph is a collection of vertices (nodes) and edges that connect these vertices. Figure 12.1 shows a graph:

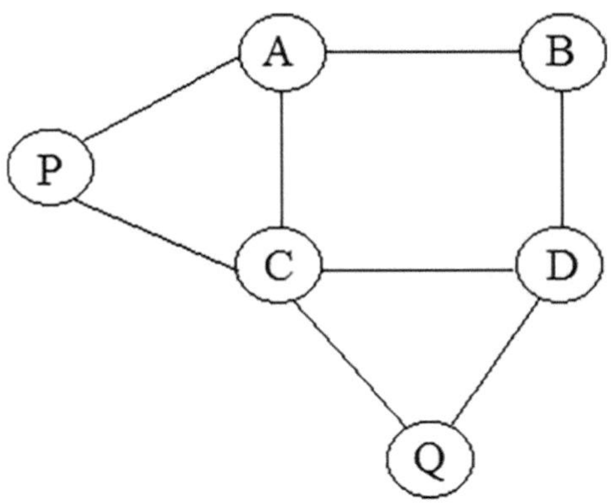

FIGURE 12.1 A graph.

Thus, a graph G can be defined as an ordered set of vertices and edges (V, E), where $V(G)$ represents the set of vertices and $E(G)$ represents the set of edges that connect these vertices. In the previous figure, $V(G) = \{A, B, C, D, P, Q\}$ represents the set of vertices and $E(G) = \{(A, B), (B, D), (D, C), (C, A), (C, Q), (Q, D), (A, P), (P, C)\}$ represents the set of edges.

Practical Application:

A simple illustration of a graph is that when we connect with our friends on social media, say Facebook, each user is a vertex, and when two users connect, it forms an edge.

There are two types of graphs:

- *Undirected graph* – In an undirected graph, the edges do not have any direction associated with them. As we can see in Figure 12.2, the two nodes *A* and *B* can be traversed in both directions, that is, from *A* to *B* or from *B* to *A*. Thus, an undirected graph does not give any information about the direction.

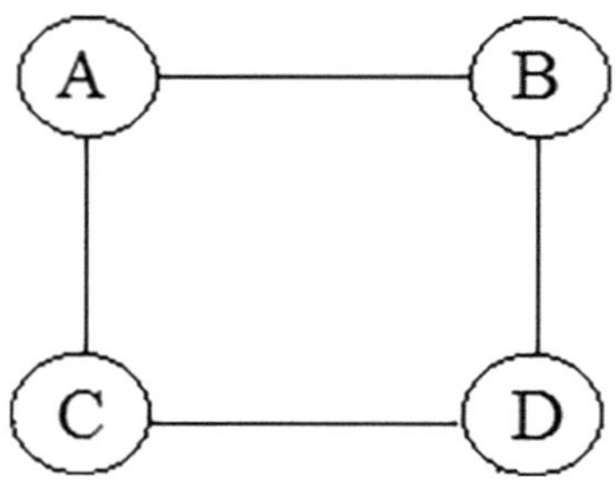

FIGURE 12.2 An undirected graph.

- *Directed graph* – In a directed graph, the edges have directions associated with them. As we can see in Figure 12.3, the two nodes *A* and *B* can be traversed in only one direction, that is, only from *A* to *B* and not from *B* to *A*. Therefore, on the edge (*A, B*), node *A* is known as the *initial node* and node *B* is known as the *final node*.

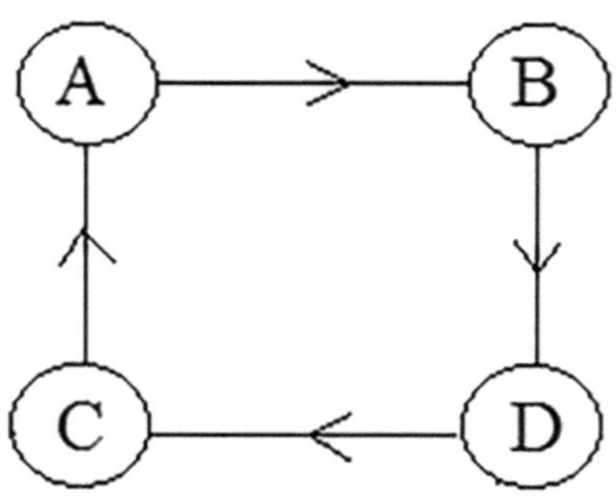

FIGURE 12.3 A directed graph.

DEFINITIONS

Let's look at the definitions of terms associated with graphs:

- *Degree of a vertex/node* – The degree of a node is the total number of edges incident to that particular node. In Figure 12.4, the degree of node *B* is three, as three edges are incident on node B.

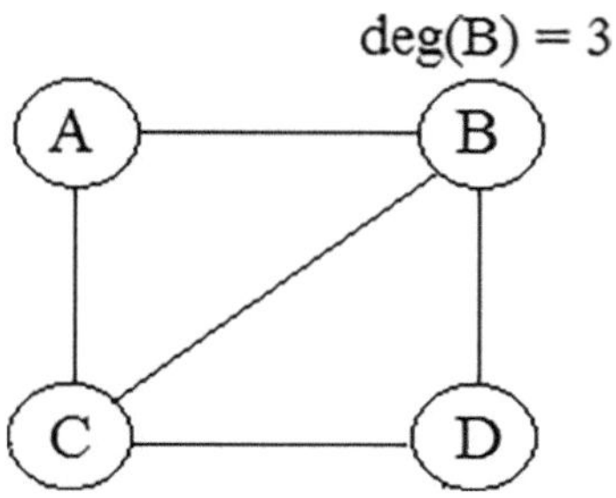

FIGURE 12.4 Graph showing the degree of node B.

- *In-degree of a node* – The in-degree of a node is equal to the number of edges arriving at that particular node, as shown in Figure 12.5.
- *Out-degree of a node* – The out-degree is equal to the number of edges leaving that particular node. This is also shown in Figure 12.5.

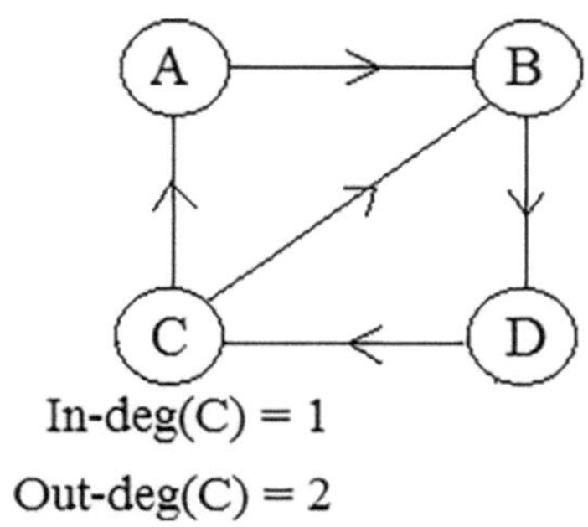

FIGURE 12.5 Graph showing the in-degree and out-degree of node C.

- *Isolated node/vertex* – A node having zero edges is known as an isolated node, as shown in Figure 12.6. The degree of such a node is zero.

FIGURE 12.6 Two isolated nodes, X and Y.

- *Pendant node/vertex* – A node having one edge is known as a pendant node, as shown in Figure 12.7. The degree of such a node is one.

FIGURE 12.7 Two pendant nodes, X and Y.

- *Adjacent nodes* – For every edge $e = (A, B)$ that connects nodes A and B, nodes A and B are said to be the adjacent nodes.
- *Parallel edges* – If there is more than one edge between the same pair of nodes, then they are known as parallel edges, as shown in Figure 12.8.

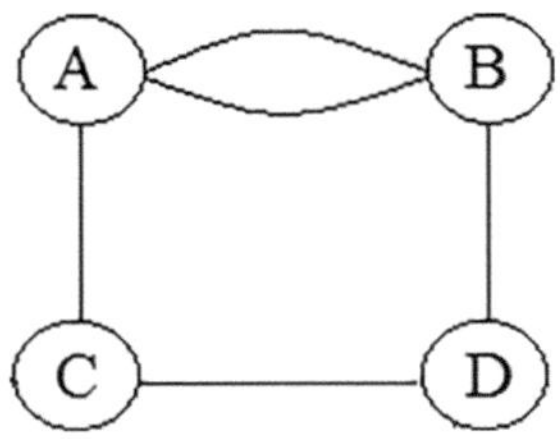

FIGURE 12.8 Parallel edges between A and B.

- *Loop* – If an edge has a starting and ending point at the same node, that is, $e = (A, A)$, then it is known as a loop, as shown in Figure 12.9.

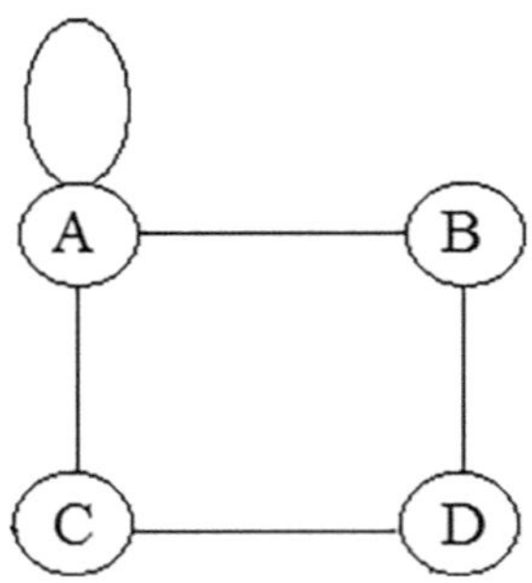

FIGURE 12.9 A loop.

- *Simple graph* – A graph G(V, E) is known as a simple graph if it does not contain any loops or parallel edges.
- *Complete graph* – A graph G(V, E) is known as a complete graph if and only if every node in the graph is connected to every other node, and there is no loop on any of the nodes. An example is shown in Figure 12.10.

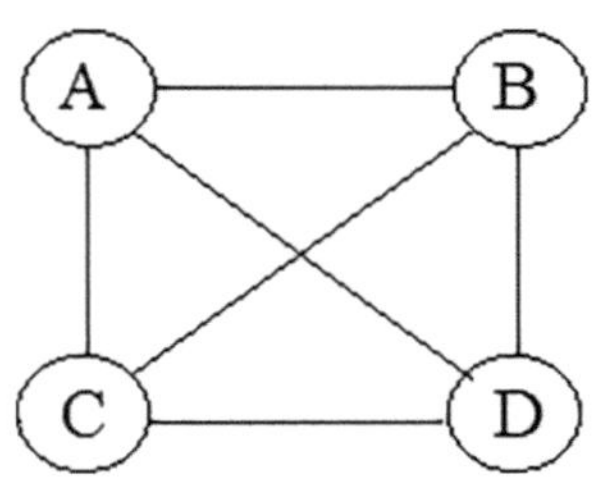

FIGURE 12.10 Complete graph.

- *Regular graph* – A regular graph is a graph in which every node has the same degree. If every node has a degree *r*, then the graph is called a regular graph of degree *r*. In Figure 12.11, all the nodes have the same degree, that is, 2; hence, it is known as a 2-regular graph.

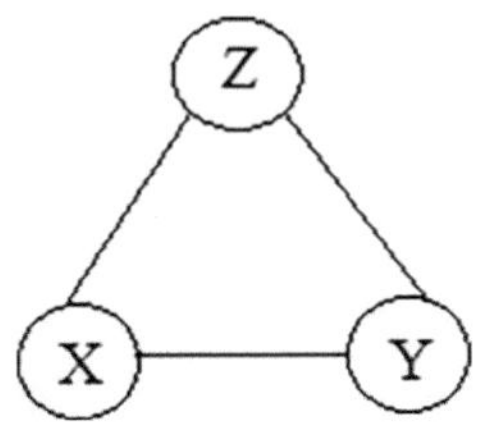

FIGURE 12.11 A 2-regular graph.

- *Multi-graph* – A graph *G*(*V*, *E*) is known as a multi-graph if it contains either a loop, parallel edges, or both, as shown in Figure 12.12.

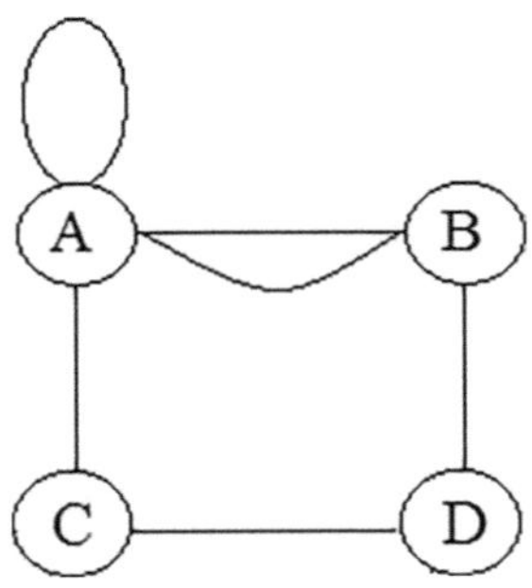

FIGURE 12.12 Multi-graph.

- *Cycle* – This is a path containing one or more edges that starts from a particular node and also terminates at the same node.
- *Cyclic graph* – A graph that has cycles in it is known as a cyclic graph.
- *Acyclic graph* – A graph without any cycles is known as an acyclic graph.
- *Connected graph* – A graph *G*(*V*, *E*) is known as a connected graph if there is a path from any node in the graph to another node in the graph, such that for every pair of distinct nodes, there must be a path. However, unlike complete graph, not all vertices need to be directly linked, each vertex should be reachable from every other vertex may be through intermediate vertices, An example is shown in Figure 12.13.

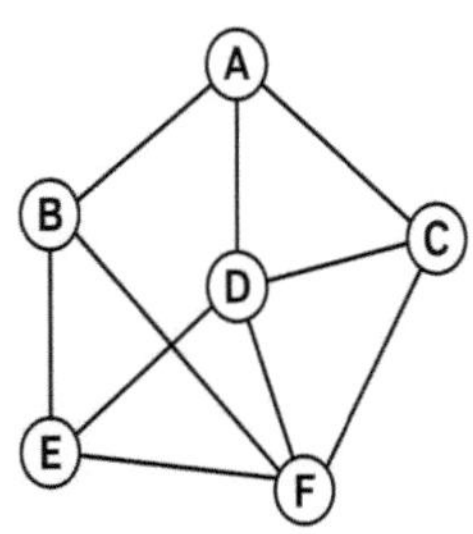

FIGURE 12.13 Connected graph.

- *Strongly connected graph* – A directed graph is said to be a strongly connected graph if there exists a dedicated path between every pair of nodes in the graph. For example, if there are two nodes, say *P* and *Q*, and there is a dedicated path from *P* to *Q*, then there must be a path from *Q* to *P*. An example is shown in Figure 12.14.

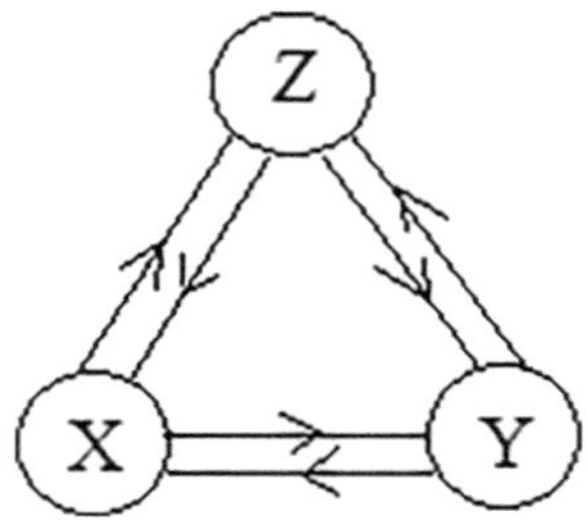

FIGURE 12.14 Strongly connected graph.

- *Size of a graph* – The size of a graph is equal to the total number of edges present in the graph.
- *Weighted graph* – A graph *G*(*V, E*) is said to be a weighted graph if all the edges in the graph are assigned some data. This data indicates the cost of traversing the edge. An example is shown in Figure 12.15.

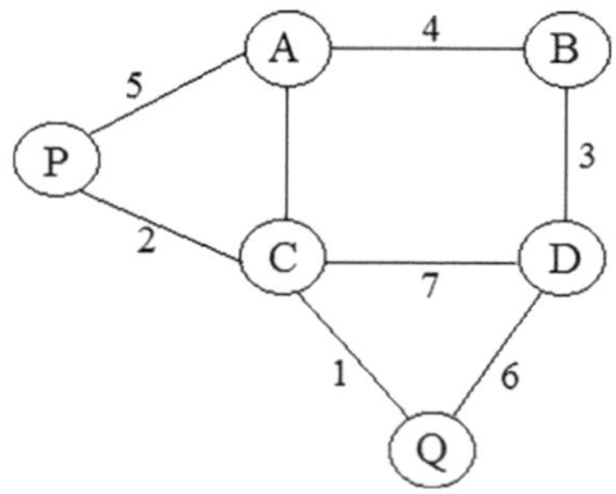

FIGURE 12.15 Weighted graph.

GRAPH REPRESENTATION

Graphs can be represented in a computer's memory in either of the following ways:

- Sequential representation using an adjacency matrix
- Linked representation using an adjacency list

Now, let us discuss both in detail.

Adjacency Matrix Representation

An *adjacency matrix* is used to represent the information of the nodes that are adjacent to one another. The two nodes will only be adjacent when there is an edge connecting those nodes. For any graph *G* having *n* nodes, the dimension of the adjacency matrix will be $(n \times n)$.

Let $G(V, E)$ be a graph having vertices $V = \{V_1, V_2, V_3 \ldots, V_n\}$, so then the adjacency matrix representation $(n \times n)$ will be given by:

$$a_{ij} = \begin{cases} 1 & \text{if there is an edge from } V_i \text{ to } V_j. \\ 0 & \text{otherwise} \end{cases}$$

The adjacency matrix is also known as a bit matrix or Boolean matrix, since it contains only 0s and 1s. Now, let us look at a few examples to discuss and understand it more clearly:

- *Example* 1 – Consider the directed graph shown in Figure 12.16 and find its adjacency matrix:

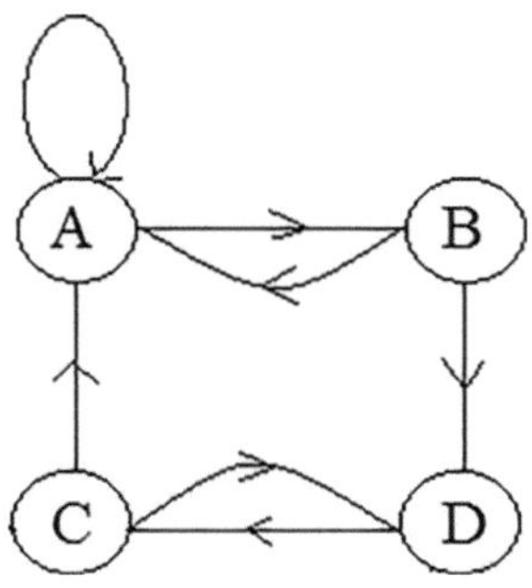

FIGURE 12.16 A directed graph.

The adjacency matrix for the graph will be:

	A	B	C	D
A	1	1	0	0
B	1	0	0	1
C	1	0	0	1
D	0	0	1	0

- *Example* 2 – Now, consider the undirected graph shown in Figure 12.17 and find its matrix:

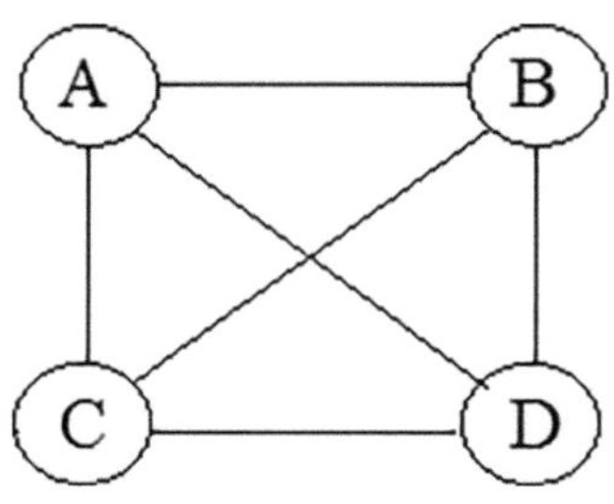

FIGURE 12.17 An undirected graph.

The adjacency matrix for the graph will be:

	A	B	C	D
A	0	1	1	1
B	1	0	1	1
C	1	1	0	1
D	1	1	1	0

- *Example* 3 – Now, consider the weighted graph shown in Figure 12.18 and find its adjacency matrix:

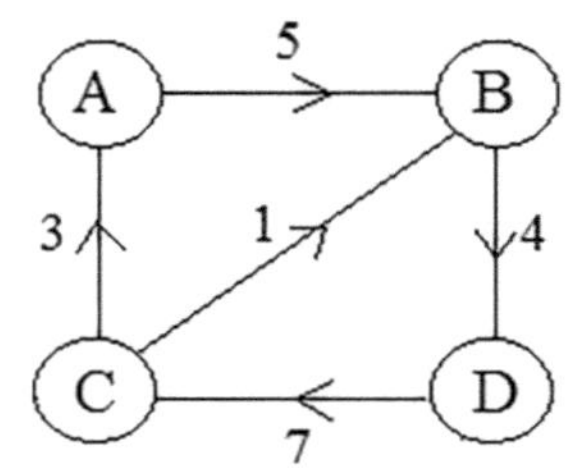

FIGURE 12.18. A directed weighted graph.

The adjacency matrix for the graph will be:

	A	B	C	D
A	0	5	0	0
B	0	0	0	4
C	3	1	0	0
D	0	0	7	0

- *Example* 4 – Consider the undirected multi-graph shown in Figure 12.19 and find its adjacency matrix:

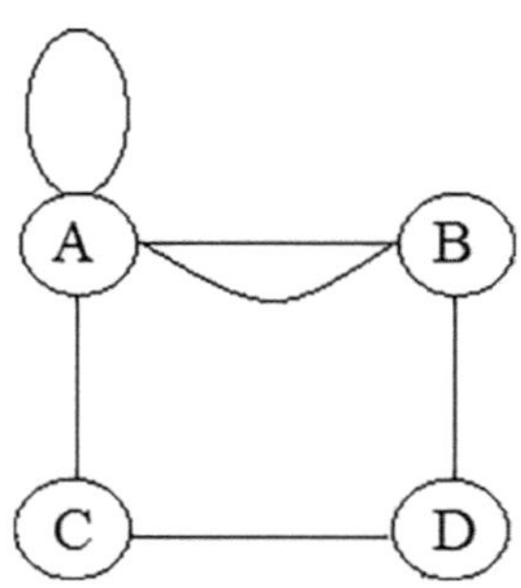

FIGURE 12.19 An undirected multi-graph.

The adjacency matrix for the graph will be:

	A	B	C	D
A	1	2	1	0
B	2	0	0	1
C	1	0	0	1
D	0	1	1	0

From the above examples, we conclude that:

- The memory space needed to represent a graph using its adjacency matrix is n^2 bits.
- The adjacency matrix for an undirected graph is always symmetric.
- The adjacency matrix for a directed graph need not be symmetric.
- The adjacency matrix for a simple graph having no loops or parallel edges will always contain 0s on the diagonal.
- The adjacency matrix for a weighted graph will always contain the weights of the edges connecting the nodes instead of 0 and 1.
- The adjacency matrix for an undirected multi-graph will contain the number of edges connecting the vertices instead of 1.

Adjacency List Representation

The adjacency matrix representation has some major drawbacks. First, it is very difficult to insert and delete the nodes in/from the graph as the size of the matrix needs to be changed accordingly, which is a very time-consuming process. Also, sometimes the matrix may contain many zeros (sparse matrix). Hence, it is not a healthy representation. Therefore, adjacency list representation is preferred for representing sparse graphs in memory.

In this representation, every node is linked to a list of all the other nodes that are adjacent to it. Adjacency list representation makes it easier to add or delete nodes. Also, it shows the adjacent nodes of a particular node. Now, let us look at a few examples to discuss and understand it more clearly:

- *Example* 1 – Consider the undirected graph shown in Figure 12.20 and find its adjacency list representation:

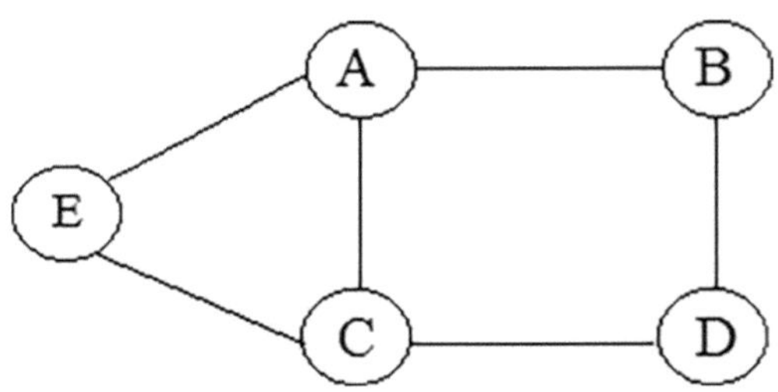

FIGURE 12.20 An undirected graph.

The adjacency list representation of the graph will be:

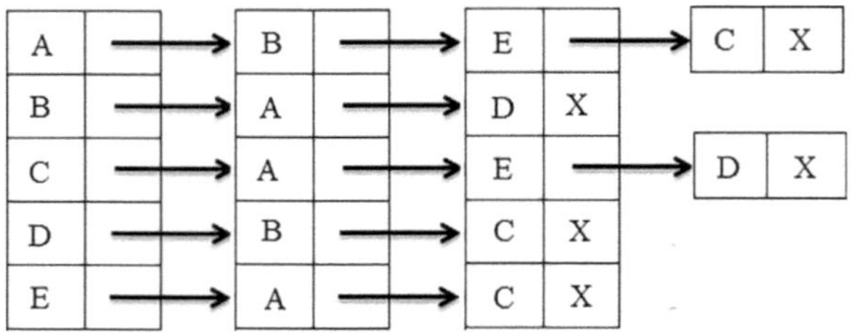

- *Example* 2 – Consider the directed graph shown in Figure 12.21 and find its adjacency list representation:

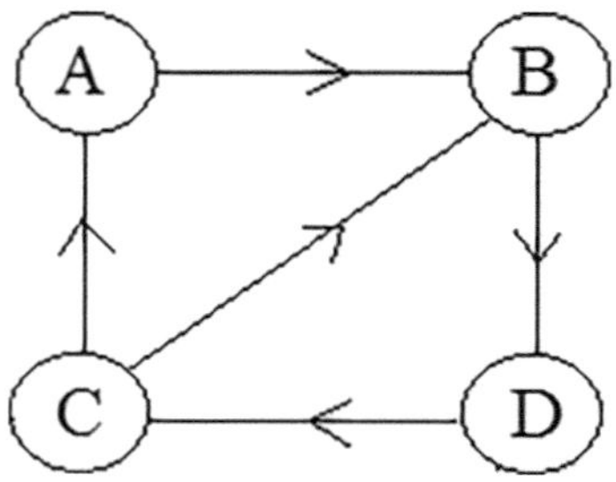

FIGURE 12.21 A directed graph.

The adjacency list representation of the graph will be:

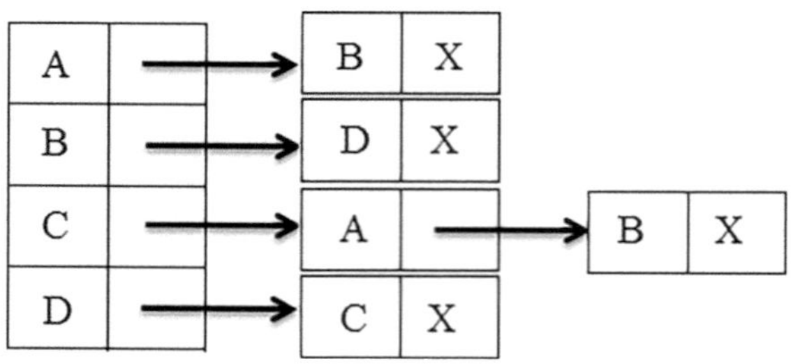

- *Example* 3 – Now, consider the weighted graph shown in Figure 12.22 and find its adjacency list representation:

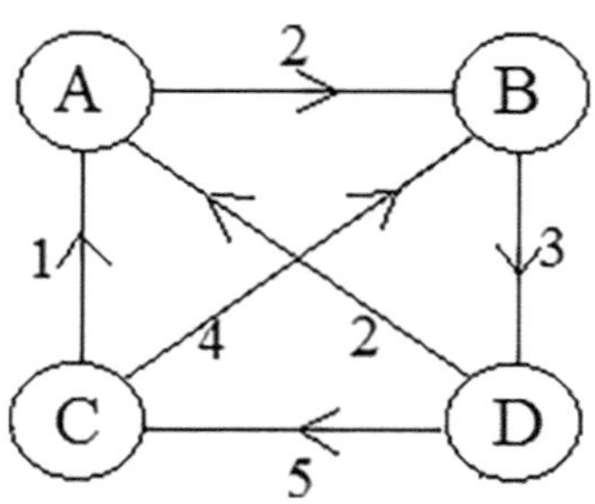

FIGURE 12.22 A directed weighted graph.

The adjacency list representation of the graph will be:

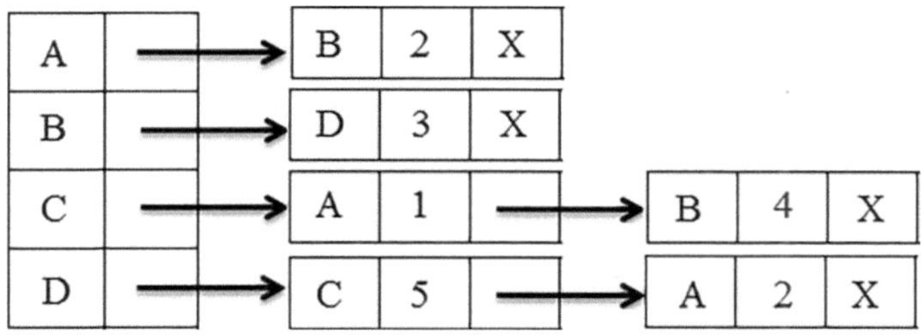

GRAPH TRAVERSAL TECHNIQUES

In this section, we will discuss various techniques to traverse a graph. As we know, a graph is a collection of nodes and edges. Thus, traversing a graph is the process of visiting each node and edge in some systematic approach. There are two types of standard graph traversal techniques:

- Breadth-first search
- Depth-first search

Now, we will discuss both of these techniques in detail.

Breadth-First Search

Breadth-first search (*BFS*) is a traversal technique that uses a queue as an auxiliary data structure for traversing all member nodes of a graph. In this technique, first, we will select any node in the graph as a starting node, and then we will take all the nodes adjacent to the starting node. We will maintain the same approach for all the other nodes. We will also maintain the status of all the traversed/visited nodes in a queue so that no nodes are traversed again. Now, let us take the graph shown in Figure 12.23 and apply BFS to traverse the graph.

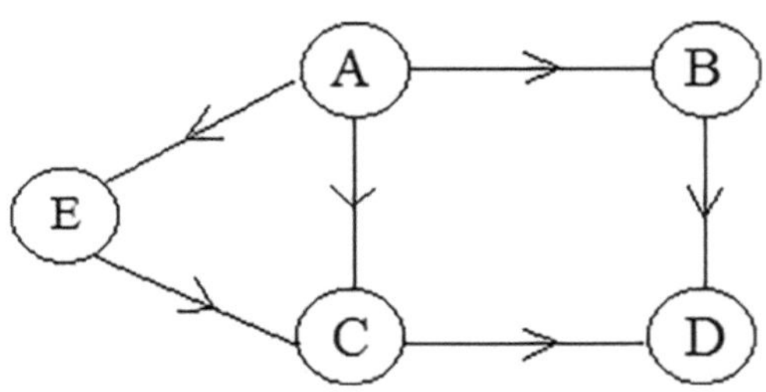

FIGURE 12.23 A sample graph

We will start the traversal of the graph by taking node *A* as the starting node of the above sample graph. Then, we will traverse all the nodes adjacent to the starting node *A*. As we can see, *B*, *C*, and *E* are the adjacent nodes of *A*. We will traverse these nodes in any order, say *E*, *C*, and *B*, so the traversal is:

A, E, C, B

Now, we will traverse all the nodes adjacent to *E*. Node *C* is adjacent to node *E*, but node *C* has already been traversed, so we will ignore it, and we will move to the next step. Now, we will traverse all the nodes adjacent to node *C*. As we can see, *D* is the adjacent node of *C*. So we will traverse node *D*, and the traversal is:

A, E, C, B, D

We can see that all the nodes have been traversed, and hence, this was a BFS traversal by taking node *A* as a starting node.

Now, we will implement the BFS traversal technique with the help of a queue. In this, we will maintain an array that will store all the adjacent unvisited neighbor nodes of a given node under consideration. Initially, the front and rear are set to −1. We will also maintain the status of the visited nodes in a Boolean array, which will have the value 1 if the node is visited and 0 if it is not visited:

1. First, we will enqueue/insert the starting node into the queue.
2. Second, the first node/element in the queue is deleted from the queue, and all the adjacent unvisited nodes are inserted into the queue. This is repeated until the queue becomes empty.

For example, let's consider the sample graph shown in Figure 12.24 and traverse the graph using the BFS technique.

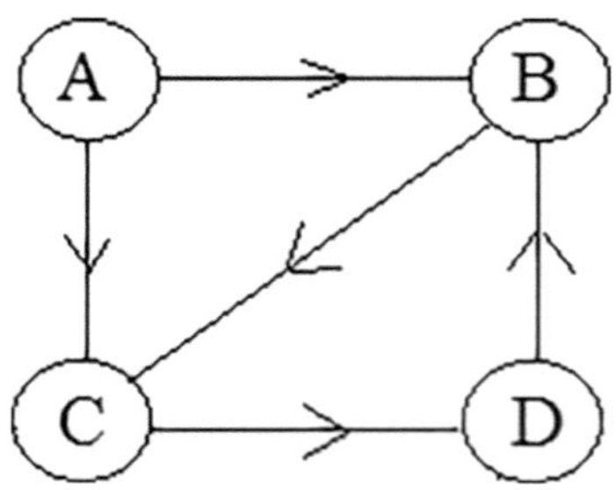

FIGURE 12.24 A sample graph

The appropriate adjacency list representation of a graph is given as follows:

Node	Adjacency List
A	B, C
B	C
C	D
D	B

In this example, we are taking *A* as the starting node:

1. First, node *A* is inserted into the queue:

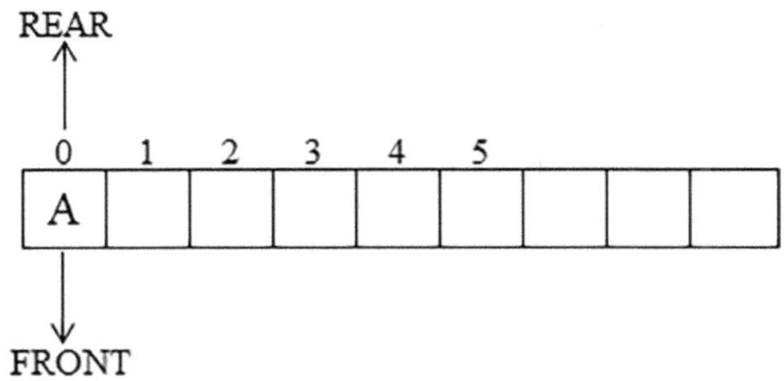

2. Next, node *A* is deleted from the queue, and *FRONT* is incremented by 1. Now, insert all the nodes adjacent to *A*, which are nodes *B* and *C*, by incrementing *REAR*. Node *A* has been traversed.

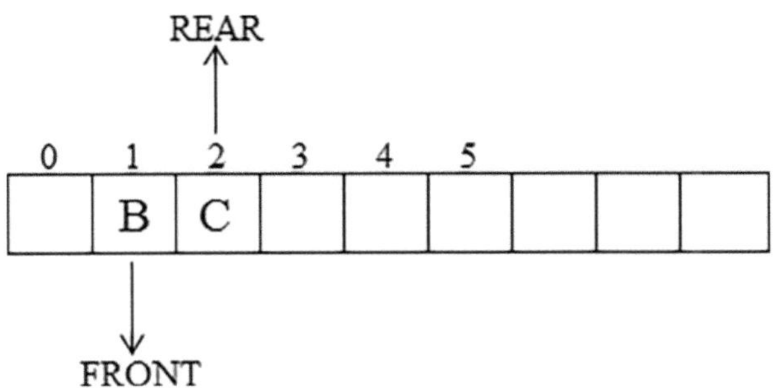

3. Similarly, node *B* is deleted from the queue, and *FRONT* is incremented by 1. Now, we insert all the nodes adjacent to *B*, which is node *C*, by incrementing *REAR*, but *C* has already been inserted into the queue. In this case, node *C* is also deleted by incrementing *FRONT* by 1, and the node adjacent to *C* (that is, *D*) is inserted into the queue. Therefore, nodes *A*, *B*, and *C* are traversed:

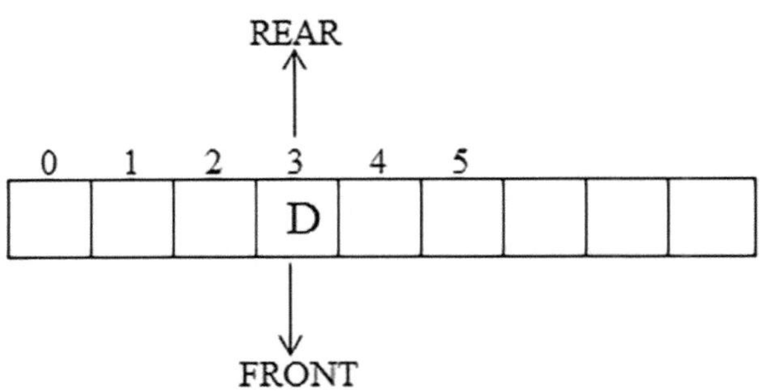

4. Now, we will again delete the *FRONT* element from the queue, which is *D*. We will insert the adjacent node of *D*, that is, *B*, but it is already traversed. Finally, as we delete the front element *D*, we notice that *FRONT* > *REAR*, which is not possible. Hence, we have traversed all the nodes in the graph:

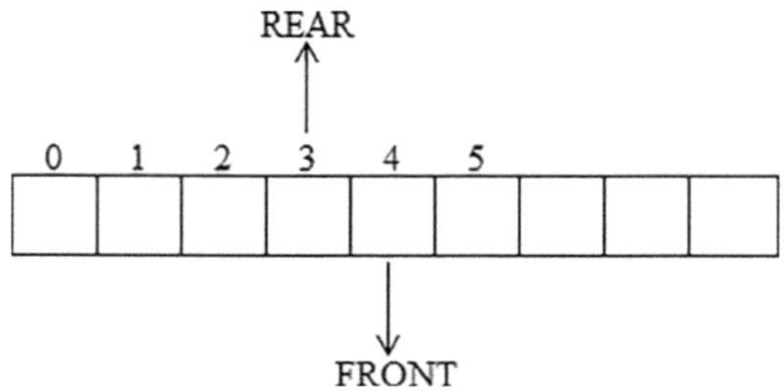

Therefore, the BFS traversal of the graph is given as:

A, B, C, D

Now, let us look at the function for a BFS traversal:

```
Breadth_First_Search(int node)
{
int i , front, rear;
int queue[SIZE];
  front=rear= -1;
printf("%d", node);
visited[node]= 1;
  rear++;
  front++;
  queue[rear]= node ;
  while(front<=rear)
{
node = queue[front];
    front++;
for(i=1; i<=n; i++)
   {
      if(adjacent[node][i]==1) && visited[node]==0)
      {
printf("%d", node);
visited[i]=1;
        rear++;
        queue[rear]= i ;
}
   }
 }
}
```

Depth-First Search

Depth-first search (*DFS*) is another traversal technique that uses the stack as an auxiliary data structure for traversing all the member nodes of a graph. In this technique, we first select any node in the graph as a starting node, and then we travel along a path that begins from the starting node. We will visit the adjacent node of the starting node, the adjacent node of the previous node, and so on. We will maintain the same approach for all the other nodes. Now, let us take the graph shown in Figure 12.25 and apply DFS to traverse the graph:

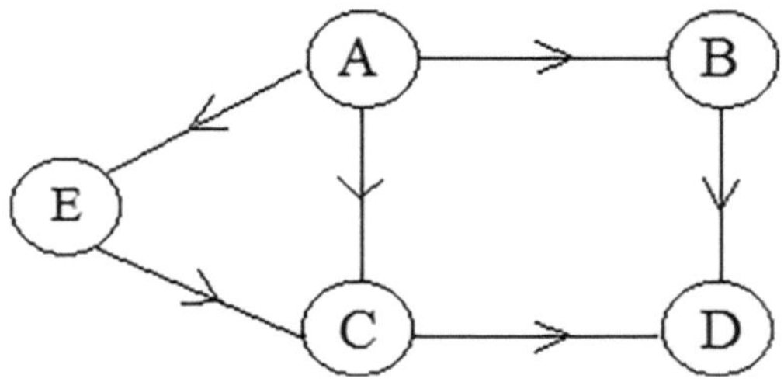

FIGURE 12.25 A sample graph.

We will start the traversal of the graph by taking node *A* as the starting node. Then, we will traverse any of the nodes adjacent to the starting node *A*. As we can see, *B*, *C*, and *E* are the adjacent nodes of *A*. If we traverse node *E*, then we will traverse the node adjacent to *E*, that is, *C*. After traversing *C*, we will traverse the node adjacent to *C*, which is D. Now, there is no adjacent node to *D*; hence, we have reached a dead end. Thus, the traversal until now is:

A, E, C, D

Because of the dead end, we will move backward. Now, we reach node *C*. We will check whether there is any other node adjacent to *C*. There is no such node, and thus, we again move backward. Now, we reach *E*. We will again check whether there is any other node adjacent to *E*. There is no such node, and thus, we again move backward. Now, we reach *A*. We will check whether there is any other node adjacent to *A*. There are two nodes, *B* and *C*, adjacent to node *A*. As *C* is already traversed, it will be ignored. Now, we will traverse node *B*. After traversing *B*, we will traverse the node adjacent to *B*, which is *D*, but *D* is already traversed. Thus, we can't move backward or forward. Thus, we have completed the traversal. The final traversal is given as:

A, E, C, D, B

Now, we will implement the DFS traversal technique with the help of a stack. In this, we will maintain an array that will store all the adjacent unvisited neighbor nodes of a given node. Initially, the top is set to −1. We will also maintain the status of the visited nodes in a Boolean array, which will have the value 1 if the node is visited and 0 if it is not visited:

1. First, we will push the starting node onto the stack.
2. Second, the topmost node/element is popped out from the stack and is traversed. If it is already traversed, then we will ignore it.
3. Third, all the adjacent unvisited nodes of the popped node/element are pushed onto the stack. This process is repeated until the queue becomes empty. The steps are repeated until the stack becomes empty.

For example, let's consider the sample graph shown in Figure 12.26 and traverse the graph using the DFS technique:

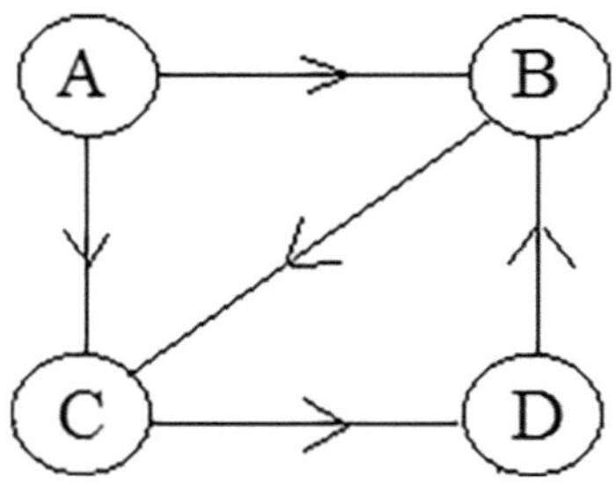

FIGURE 12.26 A sample graph.

In this example, we are taking *A* as the starting node:

1. Push *A* onto the stack:

A

2. Now, pop the topmost element from the stack, that is, *A*. Thus, *A* is traversed. Now, push all the nodes adjacent to *A*, that is, push *B* and *C*:

B, C

3. Pop the topmost element from the stack, that is, *C*. Thus, *C* is also traversed. Now, push all the nodes adjacent to *C*, that is, push *D*:

B, D

4. Now, again, pop the topmost element from the stack, that is, *D*. Thus, *D* is also traversed. Now, push all the nodes adjacent to *D*, that is, push *B*. As *B* is already in the stack, however, no push is performed. Thus, the stack becomes:

B

5. Again, pop the topmost element from the stack, that is, *B*. Thus, *B* is also traversed. Now, push all the nodes adjacent to *B*, that is, push *C*, but *C* is already traversed; hence, the stack becomes empty.

   Therefore, the DFS traversal of the graph is given as follows:

   A, C, D, B

Now, let us look at the function for the DFS traversal:

```
Depth_First_Search(int node)
{
inti, stack[SIZE], top = -1, pop ;
   top++ ;
   stack[top] = node ;
   while(top >= 0)
  {
    pop = stack[top] ;
top-- ;
    if(visited[pop] == 0)
   {
printf("%d", pop) ;
visited[pop] = 1;
     }
    else
continue
 for(i=n; i>=1 ;i--)
{
if(adjacent[pop][node] == 1 && visited[node] == 0)
{
     top++  ;
stack[top] = node ;
}
}
 }
}
```

Memory Aid:

To remember which of the data structures are used in implementing a BFS and a DFS, we can use this memory aid. BFS is implemented using a queue data structure, and DFS is implemented using a stack data structure, and it can be remembered using alphabetical order. B (BFS) and Q (queue) come before D (DFS) and S (stack) in alphabetical order.

TOPOLOGICAL SORT

Topological sort is a procedure to determine the linear ordering of the nodes of a *directed acyclic graph* (*DAG*) in which each node comes before all those nodes that have zero predecessors. A topological sort of a DAG is a linear ordering of the vertices of a graph $G(V, E)$ such that if (a, b) is an edge, then a must appear before b in the topological ordering. The main idea behind this is that in a graph, if a vertex has an in-degree of 0, then that vertex should be selected as the first element in the topological order. Also, a topological sort is possible only in DAGs. An acyclic graph is one that does not have any cycles in it. Topological sorting is widely used in scheduling tasks, applications, and so on. Now, let us look at the algorithm of topological sorting:

```
Step 1: START
Step 2: Find the in-degree of every node.
Step 3: Insert all the nodes/elements having in-degree zero in the
        queue.
Step 4: Repeat Steps 5 and 6 until the queue becomes empty.
Step 5: Delete the first node from the queue by incrementing FRONT
        by 1.
Step 6: Repeat for each neighbor P of node N -
        a) Delete the edge from P to M by decreasing the in-degree by 1.
        b) If in-degree of P is zero, then add P to the rear of the queue.
Step 7: END
```

For example, let's consider the DAG shown in Figure 12.27 and find its topological sort:

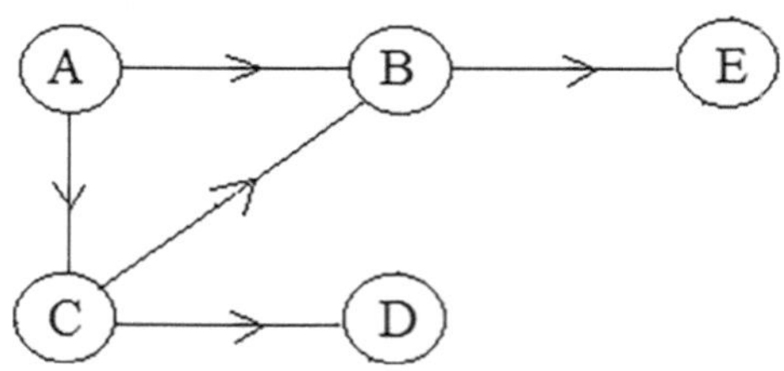

FIGURE 12.27 A DAG.

The appropriate adjacency list representation of the previous graph is given as follows:

Node	Adjacency List
A	B, C
B	E
C	B, D
D	–
E	–

1. First, we find the in-degree of all the nodes:
   - In-degree $A - 0$
   - In-degree $B - 2$
   - In-degree $C - 1$
   - In-degree $D - 1$
   - In-degree $E - 1$

   Now, we have node A with an in-degree of 0; thus, A will be added to the queue.
2. Now, we insert node A into the queue:
   $FRONT = 1$, $REAR = 1$, and $QUEUE = A$
3. Now, we delete node A from the queue. We also delete all the edges going from A:
   $FRONT = 0$, $REAR = 0$, and $TOPOLOGICAL\ SORT = A$
   Thus, the graph becomes:

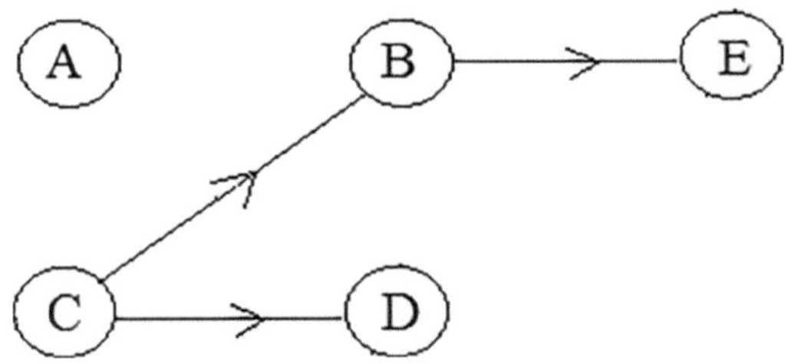

   Now, the in-degrees of all the nodes are as follows:
   - In-degree $B - 1$
   - In-degree $C - 0$
   - In-degree $D - 1$
   - In-degree $E - 1$

   Now, we have node C with an in-degree of 0; thus, C will be added to the queue.
4. Next, insert node C into the queue:
   $FRONT = 1$, $REAR = 1$, and $QUEUE = C$
5. Now, delete node C from the queue. Also, delete all the edges going from C:
   $FRONT = 0$, $REAR = 0$, and $TOPOLOGICAL\ SORT = A, C$
   Thus, the graph becomes:

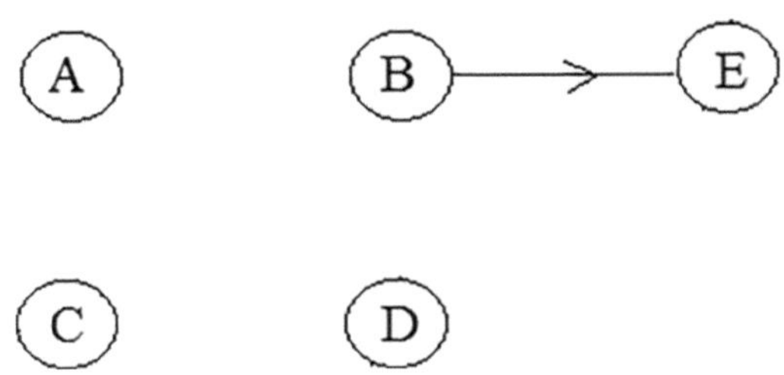

Now, the in-degrees of all the nodes are as follows:

- In-degree *B* – 0
- In-degree *D* – 0
- In-degree *E* – 1

Now, we have two nodes, *B* and *D*, with an in-degree of 0; thus, *B* and *D* will be added to the queue.

6. Next, insert nodes *B* and *D* into the queue:
   FRONT = 1, *REAR* = 2, and *QUEUE* = *B*, *D*
7. Now, delete node *B* from the queue. Also, delete all the edges going from *B*. There will be no change in the in-degree of the nodes:
   FRONT = 1, *REAR* = 1, *TOPOLOGICAL SORT* = *A*, *C*, *B*, and *QUEUE* = *D*
8. Next, delete node *D* from the queue. Also, delete all the edges going from *D*:
   FRONT = 0, *REAR* = 0, and *TOPOLOGICAL SORT* = *A*, *C*, *B*, *D*
   Thus, the graph becomes:

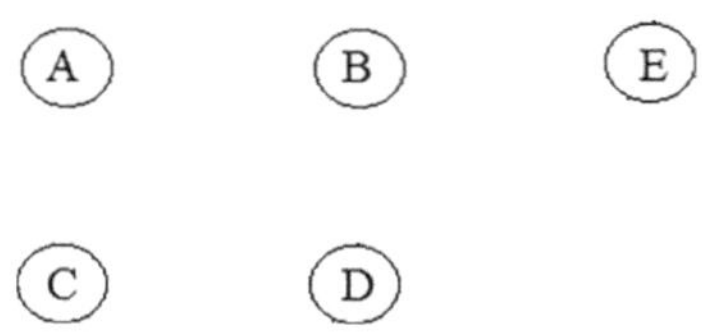

The in-degree of the node is as follows:

- In-degree *E* – 0

Now, we have node *E* with an in-degree of 0. Thus, *E* will be added to the queue.

9. Next, insert node *E* into the queue:
   FRONT = 1, *REAR* = 1, and *QUEUE* = *E*
10. Now, delete node *E* from the queue. Also, delete all the edges going from *E*:
    FRONT = 0, *REAR* = 0, and *TOPOLOGICAL SORT* = *A*, *C*, *B*, *D*, *E*
    Now, we have no nodes left in the graph. Thus, the topological sort of the graph will be:

A, C, B, D, E

MINIMUM SPANNING TREE

A *spanning tree* of an undirected and connected graph *G* is a subgraph that contains all the vertices and edges that connect these vertices and is a tree. The weights/costs can be assigned to the edges, and these weights/costs can be used to calculate the weight/cost of the spanning tree by calculating the sum of the weights/costs of each edge. A graph can have many spanning trees. Thus, a *minimum spanning tree* (*MST*) is defined as a spanning tree that has weights/costs associated with the edges such that the total weight/cost of the spanning tree is at a minimum. Although there are various approaches for determining an MST, the two most popular approaches for determining a minimum cost spanning tree of a graph are:

- Prim's algorithm
- Kruskal's algorithm

Now, let us discuss both of them in detail.

Prim's Algorithm

Prim's algorithm is used to build a minimum cost spanning tree. This algorithm works in such a way that it builds a tree edge by edge. The next edge to be included is chosen according to some criterion. The steps involved in Prim's algorithm are as follows:

1. Select a starting vertex/node and add it to the spanning tree.
2. During each iteration, select a vertex/node in such a way that the edge connecting vertex V_i to another vertex V_j has the minimum cost/weight assigned to it. Remember, the edge forming a cycle must not be added.
3. End the process when $(n-1)$ number of edges have been inserted into the tree.

Frequently Asked Questions

Q1. Consider the following graph and construct a minimum spanning tree using Prim's algorithm:

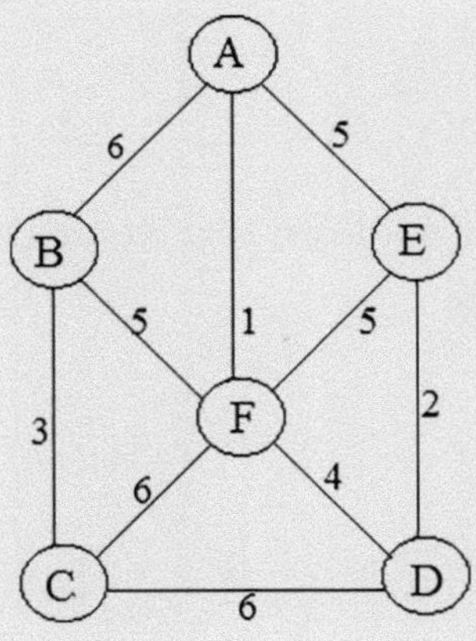

Answer.

1. The starting node is *F*:

2. The lowest weighted/cost edge is (*F*, *A*), that is, 1. Hence, it is added to the tree:

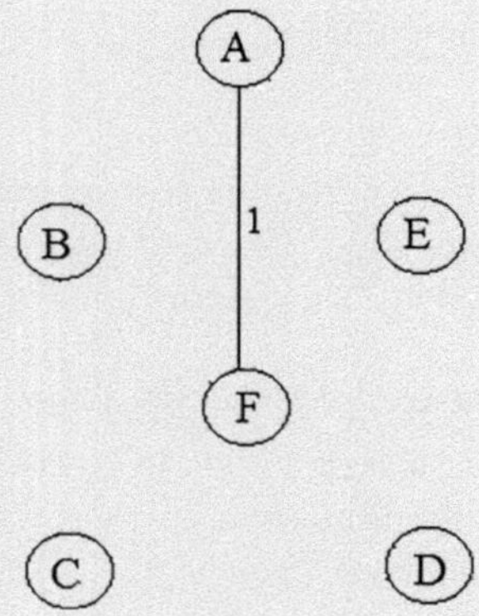

3. Now, the lowest weighted/cost edge is (*F*, *D*), that is, 4. Hence, it is added to the tree:

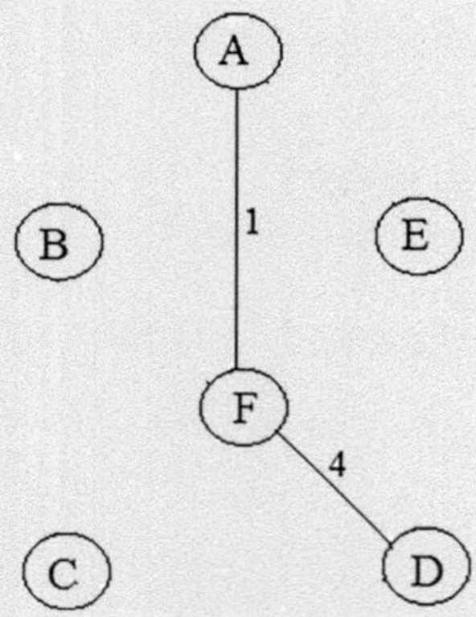

4. Step 4: Next, the lowest weighted/cost edge is (*D*, *E*), that is, 2. Hence, it is added to the tree:

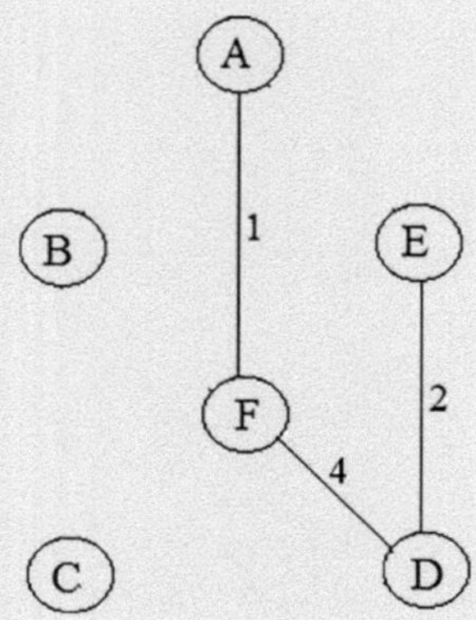

5. Step 5: Next, the lowest weighted/cost edge is (*F*, *B*), that is, 5. Hence, it is added to the tree:

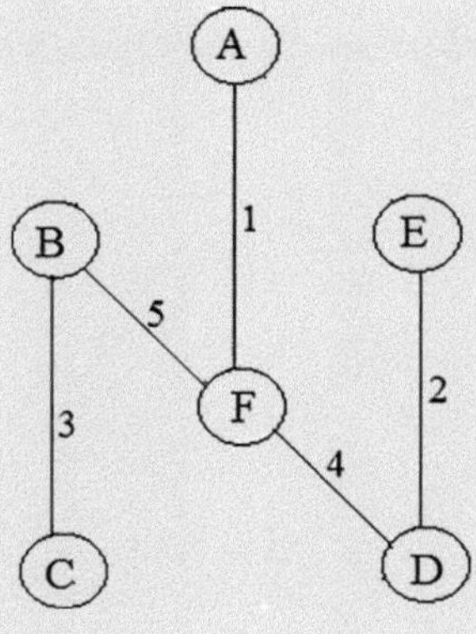

6. Finally, the minimum spanning tree is constructed by adding (*B*,*C*) edge:

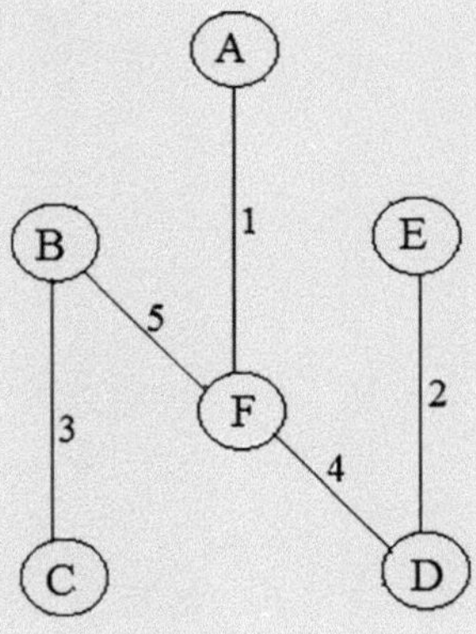

Kruskal's Algorithm

Kruskal's algorithm is another approach for determining the minimum cost spanning tree of a graph. In this approach, the tree is also built edge by edge. The next edge to be included is chosen according to some criterion. The steps involved in Kruskal's algorithm are as follows:

1. The weights/costs assigned to the edges are sorted in ascending order.
2. In this step, the lowest weighted/cost edge is added to the tree. Remember, the edge forming a cycle must not be added.
3. End the process when $(n-1)$ number of edges have been inserted into the tree.

Frequently Asked Questions

Q2. Consider the following graph and construct a minimum spanning tree using Kruskal's algorithm:

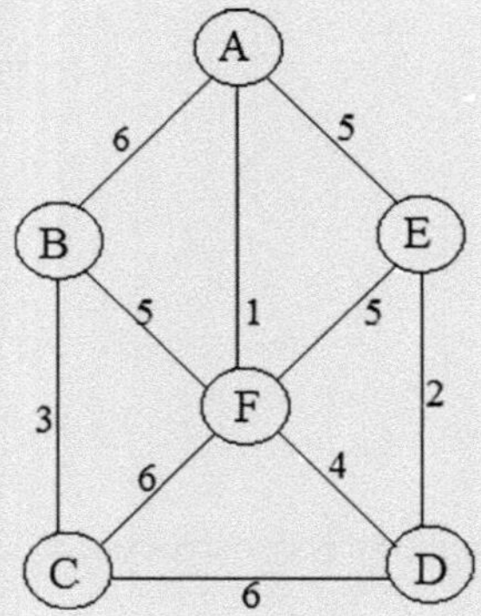

Answer.

1. Initially, the tree is given as:

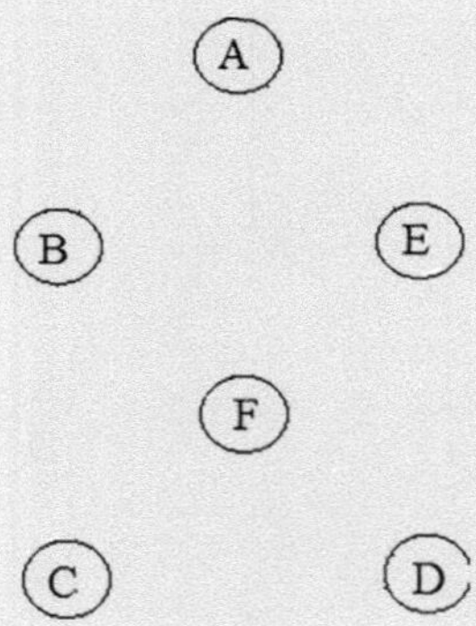

2. Choose edge (*F*, *A*):

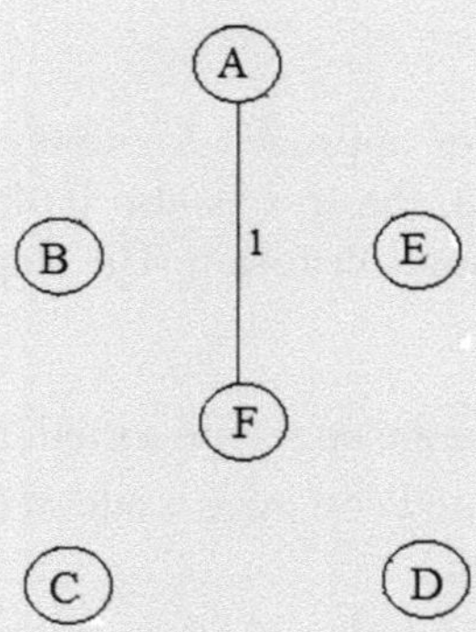

3. Choose edge (*D*, *E*):

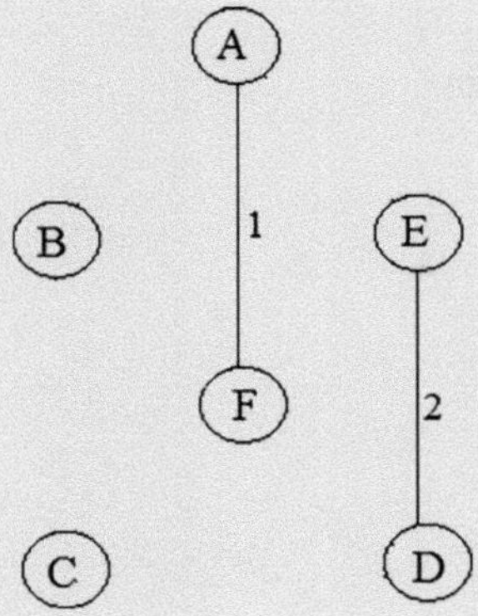

4. Choose edge (*B*, *C*):

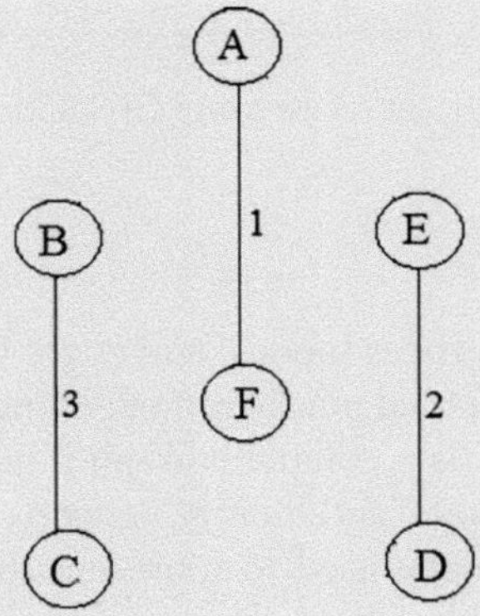

5. Choose edge (*F*, *D*):

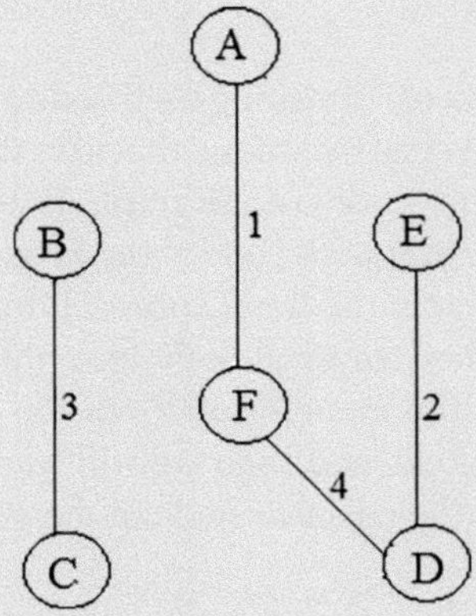

6. Choose edge (*F*, *B*):

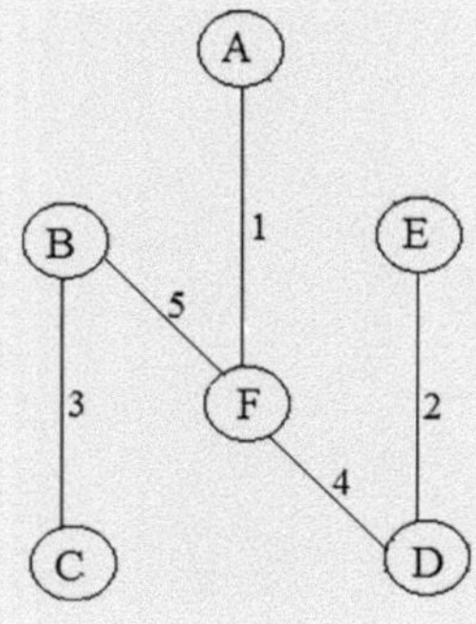

Practical Application:

Graphs are used to find the shortest route using GPS, Google Maps, and Yahoo Maps.

SUMMARY

- A graph is a collection of vertices (nodes) and edges that connect these vertices.
- The degree of a node is the total number of edges incident on that particular node.
- A graph *G*(*V*, *E*) is known as a complete graph if and only if every node in the graph is connected to every other node, and there is no loop on any of the nodes.
- An adjacency matrix is usually used to represent the information of the nodes that are adjacent to one another. The adjacency matrix is also known as a bit matrix or Boolean matrix since it contains only 0s and 1s.
- In adjacency list representation, every node is linked to its list of all the other nodes that are adjacent to it.
- Traversing a graph is the process of visiting each node and edge in some systematic approach.
- Breadth-first search is a traversal technique that uses the queue as an auxiliary data structure for traversing all the member nodes of the graph. In this technique, first, we will select any node in the graph as a starting node, and then we will take all the nodes adjacent to the starting node. We will maintain the same approach for all the other nodes.
- Depth-first search is another traversal technique that uses the stack as an auxiliary data structure for traversing all the member nodes of the graph. First, we will select any node in the graph as a starting node, and then we will travel along a path that begins from the starting node. We will visit the adjacent node of the starting node, the adjacent node of the previous node, and so on.
- Topological sort is a procedure to determine a linear ordering of the nodes of a directed acyclic graph in which each node comes before all those nodes that have zero predecessors.
- A minimum spanning tree is defined as a spanning tree that has weights/costs associated with the edges such that the total weight/cost of the spanning tree is at a minimum.

EXERCISES

Theory Questions

1. What is a graph? Explain its features.
2. What do you understand by a complete graph?
3. What is a multi-graph?
4. How can a graph be represented in the computer's memory? Discuss.
5. Differentiate between a directed and an undirected graph with an example of each.
6. Consider the following graph, and find the following:
   a) Adjacency matrix representation
   b) Degree of each node
   c) Whether the graph is complete
   d) Pendant nodes

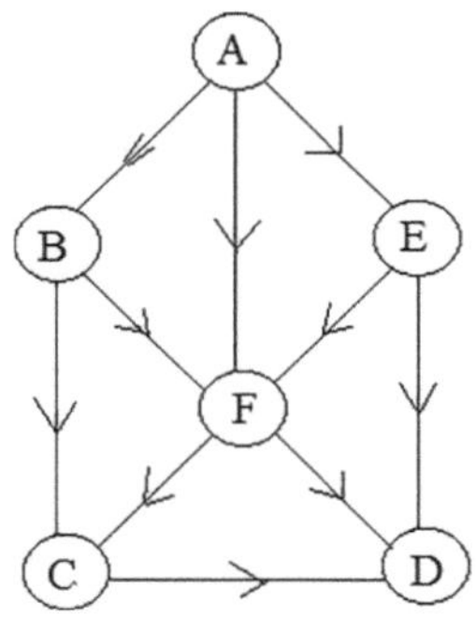

7. Explain why adjacency list representation is preferred for storing sparse matrices over adjacency matrix representation.
8. What are the different types of graph traversal techniques? Explain each of them in detail with the help of an example.
9. What do you understand by topological sort?
10. In what kind of graphs can topological sorting be used?
11. Differentiate between breadth-first search and depth-first search.
12. Consider the following graph and find out its BFS and DFS traversal:
13. What is a spanning tree?
14. Why is a minimum spanning tree called a spanning tree? Discuss.
15. Consider the given adjacency matrix and draw the directed graph:

	A	B	C	D
A	1	1	0	1
B	1	1	1	0
C	0	1	1	0
D	0	1	1	1

16. Write a short note on Prim's algorithm.
17. Explain Kruskal's algorithm.
18. List some of the real-life applications of graphs.
19. Consider the following graph and find the minimum spanning tree using
    a) Prim's algorithm
    b) Kruskal's algorithm

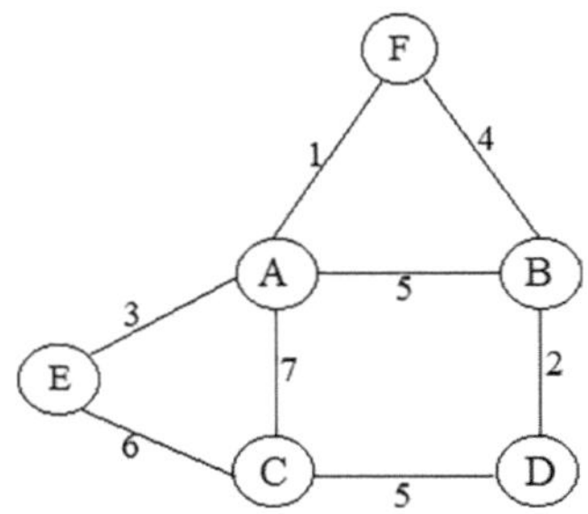

Programming Questions

1. Write a program to create and display a graph.
2. Write an algorithm to perform a topological sort on the graph.
3. Write an algorithm to find the degree of a node *N* in a graph.
4. Write a program to traverse a graph using depth-first search.
5. Write an algorithm to traverse a graph using breadth-first search.
6. Write a program to find the shortest path using Prim's algorithm.
7. Write a program to find the shortest path using Kruskal's algorithm.

Multiple Choice Questions

1. To implement breadth-first search, what type of data structure is used?
   A. Stack
   B. Queue
   C. Trees
   D. Linked list
2. A graph having multiple edges is known as a ________.
   A. Connected graph
   B. Complete graph
   C. Simple graph
   D. Multi-graph
3. What is an edge having initial and end points at the same node called?
   A. Degree
   B. Cycle
   C. Loop
   D. Parallel edge

4. What is an adjacency matrix also known as?
   A. Bit matrix
   B. Boolean matrix
   C. Both of the above
   D. None of the above
5. To implement depth-first search, what type of data structure is used?
   A. Stack
   B. Queue
   C. Trees
   D. Linked list
6. Topological sort is performed only on ________.
   A. Cyclic directed graphs
   B. Acyclic directed graphs
   C. Both of the above
   D. None of the above
7. Which one of the following nodes has a zero degree?
   A. Simple node
   B. Isolated node
   C. Pendant node
   D. None of the above
8. ________ is the total number of nodes in a graph.
   A. Degree
   B. In-degree
   C. Out-degree
   D. Size
9. A graph *G* can have many spanning trees. True or false?
   A. True
   B. False
   C. Not possible to comment
10. What is the memory use of an adjacency matrix?
    A. $O(\log n)$
    B. $O(\log n^2)$
    C. $O(n)$
    D. $O(n^2)$

INDEX

Bei Fragen zur Produktsicherheit wenden Sie sich bitte an:
If you have any questions regarding product safety,
please contact:

Walter de Gruyter GmbH
Genthiner Straße 13
10785 Berlin
productsafety@degruyterbrill.com